Population structure and human variation

T0297182

THE INTERNATIONAL BIOLOGICAL PROGRAMME

The International Biological Programme was established by the International Council of Scientific Unions in 1964 as a counterpart of the International Geophysical Year. The subject of the IBP was defined as 'The Biological Basis of Productivity and Human Welfare', and the reason for its establishment was recognition that the rapidly increasing human population called for a better understanding of the environment as a basis for the rational management of natural resources. This could be achieved only on the basis of scientific knowledge, which in many fields of biology and in many parts of the world was felt to be inadequate. At the same time it was recognized that human activities were creating rapid and comprehensive changes in the environment. Thus, in terms of human welfare, the reason for the IBP lay in its promotion of basic knowledge relevant to the needs of man.

The IBP provided the first occasion on which biologists throughout the world were challenged to work together for a common cause. It involved an integrated and concerted examination of a wide range of problems. The Programme was co-ordinated through a series of seven sections representing the major subject areas of research. Four of these sections were concerned with the study of biological productivity on land, in freshwater, and in the seas, together with the processes of photosynthesis and nitrogen fixation. Three sections were concerned with adaptability of human populations, conservation of ecosystems and the use of biological resources.

After a decade of work, the Programme terminated in June 1974 and this series of volumes brings together, in the form of syntheses, the results of national and international activities.

INTERNATIONAL BIOLOGICAL PROGRAMME 11

Population structure and human variation

EDITED BY

G. A. HARRISON

Professor of Biological Anthropology
University of Oxford

CAMBRIDGE UNIVERSITY PRESS

CAMBRIDGE
LONDON · NEW YORK · MELBOURNE

CAMBRIDGE UNIVERSITY PRESS
Cambridge, New York, Melbourne, Madrid, Cape Town, Singapore, São Paulo, Delhi

Cambridge University Press
The Edinburgh Building, Cambridge CB2 8RU, UK

Published in the United States of America by Cambridge University Press, New York

www.cambridge.org
Information on this title: www.cambridge.org/9780521112628

First published 1977
This digitally printed version 2009

A catalogue record for this publication is available from the British Library

Library of Congress Cataloguing in Publication data

Main entry under title:

Population structure and human variation.

(International biological programme; 11)

Includes index.

1. Physical anthropology – Addresses, essays, lectures.   2. Population – Addresses,
essays, lectures.   3. Human population genetics – Addresses, essays, lectures.
I. Harrison, Geoffrey Ainsworth.   II. Series.
GN62.8.P66   573   76–22987

ISBN 978-0-521-21399-8 hardback
ISBN 978-0-521-11262-8 paperback

Contents

Contents

Table des matières

Table des matières

Содержание

Содержание

x

Contenido

Contenido

Contributors

Cavalli-Sforza, L. L.   Department of Genetics, Stanford University, California 94305, USA

Edholm, O. G.   School of Environmental Studies, University College London, Gower Street, London WC1E 6BT, UK

Friedlaender, J.   Department of Anthropology, Harvard University, Cambridge, Massachusetts, USA

Harrison, G. A.   Department of Biological Anthropology, University of Oxford, 58 Banbury Road, Oxford OX1 3QX, UK

Hiernaux, J.   Equipe d'Ecologie Humaine, Laboratoire d'Anthropologie Biblogique, Université Paris VII, France

Hooper, A.   Department of Anthropology, University of Auckland, New Zealand

Hornabrook, R. W.   27 Orchard Street, Wadestown, Wellington, New Zealand

Huizinga, J.   Institute of Human Biology, State University of Utrecht, The Netherlands

Huntsman, J. W.   Department of Anthropology, University of Auckland, New Zealand

Layrisse, M.   Instituto Venezolano de Investigaciones Cientificas, Apartado 1827 Caracas, Venezuela

Moellering, R. C.   Department of Medicine, Harvard Medical School, Cambridge, Massachusetts, USA

Mourant, A. E.   Serological Population Genetics Laboratory, St Bartholomew's Hospital, London EC1A 7BE, UK

Neel, J. V.   Department of Human Genetics, University of Michigan Medical School, Ann Arbor, Michigan 48104, USA

Page, Lot B.   Department of Medicine, Harvard Medical School, Newton Wellesley Hospital, Cambridge, Massachusetts 02162, USA

Prior, I. A. M.   Epidemiology Unit, Wellington Hospital, Wellington 2, New Zealand

Rychkov, Yu. G.   Department of Anthropology, Moscow State University, Moscow, USSR

Contributors

Salmond, C. E.	Epidemiology Unit, Wellington Hospital, New Zealand
Salzano, F. M.	Departamento de Genetica, Instituto de Biociencias, Universidade Federal do Rio Grande do Sul, Porto Alegre, Brasil
Samueloff, S.	Department of Physiology, Hadassah Medical School, Hebrew University, Jerusalem, Israel
Sheremetyeva, V. A.	Department of Anthropology, Moscow State University, Moscow, USSR
Stanhope, J. M.	Epidemiology Unit, Wellington Hospital, New Zealand
Vorster, D. J. M.	National Institute for Personnel Research, Johannesburg, South Africa

Foreword

J. S. WEINER
Convenor HA Section, IBP

The investigations of human populations within the Human Adaptability Section of the IBP (1964–74) were planned on a world-wide scale. One objective was to obtain comparable data on population characteristics over a wide range of ecosystems. Another was to examine and compare different ethnic and economic groups within similar biomes. Yet another was to direct international biomedical teams to 'threatened' or disappearing groups – hunter-gatherers and simple agriculturists – to study their often highly distinctive ecological characteristics (and at the same time to provide biomedical help). Yet another objective of interest to many participating countries was the biological condition of populations undergoing migration or living under man-made urbanized conditions.

When in 1974 the IBP was brought to a close, the HA Section had largely fulfilled this global programme. Over 50 countries had mounted some 250 projects. Several thousands of scientific papers have been published along with some 30 monographs on particular national projects. Particulars of all the contributing countries, their projects, the team personnel and their publications and reports are to be found in *Human Adaptability: A History and Compendium of Research within the I.B.P.*, K. J. Collins & J. S. Weiner (Taylor & Francis, London).

Over and above the purely local interest of a national project – and this is always important – the data obtained from different HA investigations can be utilized in many different ways; the exploitation of the material will certainly go on for a long time. However, the Special Committee for the IBP decided that an immediate effort should be made to put together a significant proportion of the material in a series of readily accessible 'synthesis' volumes. Within this series of some 30 volumes, five have been planned to cover some of the major approaches mentioned above within the HA Section.

The Biology of Circumpolar People, ed. F. A. Milan
The Biology of High Altitude Peoples, ed. P. T. Baker
Worldwide Variation in Human Growth and Physique, by J. M. Tanner & P. B. Eveleth
Components of Human Physiological Function, ed. J. S. Weiner
Population Structure and Human Variation, ed. G. A. Harrison

The present volume is the second of these to appear.

The first and second volumes each comprise a comparative survey of the demographic, genetic and biomedical characteristic of communities living in circumpolar and high altitude biomes. Each is aimed at understanding how

XV

communities differing in their economic base, in population size, and in genetic make-up, come to terms with the particular stresses of life in these environments. In these two volumes, practically every aspect of population biology comes under scrutiny.

The third deals in a systematic and comprehensive manner with a fundamental characteristic of human beings the world over – namely, development and physique. In this volume, children's growth patterns in particular are examined on a world scale in relation to many factors, including climate, disease, nutrition and genetic constitution. The extensive tabulations make this an invaluable survey of base-line growth data for some hundreds of population samples.

Like the third, the fourth volume is concerned with major response systems of the body. Three physiological characteristics are examined for a wide range of communities living in a wide variety of conditions. These comprise physiological work capacity, respiratory capacity and tolerance to high temperatures. The significance and causes of variations in these physiological parameters are analysed in detail and a large amount of original data is brought together.

The last of the group, the present volume, edited by Professor G. A. Harrison, complements all the other four volumes in an interesting way. It comprises a series of case studies, each of which provides a vivid illustration of special aspects of population biological structure which enter into the comparative surveys in the other four volumes. Population structure may be taken as referring to the biological linkages and relationships – genetic, demographic, nutritional, etc. – between individuals and families within a society as well as to the aggregate of individual states of development, health, disease resistance, physiological functional capacity and genetic constitution. These attributes make up the variation and polymorphism characteristic of a population. In the adaptive sense, the survival and homeostasis of a community ultimately depend on two dimensions to which all the attributes contribute – physiological fitness in the short term and reproductive fitness over the long term.

These are the issues treated in the case studies. They focus on particular aspects of structure and functional constitution and on the forces or agencies which have affected the community in the past or are acting on it in the present.

In his introduction to this volume, Professor Harrison, provides a comprehensive guide to the particular topics of each case study and draws attention to their significance for understanding the major issues of evolutionary differentiation, genetic affinity, ecological adaptation, and the effects of migration, isolation and urbanization.

Finally, there is one feature of these contributions which adds enormously to their interest. In nearly every case the contributor provides a picture of the 'cases' as living communities, going about their everyday activities and in settings which range from the tundras of Siberia to the Pacific Islands, from the Amazon to the savannas of Africa, from small and isolated settlements to modern cities.

Preface

During the IBP many Human Adaptability (HA) projects from most partici-pating countries had a primary concern with the genetics, demography, or the health of human populations. It would be quite impossible to review with justice all these contributions in a book of reasonable length, especially as analysing components of human adaptability involves the micro-analysis of innumerable very individual situations. Populations differ in endless idio-syncratic ways, and the human characteristics which are relevant for examina-tion are almost infinite. It was, therefore, felt that the spirit and achievements of that part of the HA IBP dealing generally with population structure and function could best be dealt with through a number of case studies. The problem, of course, became 'which studies?' Any selection was bound to be invidious, particularly as in addition to scientific merit a number of other con-siderations had to be taken into account. Clearly it was necessary to choose projects which represented the whole variety of problems and approaches tackled by HA investigators interested in population biology. Then, as far as possible, an International representation was required, both in terms of the nationality of investigators and of the nations in which the work was done. Note, however, was taken of the fact that structural/functional studies were inevitably being encompassed in other IBP synthesis volumes, particularly *The Biology of High Altitude Peoples* and *The Biology of Circumpolar Peoples*. It was also felt desirable to concentrate on 'in depth' studies, especially those requiring a multi-disciplinary integrated approach. The IBP has provided a very important stimulus to such researches. These anyway were the factors considered in extending the invitations to the authors of this volume. Other equally good selections were possible, but it is felt that the present collection well exemplifies the IBP achievements in the field of human population biology. Contributors were asked to prepare their papers as broad reviews, aimed at the sort of level young research workers of the future would find helpful in understanding population biology, its objectives, methods and achievements, and its difficulties and failures, in the 1960s and early 1970s. For this policy the editor must assume total responsibility.

April 1976   G. AINSWORTH HARRISON

1. Introduction: structure and function in the biology of human populations

G. AINSWORTH HARRISON

Traditionally, physical anthropologists were concerned with describing and interpreting the patterns of geographical variety which are found to be so striking in the human species. The characteristics first examined were, naturally enough, the visually obvious ones, such as body and head shape and size, pigmentation of skin, hair and eyes and hair form; and for their detailed study a variety of increasingly refined anthropometric and anthroposcopic techniques were devised. Many of these characters show rather striking geographical patterns, with populations being differentiated on a broad continental basis. This permitted and indeed encouraged the drawing up of racial taxonomies, an activity which received a great deal of attention. The activity, however, was not merely one of classification, for on the Darwinian premise that similarity indicates phylogenetic affinity, it was taken that the taxonomies attempted to express the evolutionary history of the species. Ascertaining this history has always been and remains one of the main goals of physical anthropology.

It was well recognised that for this purpose the characters first available for examination had a number of disadvantages. Most of them, for instance, are influenced in their expression by the immediate environment in which the individual develops. Similarities and differences, therefore, are to some extent determined by non-genetic factors which are not relevant to establishing evolutionary affinities. At least as important is the fact that such genetic basis as there is to anthropometric variation, both within and between populations, is typically complex. Within any one population, and in groups that are hybrids between two or more, it generally displays, normal or Gaussian form, without the appearance of discrete classes. Such variation has so far remained intractable to precise genetic analysis and, in particular, although caused by particulate inheritance, has not in any case been reduced to the level of identification and localisation of responsible genes. This means not only that individuals of similar or identical phenotype may be genetically very different but also that, even if each phenotype is genetically unique, there may be no linear relationship on the particular scale of measurement between the magnitude of phenotypic difference and the magnitude of genotypic difference.

The situation changed radically with the discovery of an ever increasing

1

number of characters which show discrete variation within and between populations. The development of these characters is not affected by immediate environmental factors and, because the mode of inheritance has been precisely established, it is possible to identify – and to some extent locate – responsible genes. Many of these characters have been identified in blood, mainly because this tissue is easy to collect and handle, and the variants often reach polymorphic proportions within populations. They include the various blood group systems, haemoglobins, serum proteins and a rapidly increasing number of enzymes. For them, populations can be characterised for gene frequencies, and in population comparisons of these frequencies it is found, typically, that what varies is the level at which the polymorphism is set. On the whole the pattern of geographical distribution is less distinct than for many of the quantitative traits, and changes in gene frequency with distance tend to be clinal. Also, there is rarely close concordance in the distribution of the different characters. Having said that, it is nevertheless true that broad continental patterns emerge, and taxonomies of man, based on polymorphic systems, are broadly in harmony with those based on anthropometric traits.

Prior to the IBP, an extensive body of information had gradually accumulated concerning the world distribution of polymorphic systems. For some which had been known for a long time and were easy to test, e.g. the ABO blood group system, quite detailed information was already available, but for recently discovered systems there was a dearth of data. The IBP offered an excellent opportunity for extending knowledge and on the whole this was seized. Many investigators in the HA project throughout the world collected blood samples for genetic analysis and the results have been compiled by Dr A. E. Mourant and his colleagues who acted as the International Reference Centre for genetic surveys. Dr Mourant summarises this work in an extensive paper in this volume.

Although, through survey work, the broad patterns of genetic variation in man have become known, the causes for these patterns remain obscure and controversial. Essentially the issue of debate is whether genetic differentiation of populations is dominated by the stochastic processes which are bound to occur in small populations – and Mendelian populations have been small throughout most of human evolution – or whether it is dominated by natural selection, which can be seen primarily as a set of deterministic processes. Views about the relative importance of these two types of process have fluctuated over the past 50 years. Today the stochastic view tends to hold greatest sway, partly because of evidence produced by biochemists and biophysicists that many gene mutations do not affect, or affect only in a trivial way, protein structure; partly because of some refined mathematical theory, demonstrating that the genetic load in populations would be too great to permit population persistence if every polymorphic genetic locus affected fitness; and partly because there is a dearth of empirical evidence for the action of selection. There is of course a selectionist answer to each of these points. The biochemists have not yet taken into account the comparative efficiency of gene substitution on the economy of cellular biosynthesis, and in particular the relative availability of bases, and amino acids. Mathematical approaches, focussing directly on genes, overlook the fact that selection operates on the overall phenotype of individuals and that fitness is a relative thing – many phenotypes would survive if they were not in competition with other phenotypes. And the fact that, with a few notable exceptions, selection has not been demonstrated to act, certainly doesn't mean that it is not acting; it is well recognised that in natural situations, even strong selection is difficult to detect.

The position is, thus, in a state of considerable flux, but the point of importance here is that concern with the question of what are the causes of genetic variety has led biological anthropologists and human population geneticists to concentrate increasingly on unravelling the detailed genetic structure that exists within populations and population systems, since the diversity which exists between populations is the consequence of forces which operate within them. This has necessitated the development of refined and sophisticated techniques and procedures for field studies and the modern human biologist has to spend many months or years working with a comparatively small group of people, whereas his predecessor could quite quickly and easily sample the physical variety which existed in a population. The modern approach, and the types of problem, both logistic and academic, that have to be overcome are well exemplified in many of the contributions to this volume.

An important development in human population genetics, which has followed this increasing concern with within-population phenomena, is the incorporation of demographic studies with the genetic ones. It has long been recognised that the forces determining the genetic structure are, or are manifest in, demographic conditions. Population size and changes in it affect the magnitude of stochastic processes and the probabilities of inbreeding; migration and fission and fusion, influence genetically effective population sizes and are the vehicles of gene flow; and natural selection classically operates through genetically determined differences in fertility and mortality. Until recently, however, the concern with demography in genetics has largely been theoretical, but in the course of many HA IBP studies considerable attention has been given to elucidating the demographic structure of real populations. Not unexpectedly the real situations turn out to be vastly more complicated than any of the theoretical models but the use of these models has certainly added insight to our understanding of particular situations.

Another area which has attracted the concern of population geneticists is the social structure of populations. This not only influences demography but also genetic structure directly, through the prescribed mating systems. Although genes are often distributed as though mating were at random, such

3

mating is in fact rare, especially in any extended population; and the form and nature of deviations need to be known. Assortative marriage is particularly a function of social systems. These systems also frequently determine the actual definition of populations. What may, at first sight, appear to be a single Mendelian unit is often a set of multiple units separated by social breeding barriers. Then, understanding of the social anthropology of a population can be vital in recognising biological kin as compared with social kin.

These considerations of demography and the social sciences are well exemplified in the case studies reported in this book which focus primarily on genetic structure – in the Russian investigations of isolates in North Asia and the American studies of the Yanomama. But they are also evidenced, as indeed genetics itself is, in the other contributions. These can, rather broadly, be said to be concerned with function, and deal with such issues as adaptation, health and ecology.

The distinction of course between structural and functional studies is blurred. Typically, knowledge of a population's structure is critical to understanding questions of function. Indeed such knowledge can be vital in devising the most penetrating design for a functional study, as a number of the case histories show. Contrariwise, one can see the structure of any population as the historical product of the ways in which it has functioned in the past, and both types of study are ultimately concerned with explaining the nature and form of human variety. But once central attention moves from concern with genetic structure and population affinities to the ways individuals and populations adapt, or fail to adapt, to their environments, priorities for attention change. In particular, concern with the effects of variation is at least as important as concern with its causes, and interest is as much levelled at variation that may have no genetic basis (or at variation whose causation is not understood) as at variation which is clearly genetic. Interest in adaptive processes has provided a new impetus to classical anthropometric and other quantitative traits because, whatever their determination, many of them clearly affect or reflect the capacity to survive and reproduce, i.e. fitness.

There are many components to adaptation: individual adaptability in development, physiology and behaviour; genetic adaptations produced by classic Darwinian natural selection; and group adaptations, which may be biological, cultural or bio-cultural and which in man are especially important. Analysis of these components and the way they interact to determine fitness has been a central theme in many of the HA studies.

Although distinctions are again not clear-cut, a number of different approaches to investigations of adaptation are exemplified in the contributions to this book. One strategy is to single out for special attention some condition or variation in a trait which has a complex determination, and use contrasting ecological conditions to identify the relative importance of the various determinants. It can broadly be called the epidemiological approach and is well exemplified in the way the Solomon's study is presented here, with its special concern for the analysis of blood pressure determinants. Another approach to analysing components of individual adaptability and fitness is to focus on a marked environmental contrast and examine the various biological effects of this contrast. This is usually done by comparison of 'sedentes' in one environment with recent migrants to another. Such comparison affords a means of controlling for many genetic differences, though, as is well shown in the study of Tokelau migrants to New Zealand, migrants are often not fully representative of the populations from which they come. A most important problem which is only beginning to be considered by human biologists is what biological attributes predispose individuals to migrate.

While studies of recent migrants allow analysis of individual adaptability, investigations of peoples who are known to have moved in the more distant past from one environmental situation to another, provide an opportunity for determining the genetic adaptations to the new environment which have been produced by natural selection over the generations since colonisation. The approach is particularly rewarding when, as Professor Hiernaux does, one considers a number of populations in each of two major ecosystems such as the African savanna and the African rain forest. Systematic differences between the various groups in the contrasting environments are unlikely to be products of drift. Clearly, reasonably accurate dating of the times of separation are also useful.

Knowledge of the ontogenetic effects of an environmental change, obtained from the study of recent migrations, is obviously useful in interpreting the nature of the difference between long-separated peoples. In the simplest situation, subtraction of the components of individual adaptability and environmentally produced effects on fitness from a total difference should reveal the genetic components of that difference. This relationship is particularly important in considering classical anthropometric and anthroposcopic traits where, as already mentioned, there are usually both environmental and genetic components of the observed variation.

A third type of migrant study is exemplified by the investigations in Israel. Here one of the essential features of the design was to compare groups who had long been separated but who had recently returned to common environments in Israel. Thus there was the possibility of examining the effects of such genetic divergence as had occurred since the separation, in a situation which was at least partly controlled to avoid environmental effects. This type of design would seem to be particularly appropriate where investigations focus, as the Israel study did, on physiological characteristics. These characteristics are typically highly environmentally labile, and if there is any genetic basis to between-population differences it will surely be detectable only when environmental effects are standardised as much as possible. And in such standardisation one is as much concerned with an individual's previous environmental experience as with the environmental circumstances under which physiological tests are conducted.

Little is as yet known about the genetics of physiological variation within populations let alone between populations – a regrettable state which, unfortunately, the IBP did not do much to improve – but the increasing interest in issues of adaptation and ecology has led to the incorporation of ever more physiology in human biological field investigations (a development which owes a great deal to the IBP). Of the case studies reported here, a central concern with physiology is evidenced not only in the Israel study, but also in the New Guinea and central Africa investigations.

These last two studies, and the one of the Pygmies, especially exemplify a fifth type of approach to functional studies. In this some particular population or population group is selected, either because of its ethnic or environmental uniqueness, or, contrariwise, because it well represents general features of a particular economy, or stage of development, and all the biological elements which might conceivably bear on the overall determination and measurement of fitness are examined. The strategy is sometimes termed the 'total ecological approach', and the aim, apart from providing much-needed descriptive information, is to develop a holistic view both of the complex inter-relationships that exist between the various biological attributes and of the ways populations are biologically organised to exploit environmental opportunities and obviate environmental hazards.

Both of these major objectives present many technical and conceptual difficulties. Describing the total phenotypes of individuals necessarily requires multi-disciplinary operations at the moment. The techniques of physiology, nutrition, epidemiology and genetics are so specialised and different that it is impossible for one person to practise them all. And when experts in each discipline work together on the same population it is all too easy for general aims to become subordinated to disciplinary objectives. Truly interdisciplinary approaches are rare, though one can detect the beginnings of them in various of the case studies described here.

Then there is the complex problem of how the different types of data can be combined into overall estimates of the fitness of individuals. It is this fitness which is of primary ecological concern, and the genetic component of its variation which is of essential evolutionary concern. But integrating the functional effects of nutritional, physiological, anthropometric and polymorphic variety raises many difficulties, including ones of scale. One scale, which is comparatively easy to use is morbidity experience, i.e. a direct test of somatic fitness, but in traditional societies without written records it is almost impossible to do more than ascertain morbidity patterns at the time of the study itself. A better scale, certainly from the evolutionary point of view, is the demographic one of comparative fertility and mortality. By observation and questionnaire it is possible to gain information on the reproductive success not only of individuals themselves but also of their relatives, and clearly this can span some considerable time period. It is an approach which is mentioned in a number of studies and should surely be developed. Whether or not there are genetic components, it would be interesting to know the extent to which variation in physiological fitness affected reproductive fitness in a population. One would surely expect such an affect, especially in man, where it could be expressed indirectly through status and general socio-economic conditions. But studies of this kind require very patient and careful collection of the demographic data and are extremely time consuming in a traditional society where even a subject's age can usually only be estimated at all accurately after a great deal of questioning. The approach also requires the physiological testing of all categories of subjects: women as much as, if not more than, men; the aged as well as the young; and the chronically sick as well as the healthy. Often the physiological tests themselves preclude this, and there is an urgent need to devise simple tests of physiological fitness which can then be applied to large numbers of subjects and, without risk, to the sick. All too often, from an ecological viewpoint, physiologists in their disciplinary concern for elucidating physiological mechanisms, have confined their attention to healthy young adult males.

It is evident that demographic and ethnological data are as important in functional studies as in structural ones, and most of the case studies reported here recognise this. Even, however, when these data are not available, there is still great value in collecting biological data on the same individuals. This is particularly well shown in the New Guinea project, where not only have important inter-relationships between different physiological characteristics become evident from having such data but also a variety of associations between the genetic polymorphic variety and biochemical concentrations and disease states have been revealed. It is of course by such association that natural selection may be acting on gene marker systems, though the final proof of this requires the demonstration that the association affects demographic performance. Certainly it would seem from the New Guinea work that much further work is urgently required, examining the possible relationships between polymorphic variety and quantitative variables which affect fitness in societies still exposed to the sorts of natural environment which have prevailed throughout most of human evolution. It is here that ecological and genetic-structure studies tie up and where the controversy over neutral or adaptive genes may be resolved.

On the other major concern of the 'total ecological approach' – the way in which human populations as whole populations are biologically organised – a start is only beginning to be made. This problem, perhaps more than any other, demands a bio-social approach and is awaiting the development of a substantive and coherent body of ecological theory. It can be tackled through

7

G. A. Harrison

systems theory, applied, for example, to patterns of community energetics, but no examples of it are available for inclusion in this book. It is, however, evinced in some of the high-altitude IBP studies, which are reported on elsewhere and has become a central concern in much post-IBP work.

It has been customary for human biologists to confine their attention to aspects of physical variation. It is, however, in mental traits that man shows his most remarkable characteristics and behaviour is often the main component of adaptation. Likewise, mental health is a critical aspect of fitness. It is therefore gratifying that studies of behaviour have been included in some IBP studies, as is well evidenced in the final contribution in this book on South African populations. This study also represents, as does work in the Solomon Islands and New Zealand, another pioneering development in human biology field work – concern with the impacts of modern advanced technology and living conditions. Traditionally, human population biologists have worked with societies which have simple economies and are exposed to the rigours of natural environments, and, as has already been argued, there is a good case for this, since these are the environments under which most of human evolution occurred. But ever increasingly people are living in essentially man-made environments, particularly urban ones, and the effects of these conditions, which from the evolutionary point of view are very new, are in great need of investigation in both the developed and developing world. This need is as much practical as academic. It is a further tribute to the IBP that it played a role, both directly and indirectly, in initiating studies of urban biology, and it may not now be long before we know as much about the biological structure and function of urban populations as of more exotic ones.

2. The genetic markers of the blood

The initial concept of the HA section of IBP implied a study of the interaction, in human populations, between the genotype, in its broadest sense, and the environment. This, in turn, called for a study of gene distribution in human populations.

The most obvious human genetic systems having a possible effect on adaptation to the environment are those which directly cause diseases such as haemophilia and phenylketonuria, and those determining such visible characters as physique and skin colour which demonstrably interact with the environment.

Relatively rare diseases of genetic origin were regarded as lying outside the scope of IBP. The external characters of the body, on the other hand, are of the greatest concern to IBP and have been studied extensively as part of the Programme. However, despite the rapid and almost explosive advance of human genetics in recent years, the analysis of these characters in terms of genes has advanced very little. It has long been recognised that each character is the result of the combined action of numerous genes belonging to several separate allelic systems. There has, however, been very little advance in the identification of the separate genes and systems of genes involved. Part of the difficulty has lain in the complexity of the interactions between the genes themselves, and part in the unknown extent to which the environment interacts with the genotype to determine the final phenotype.

Thus the study of these characters, which has formed a large part of the HA investigations, has been carried out almost exclusively on a phenotypic basis.

Meanwhile very rapid advances have been taking place in the study of the so-called marker genes. With the possible exception of the vast number of genes each responsible for some very rare hereditary disease, by far the greatest number of known human genes are those which determine sets of biochemical characters most easily studied in the blood, though most of the gene products involved are present in other tissues as well.

The marker genes nearly all fall into four main classes, determining respectively the blood groups or red-cell agglutinogens, the plasma proteins, the red cell enzymes and haemoglobins, and the histocompatibility antigens.

The blood groups have been known since 1900; their study has been fostered by their vital importance in transfusion therapy and the prophylaxis and treatment of haemolytic disease of the newborn. A dozen or more major genetic systems of blood groups are known, determining well-defined poly-

9

morphisms characterised by two or more genes each having frequencies of several per cent, frequencies which in nearly all cases have been shown to vary significantly between different ethnic groups. There are, in addition, a very large number of systems within each of which nearly all individuals are of a single homozygous type, but in which there exist rare persons of a different hereditary type. Techniques of study are now highly standardized and reagents widely available, so that the blood groups remain even now the most widely used of marker systems in family studies, forensic paternity testing, and population investigations. Unfortunately, apart from their negative function of causing haemolytic disease of the newborn, very little is known of the normal role in the body of the blood group antigens and the nature of any possible interaction with the environment, and it has been one of the objects of HA work to search for such interactions.

Large numbers of the proteins of the plasma, hitherto characterised by their broad physical and chemical properties, are now recognised by more sophisticated tests as the products of a number of distinct systems of allelic genes. The functions of most of these proteins in the body's economy are known and in some cases differential physiological effects of different alleles have been ascertained. There are about a dozen useful plasma protein polymorphisms; by far the most valuable of these is that of the Gm groups, detected by antigenic differences arising from variations of the amino-acid sequence in immunoglobulin molecules.

Gm investigations require a rather laborious type of immunological test; tests on populations have been fairly numerous but far less in number than for some of the blood group systems. It is clear, however, that if Gm tests were more widely used they would be of greater value in discriminating between populations, and in showing relationships between them, than any one of the classical blood group systems.

The principal protein of the red cells is haemoglobin, but they also contain a very large number of distinct enzymes. Haemoglobin polymorphisms are of great importance in certain regions, especially Africa. More and more of the enzymes are being shown to be polymorphic; polymorphisms are detected mainly by means of different migration speeds during gel electrophoresis. For many of the enzyme systems there are substantial allele frequency differences between ethnic groups but, chiefly because characteristic world patterns are only beginning to be recognisable, the enzymes have not yet contributed greatly to the taxonomic side of population studies. However, since the functions of the enzymes are well understood and are vitally important in the functioning of the body, and since in many cases the different gene products differ measurably in their chemical activity, there is every prospect of these polymorphisms proving important in giving rise to differential responses to the environment, as has already been proved for glucose-6-phosphate dehydrogenase.

10

The history of the histocompatibility antigens is instructive. They had for many years been studied by means of crude agglutination tests on white cells and platelets, and it was shown that they were genetically determined. Gradually it was established that they are present in most tissues, and that they are largely responsible for the compatibility or otherwise of tissue and organ grafts. By about 1970 their complex genetics were largely understood, but it was not until 1972 (see Dausset & Colombani, 1973) that their great potential value in population studies was realised. If availability of reagents and the considerable skill needed to use them were sufficiently widespread, they would rival the Gm system in taxonomic discriminatory power. But it was only even more recently, in 1973, that it was established that, of all the genetic markers of the blood, they are the ones which show the most marked statistical associations with particular diseases. These are mainly diseases which had previously been suspected of having an auto-immune aetiology. It is possible that closely linked genes for the histocompatibility antigens and for susceptibility are involved, or that a particular antigen inhibits the production of an antibody to a virus of very slow incubation possessing similar antigenic features. Thus in any sequel to the HA work, such as MAB (Man and Biosphere), an attempt should be made to include histocompatibility antigens in all blood studies where an interaction with the environment is in question.

A number of other inherited biochemical carriers have from time to time been included in population surveys. These include the salivary secretion of the ABH red-cell antigens, the tasting of phenylthiocarbamide, and the rate of acetylation of isoniazid and other drugs.

The planning of population surveys

During the planning stage of IBP, it was realised that valuable contributions to a knowledge of blood group distributions could be made at widely differing levels of collaboration and sophistication.

In some countries, where even the basic distribution of the ABO and Rh blood groups had not yet been firmly established, even solitary workers could produce important new data. Elsewhere, varying degrees of elaboration would be possible, culminating in multi-disciplinary surveys carried out by large teams, often international, with access to a wide range of specialist laboratories.

It was therefore decided that virtually any survey of the incidence of the hereditary blood factors would be acceptable as a contribution to HA, though the adaptability aspect was to be stressed wherever possible.

As a result many gaps in the world distribution map of particular blood factors have been filled with new data. In other cases the stimulus of IBP has led to the publication, in an accessible collected form, of the results of work previously carried out but either unpublished or published in journals of limited circulation. In the still more elaborate surveys much light has been thrown on the genetic structure of particular populations or groups of populations, as well as on more remote ancestral relationships between populations.

Moreover, however thoroughly the observations have been analysed by the original authors, there is no doubt that in a world context much more information is capable of being extracted from most of them, and it is hoped that all authors will agree to deposit the records of their work in a form which will permit this.

Because of the very varied kinds of survey carried out in different countries or regions, or even in different projects within a single country, it has been difficult to summarise results on any uniform plan. In some regions little more can be attempted than catalogues of results, yet the latter, as stated above, are likely to form an indispensable part of comparative regional studies. In other regions the IBP results are so comprehensive both in populations and areas studied, and in the range of investigations carried out, as to permit a more unified account to be written in this chapter. In several such cases a much longer and more detailed account will appear in one of the other final IBP volumes.

Some hundreds of papers, mostly with several authors, have been written on work done under IBP auspices on the frequencies of the marker genes, and their implications. In a general summary it is impossible to mention every author, or even every paper, but an attempt has been made to include in the Bibliography every relevant paper that has been submitted to the IBP Central Office, or has been traced by other means. It is possible, however, that the results of some completed investigations have failed to reach me, and I apologise for any omissions of such work. On the other hand, some data have been included which are the results of recent work not sponsored by IBP but are essential to an overall view of the peoples of a particular region. This applies especially, but not solely, to the Scandinavian region.

Genetic distances between populations

One of the principal aims of the genetic study of populations is the taxonomic one of ascertaining how closely they are related. The other, implied by the term 'human adaptability', is to detect selective effects of the environment on gene frequencies. Both of these aims are served by population distance statistics which, by taking into account all the available evidence, generate a series of quantities which express in a relative sense the distance between every possible pair among the set of populations under consideration.

If it is assumed that gene frequencies have varied only in a random manner ('genetic drift') then, subject only to random errors of a known order of magnitude, distance statistics express true taxonomic distances. It is not proposed to discuss here the relative value of different ways of expressing population distances. An admirable critical summary is given by Hiernaux (1964). However, one additional method, that of Cavalli-Sforza & Edwards (1967), calls for special mention, since it assumes that distances are directly related to the length of time since particular populations have separated, and the results are expressed as a branching tree. If indeed only random processes are involved, and provided that a sufficient number of genetic systems have been taken into consideration, the result should give a close approximation to the historical sequence of events. The results will, however, be distorted if there has been remixing of once-separated populations, or if different populations have, because of varied environments, been subject to different intensities of natural selection. A further complication, discussed by Fitch & Neel (1969) in relation to their work on South American Indians, is the effect of mating patterns on apparent gene frequencies based on limited samples.

Much work has been on the mathematics of mixing, but almost solely in cases, such as that of American Negroids and Caucasoids, where fairly precise knowledge of the original gene frequencies of the participating populations can be assumed.

The detection of natural selection demands, in each separate case, the formulation of a hypothesis, usually based on a comparison of a small number of populations. If gene frequencies in these agree well with the hypothesis, the latter is then tested on other pairs or sets of populations. The best established example of this kind is the relation between the possession of the haemoglobin S gene and reduced susceptibility to falciparum malaria (Allison, 1954).

Europe

Genetic surveys of European populations were continually being carried out at the time of the IBP, but relatively few of them were officially incorporated in the Programme. In some countries it is clear that the advent of IBP gave a strong stimulus to the initiation of population surveys and so to making the genetic map more complete than heretofore, but only in the case of the Circumpolar project was there a co-ordinated underlying international scheme. The results of this project are described in a separate section. For the rest of Europe little more can be done here than to list the various surveys and draw attention to any outstanding results. For several countries surveys are known to have been in progress but results have not been received in time for mention here. The total effect of all the work is nevertheless considerable but it will be seen clearly only when an updating can be made of the work of Mourant, Kopeć & Domaniewska-Sobczak (1976), which work does not include systematically results published since 1969.

In West Germany (BRD), Professor E. F. Vogel and his colleagues

13

(BRD/HA/2) have continued their work locally as well as internationally on associations between blood polymorphisms and diseases, especially infectious diseases. This very important but broad and contentious subject cannot adequately be discussed here.

In Czechoslovakia (Czechoslovakia/HA/1) extensive surveys have been done but results are not yet available.

Results of surveys in East Germany (DDR) on frequencies of marker genes in healthy and diseased persons have been summarised by Grimm (1973).

In Finland (HA/1 and HA/2), efforts have been directed mainly towards the Lapps, as part of the Circumpolar project, but Eriksson and others have also continued their very detailed work on gene distribution and population structure in the Åland Islands.

In Greece, Valaoras (Greece/HA/6) has studied ABO blood group distribution, especially in Army recruits from different parts of the country, while Motulsky (USA/HA/15) in collaboration with local workers has studied the distribution of glucose-6-phosphate dehydrogenase deficiency.

A very extensive survey of the population of Ireland (UK/HA/19) has included a wide range of genetic markers. Preliminary results of work on blood groups and other markers (Mourant *et al.*, 1976; Tills, 1976; Tills, Teesdale & Mourant, 1976) have been published and an extremely full analysis has been made by Tills (1975).

In Italy (HA/5) work has been concentrated on a study of population heterogeneity in Sardinia. The isolation of the villages gives rise to a general heterogeneity but upon this is superimposed a systematic difference between highland and lowland populations, attributed to the effects of past malarial endemicity in the lowlands.

In Poland, IBP genetic studies have concentrated on the hereditary aspects of growth and development, and of blood pressure, rather than on discrete genetic markers.

In Romania (HA/1, 2, 3), where much work has been done on blood group distribution, this has largely been treated retrospectively, but only recent references are included in the present Bibliography.

In Spain (HA/3) work has been done on a number of provincial populations, including the inhabitants of Gran Canaria whose essentially west European genetic constitution has been confirmed.

Asia

Studies carried out under IBP auspices have added greatly to our knowledge of the distribution of the hereditary blood factors throughout Asia, but apart from the circumpolar region, considered on pp. 26–29, and the comprehensive study of Jews in Israel, there are no over-riding themes, and results must in most cases be summarised separately country by country.

Kurdish and Yemenite Jews, Kurds and Arabs

A joint multi-disciplinary survey (Israel, HA/5, HA/7; UK, HA/8) of Jews in Israel, under the direction of Professor S. Samueloff and Dr O. G. Edholm, included studies of inherited blood factors. For comparative purposes Professor H. Lehmann undertook a study of the Kurds of Iran among whom many of the Kurdish Jews formerly lived. The ideal comparative population for the Yemenite Jews would have been the Yemenite Arabs. The latter, however, have been tested for only a small range of factors whereas the southern Arabs (Marengo-Rowe *et al.*, 1974) have been examined for a much wider range and hence can be compared more fully.

Without elaborate calculations it is clear that the Yemenite and Kurdish Jewish populations differ considerably from one another in their gene frequencies, while there are considerable resemblances between the Kurdish Jews and the Iranian Kurds on the one hand, and between the Yemenite Jews and the southern Arabs on the other. The major unexplained difference between the Kurdish Jews and the indigenous Kurds is the extremely high frequency of glucose-6-phosphate dehydrogenase deficiency in the former. It is most improbable that this somewhat disadvantageous character would have persisted unless there had been some selective environmental influence favouring it; since the Kurds and the Jews have lived for thousands of years in a common physical environment the influence is likely to have been a cultural, perhaps a hygienic, one.

India

The only project officially sponsored by India which has yielded frequency data of blood-group or other marker genes is HA/4, 'Genetic survey of selected castes and tribes', co-ordinated by Dr L. D. Sanghvi. In addition, a large amount of work in this field, not sponsored by IBP, has been and is going on in India, both on the tribal and the settled populations (see Sanghvi *et al.*, 1974). The most urgent anthropological work in India, as elsewhere, is the recording of the physical and cultural anthropology of the tribal peoples before they disappear as separate entities, and the HA teams have concentrated on tribal peoples in Tamilnadu, Gujarat and Maharashtra. In the case of Tamilnadu each population was sampled in about five localities and the results for each locality are tabulated separately. A full analysis by localities would probably reveal significant differences between the local sections of any one tribe and it is hoped that analyses at this level will be carried out.

Perhaps the most surprising feature of the results is the general genetic resemblance between the tribes in any one region, though with some marked exceptions. In Tamilnadu the Ayer have, in comparison with surrounding populations, a particularly high total frequency of *A*, though this does not

15

approach that previously found in the Paniyans. The Kallar have a remarkably high frequency of *B* for southern India. The low frequency or absence of haemoglobin S in most of Tamilnadu is in contrast to the high frequency in the Irula; it has long been known that the latter and several neighbouring tribes show this characteristic. Frequencies of the blood group genes of other systems show no striking departures from the previously known regional pattern.

In Maharashtra rather high frequencies of *B* and *cde* in the Son Koli suggest northern affinities. In Gujarat, the Waghri similarly show high frequencies of *B* and of *cde*. Like the Son Koli they also have a rather low ABH secretor frequency. The almost constant and rather high frequency of haemoglobin S in the tribes of Gujarat is to be noted. It is presumably related to the prevalence of falciparum malaria in the past, if not at present.

In both Gujarat and Maharashtra the frequency of the *taster* gene is rather low by world standards, as has previously been reported for these and some other parts of India.

Polymorphisms and diseases in India and Thailand

Vogel, Chakravartti and others (BRD/HA/2) have carried out extensive surveys of the relative incidence of the ABO blood groups in cases of smallpox in India and of leprosy in India (West Bengal and Bihar) and Thailand.

The smallpox results taken by themselves show significantly greater severity and mortality in groups A and AB than in groups O and B. Other authors, both in India and elsewhere, have failed to find such an association; it is impossible to discuss the matter fully here, but Vogel and co-authors claim that the very significant associations which they find are because they, but not other workers, have tested unvaccinated populations. Their case is a strong one and supports their further claim that smallpox mortality has been a major selective factor in determining the present distribution of the ABO blood groups. The problem is one of great importance and demands further work in those few parts of the world where smallpox is still endemic.

In the case of leprosy, previous work in various regions showed a slight association of the disease with blood group O when previous observations were combined. The present results in India and Thailand tend to confirm this association, but it is not a strong one. A non-significant association of the non-lepromatous type of leprosy with group A is also supported by data from other continents. Despite the inconclusiveness of these results the problem of association is an important one because of the very wide present distribution of leprosy, and the even wider past distribution, which may have an important bearing on the history of ABO blood group distribution; further data are therefore much to be desired. No significant associations with the occurrence or type of leprosy were found for haptoglobin, caeruloplasmin, transferrin,

β_2-glycoprotein I, or protease inhibitor types, but a highly significant association of the Gc 1-1 phenotype was found with leprosy in general, and this is supported by data from elsewhere. This clearly demands further research, not only from the point of view of leprosy epidemiology but because it relates to the possible function of the Gc proteins.

Burma

The recent work of Mya Tu and his colleagues has added enormously to our previously scanty knowledge of the distribution of blood groups in Burma. There is now seen to be an extremely wide range of frequencies of the genes of the ABO system. The A_2 gene, previously thought to be very rare or absent in Burma, is in fact present in several populations, with frequencies of the order of 1 %, and with the very high frequencies of 17 % in the Nagas near the Indian border and 7 % in the neighbouring Shans at Khamti. Total frequencies of the A gene average about 20 %, as do those of the B gene, but whereas the Nagas, already mentioned, have 42 % of A and only 7 % of B, a small combined sample of the Tarons and Htalus, closely related and somewhat mixed with one another, living near the northern tip of the country, have only 1 % of A genes, and the world record frequency of 52 % of B, in confirmation of similar figures, previously unsupported, given by von Eickstedt (1944) for Upper Burma. Similar but less extreme frequencies are found in the Semai Senoi of Malaya and the Veddas of Ceylon.

The M gene frequency averages about 70 %, considerably higher than in India, but the S gene (mostly in the combination MS) falls markedly below the Indian level to an average of 10 %.

Rh negatives of genotype cde/cde are recorded only for the Burmese in the restricted sense, for the Tavoyans and the Mons. The CDe allele has in general a high frequency, as in most of Asia, while cDE, usually in Asia a marker of northern affinities, reaches 24 % in the Nagas, but nearly 28 % in the central Pa-Os. Other work on the Burmese *sensu stricto* by Ikin *et al.* (1969), not planned as part of IBP, shows them to possess the Mongoloid Di^a gene, and to have 2 % of Lu^a and 4 % of K genes, perhaps indicative of European admixture. They have also a frequency of 7 % of the HbE gene.

Malaysia

West Malaysia, formerly known as Malaya, is the continental portion of Malaysia, and is inhabited by a considerable variety of populations. The undeveloped centre of the country is occupied by three aboriginal groups, in the north the Senoi (Semai Senoi and Temiar Senoi), in the east the Negritos, and in the south the Aboriginal Malays. In the peripheral parts of the country the autochthonous population consists of the Malays, but there are also very

17

numerous Chinese and Indians who have mostly remained culturally and genetically distinct. Blood samples have been tested, as part of the Programme, from most of these populations, adding substantially to what was previously known.

Three IBP projects have yielded genetic data. HA/1, for which Mr A. G. Fix is responsible, is a detailed study of population structure and genetic distances among the geographically isolated units of the Semai Senoi. As, however, most of the laboratory tests were done by Dr Lie-Injo Luan Eng, who is also responsible for HA/4, the two projects will be considered together. Dr Lie-Injo has been responsible for a very large body of important investigations on the blood of the peoples of Malaya and surrounding countries, beginning many years before IBP and including intensive studies of many rare and important genetic variants of haemoglobin and numerous other blood proteins. Here we shall consider mainly those results referring to whole populations, and published from 1965 onwards.

Project HA/5 under Dr S. S. Dhaliwal comprises studies of Malays, Indians and Chinese. Tests were mostly for the serum groups, red-cell enzymes and haemoglobin variants.

Owing to the high frequency of hereditary ovalocytosis in some populations, its presence or absence may be regarded as constituting a useful genetic polymorphism in this region. It was unfortunately not possible to ascertain whether or not the type found is that which is linked with the Rh blood groups.

The new data on the settled populations of Malays, Chinese and Indians fit well into the known regional pattern, but in the absence of any extensive data from China itself it is useful to have these data for the Chinese to supplement the few already available.

The data on the aboriginal peoples, on the other hand, break much new ground. Those on their serum groups, red-cell enzymes and other factors should be studied in conjunction with the results of the blood group studies carried out by Polunin & Sneath (1953). It is unfortunate that only the Semai Senoi were studied in large numbers in the field but, despite this, certain patterns are evident. The Semai Senoi are once again seen to have a high frequency of *B* and a very low one of *A*, like the Tarons and Htalus of Burma, and the Veddas, already mentioned. They stand out in certain other respects. In the Gm system they have virtually only one allele, *Gm* [1,3,5,13,14]; they have the high frequency of 11% of the gene for ovalocytosis, which may be the highest in the world, and they have a very high frequency of haemoglobin E. Neither haemoglobin E nor ovalocytosis, alone or in combination, seems to cause any disability. They have also an unusually high frequency of $PGM_1{}^2$. It will be of interest to look for similar features in other populations in Asia with high *B* and low *A* frequencies.

Japan

From the earliest days of anthropological blood group research the Japanese population has been closely studied and there are few countries for which we have a fuller knowledge of the ABO blood groups. The IBP was made the occasion for a survey of the distribution of more recently discovered factors, especially of the serum groups and red cell isoenzymes, mostly under project HA/3 for which Dr E. Matsunaga has been primarily responsible. A great many workers have, however, taken part and the names of those who have published papers will be found in the Bibliography. A few blood group data published by Yanase *et al.* (1973), from a number of isolates, come under the heading of HA/1. The papers mentioned above add a considerable amount of important local detail to our knowledge of the distribution of the blood groups *sensu stricto* but they do not substantially alter the picture already established. They do, however, add greatly to available information on the distribution of the other hereditary blood factors.

The Ainu, already known to differ physically, and in their blood groups, from the Japanese, are found to differ in other respects, notably with regard to their haptoglobins and their Gm groups.

It has been suggested that the Okinawa (Ryukyu) islanders are related to the Ainu, but they differ relatively little from the Japanese, and fit much better than the Ainu into the general far eastern picture.

The IBP has also been the occasion for some detailed surveys of the population structure of a number of isolates, and for studies of the inheritance of a number of blood factors in Japanese families.

Korea

A considerable amount of new information has recently been published on the genetics of the Koreans. Only one investigation under IBP auspices has, however, been notified, that by Kang *et al.* on glucose-6-phosphate dehydrogenase (G6PDH) and the acetylator system.

The gene frequency of G6PD deficiency is 2.6% which though low is considerably above the 0.2% found in Japan. The frequency of the rapid acetylator gene Ac^R is 66%, which is high by world standards but in close accordance with previous observations on Koreans in Japan, as well as being near those found in the Japanese, including Ryukyuans and Ainu.

Africa

It is now generally accepted that it was in Africa that man evolved from his pre-human ancestors, and it is probable that here too a great many of the important steps in the subsequent evolution of man himself took place. This

19

gives to the genetics of African man a unique importance. Much work has been done on the distribution of the inherited blood factors in African populations, but because of the political fragmentation of the continent, and the objective difficulties of travel and of transport of biological specimens, little of the work has been planned on a continent-wide basis, and large areas and large and important populations have remained almost totally unexamined. Only a few African countries took part officially in HA projects, though many more gave invaluable help to European and American HA teams.

Much genetic survey work has been offered in retrospect as a contribution to HA. Most of this, and a substantial part of that carried out during the course of IBP, has consisted mainly or entirely of blood tests on individual populations, though several important multi-disciplinary studies have also been carried out. Only in a few cases, however, has it been possible to relate the results of the genetic work closely to those of the other studies. Thus a vital step in HA work, regarded on a continental basis, has consisted in the preparation and publication of extensive tables of genetic results. (Hiernaux, 1964; Tobias, 1966, 1974; Mourant *et al.*, 1976).

For many reasons much work remained unpublished at the time of writing, but much of it appeared likely to be published in 1975. Only as this chapter is being written has a further step been reached – that of submitting both published and unpublished results to computer-based genetic distance and related statistics such as the evolutionary tree construction of Cavalli-Sforza & Edwards (1967).

Within subsaharan Africa attention has been paid especially to an attempt to reconstruct the differentiation and subsequent evolution of the Khoisan (click-speaking), Pygmy, and Negro peoples.

Though not listed as an HA project, the work of T. Jenkins and of J. S. Weiner and their colleagues on the Khoisan peoples is of outstanding importance for an understanding of the whole genetical anthropology of Africa.

These peoples and especially the San (Bushmen) represent an extreme variant of the general Negroid type, exposed at present to a harsh environment and showing specialised anatomical and physiological characters which must to some degree represent an adaptation to that environment. Yet some of the general African marker genes are present in particularly high frequencies in the Khoisan: A_{bantu}, *Henshaw*, *cDe*, *V*. In particular, they possess over 90 % of the *cDe* (or R_0) complex. Similar frequencies are found in the click-speaking Hadza of Tanzania, in some Pygmies of Central Africa and, rather surprisingly, in the Northern Nilotes of the Sudan. These facts, together with some archaeological evidence, suggest that the Khoisan peoples represent one, and the Negro peoples another, differentiate from a more generalised previous African type.

Two Tanzanian tribes with clicks in their languages, the Hadza and the Sandawe, have been subjected to multi-disciplinary studies. Genetically the

Sandawe differ little from neighbouring Bantu-speaking peoples, whereas the Hadza are of a more extreme type, especially in their high *cDe* frequency.

The spread of the Bantu-speaking peoples, iron workers and cultivators, from a focus somewhere in central Africa to most parts of central and southern Africa, has taken place mainly since the beginning of the Christian era. According to Brain (1953) and to Singer (1962) they came as a series of waves – one wave which reached South Africa brought neither the haemoglobin *S* gene nor short-horned zebu cattle, whereas a subsequent one, bringing both, stopped at the Zambezi River but, like a tidal wave, overflowed, perhaps as a concourse of slaves, into Madagascar.

The early ancestors of the present peoples of north Africa appear to have been Caucasoids from the Near East who entered in Mesolithic times. Culturally related peoples from the same source also entered East Africa, but whereas in the north the Sahara Desert kept the two races largely separate, in the east there appears to have been more mixing and to this day there exists in East Africa a mosaic of peoples and languages which has by no means been resolved.

The team of Ruffié (France HA/8.1) has carried out multi-disciplinary studies on the peoples of the Sahara, both the dominant groups who, physically and genetically, are essentially Caucasoid, and the somewhat mixed Negroid Haratines.

Most of the tribes of Ethiopia show, from the frequencies of their blood factors, that they were originally a mixture, in about equal proportions, of Caucasoid and Negroid, but they are probably a very ancient mixture. One of the areas inhabited by the Amhara is in the Simien Mountains where there are villages at widely differing altitudes. There they have been the subjects of an intensive series of multi-disciplinary studies. Anatomically and physiologically there are marked differences between those living at 1500 and at 3000 m. Blood-factor gene frequencies do not differ significantly except in the case of the alleles of 6-phosphogluconate dehydrogenase where the frequency of *PGD^c* is 5.3% at 1500 m, and 13.5% at 3000 m (Harrison *et al.*, 1969). There is, however, no indication of any systematic relation to altitude in the frequencies of this gene outside Ethiopia; the difference is probably to be regarded as an effect of sampling or genetic drift, but attention should be paid in future studies to possible environmental effects of this system.

Northern and Central America

Most of the HA genetic population surveys by workers from the United States and Canada have been carried out as part of the Circumpolar project (USA/HA/9 and 13, Canada/HA/1) or in South America, and are described in separate sections. The studies of the French team (France/HA/8.3), centred

21

on South American Indian populations, were extended to the island of Dominica, as well as to Yucatan, Mexico. No HA studies appear to have been sponsored officially by Mexico.

South American Indians

South America has, genetically speaking, been the most intensively cultivated of all fields by IBP workers. Central to the whole of the work, both geographically and scientifically, has been that of J. V. Neel and his numerous colleagues in Brazil, but they have also collaborated closely with F. M. Salzano and his team in Brazil, and with M. Layrisse and his group in and on the borders of Venezuela. J. Ruffié and his group have worked on the Altiplano in Bolivia, and in French Guiana. Other IBP genetic workers have been the groups of R. Cruz-Coke in Chile and of G. Morpurgo in Peru (Italy HA/4). Numerous other investigators not affiliated to IBP have also been at work in this subcontinent, including G. A. Matson, who has carried out very full investigations of numerous Amerindian populations in several different countries. Much work, not summarised here, has been done in South America on the genetics of various diseases.

The planning of the work of the American and Brazilian teams was largely inspired by the theories and computer-based phylogenetic techniques of Cavalli-Sforza & Edwards (1967) and of Fitch & Margoliash (1967), which assume a series of dichotomies of a single original population, followed by separate evolution of each division. The system was applied to previously available data on the Amerindian tribes of South and Central America (Fitch & Neel, 1969) and the work was so planned as to try to place within this scheme the tribes to be studied under the IBP, but, more important, to study the evolution of large but limited tribal groupings with known histories of dichotomy, and to study within these, in the greatest detail, with full genealogies, the actual process of evolution as shown by the widest possible range of marker genes.

The initial major attack was concentrated on the Xavante Indians of the Mato Grosso of Brazil, a large and scattered tribe hitherto very little exposed to 'acculturation' by peoples of European origin.

If any one part of the investigation was central to the whole, it was the very full testing for the blood markers and other inherited biochemical characters, but there was a most carefully integrated combination of studies of social anthropology (especially mating customs), anthropometry and observations of non-metrical somatic characters, genealogy, and medical and pathological examination involving both congenital and acquired diseases. Sera were examined for antibodies to a wide range of micro-organisms.

As stated by the investigators (Neel & Salzano, 1967), these investigations represent an 'effort to understand and quantify the breeding structure and the important biological determinants in the survival and reproduction of certain primitive peoples . . . there is the further possibility that the investigation of these more simply structured communities may point the way towards factors of importance in more complex communities'. It is impossible to set out here the very extensive conclusions derived from this major series of investigations; but it is important that its lessons of multi-disciplinary integration should be borne in mind by future planners.

This series of investigations was followed by a somewhat similar series on the Yanomama Indians of southeastern Venezuela and northern Brazil by the American and Brazilian teams, in collaboration with that of Layrisse of Venezuela who had already made a major survey of the Amerindian tribes of that country. The Yanomama present a large number of topographical and social units, those in the central area having had little contact with other peoples, but those on the periphery having met a variety of other Amerindian tribes.

In all, over 30 polymorphisms were studied in 37 villages. Though this was, like that of the Xavante, a multi-disciplinary study, the stress in this case was relatively more on genetic structure and demography and especially on comparison between the numerous villages. At an early stage the investigators (Arends *et al.*, 1967) claimed that 'the Yanomama are the best approximation to what one might expect in an expanding Indian tribe in pre-Columbian times thus far studied in depth', and very wide ranges of gene frequencies of the various markers were indeed found in the different villages. Among these several new alleles were discovered.

The American and Venezuelan teams also investigated the Makiritare Indians of Venezuela, and the Brazilian and American teams, in collaboration with R. Morena of Paraguay and M. Palatnik of Argentina, examined the Maca Indians of Paraguay, while the Brazilian team carried out work on the Brazilian Cayapo Indians.

The French team directed by J. Ruffié has carried out extensive surveys of Amerindian populations in Bolivia and Peru, especially in the inter-Andine Corridor. These have covered frequency studies of a wide range of hereditary blood factors, together with physiological investigations directed to the understanding of the adaptation of populations to this high-altitude region. Special attention has been paid to the Quechua and Aymara tribes which appear to have lived in the region for a very long time.

The Quechua and other tribes of Peru have been studied also by an Italian team (HA/4), and the Aymara by an American one (HA/26). The abovementioned French team has studied also the hereditary blood factors of several Amerindian tribes in French Guyana. The results of multi-disciplinary investigations by this team in Central and South America have been issued as a large volume of collected reprints.

As already indicated, large contributions to our knowledge of gene distribution in the Indians of South America have been made by other teams working outside the IBP, notably that directed by G. A. Matson which has carried out surveys for a wide range of blood factors in numerous tribes living in a large proportion of the States of South America. All available data, whether or not sponsored by IBP, up to 1969, have been tabulated and, for many systems, mapped, by Mourant *et al.* (1976). It is now possible to see consistent continent-wide trends and apparent clines for a considerable number of genes. This is notably the case for the *M* gene which reaches frequencies around 90% in eastern Ecuador and north-eastern Peru, and over 80% over a large area lying mainly to the east of the northern third of the Andes. From here, frequencies diminish southward and eastward to levels of 54 to 55% near the southern half of the east coast. On the west coast, south of Ecuador, frequencies are nearly everywhere between 60 and 70%.

Total frequencies of the *E* gene of the Rh system (i.e. *CDE+cDE*) are notably high among South American Indians in general, averaging over 40% They tend to be relatively high in the south and west and lower in the north and east.

In their separate areas the original authors, and especially the USA team, have carried out sophisticated attempts to correlate the frequencies of the marker genes with ethnography and linguistics, but the opportunity now arises, by considering the data for the whole of South America, to do this on a much wider scale than has yet been attempted. In view also of the great variety of environments presented by the subcontinent, correlation of gene frequencies with environmental factors may now also be possible.

While most IBP investigations have been concentrated on the Amerindian population, Caucasoid, Negroid and mixed populations have also been subjects of study, especially in Brazil by the team of N. Freire-Maia (HA/7, 8, 16), which has concentrated largely on population structure and the effects of inbreeding. Both this group and that from the USA have investigated possible associations between the phenylthiocarbamide taster types and the incidence of diseases. A Netherlands team (HA/3) has studied blood factor distribution in the Negroid population of Surinam, with particular reference to the typical African marker genes.

Australia

The main contribution of Australian workers to HA has been through work in Papua and New Guinea. In Australia itself, studies of the frequencies of hereditary blood factors in aboriginal populations have continued during the period of IBP, but only a few of these, in the Northern Territory, have been associated officially with the Programme and are listed in the Bibliography. It is difficult to discuss these results on their own, but they have been considered elsewhere in relation to the whole Australian blood group picture (Mourant *et al.*, 1976).

Melanesia

New Guinea

The island of New Guinea comprises Papua in the southeast, the mainland part of the Territory of Papua and New Guinea in the northeast, and Irian Barat – formerly Netherlands New Guinea – in the west.

As part of a comprehensive inter-disciplinary study, a very extensive genetic survey of the peoples of the Territory of Papua and New Guinea has been undertaken by Dr R. W. Hornabrook and the staff of the Institute of Human Biology, who did the planning and collecting, the staff of the Serological Population Genetics Laboratory, London, who did most of the tests, and Dr P. Booth who also did numerous tests and has analysed the data (Booth, 1974). Most of this summary refers to the above survey, but account has been taken also of other recent work in the Territory as well as in Papua and West Irian. Mention should be made of a previous genetic distance survey of the peoples of the Western Highlands of Papua New Guinea (Sinnett *et al.*, 1970).

The population of the island falls into two major divisions: peoples with a mainly coastal distribution, speaking languages of the Austronesian group; and those, mainly of the mountainous interior, speaking a great variety of local non-Austronesian languages. Booth has shown, mainly by calculating genetic distances, that there is a close correspondence between language and genetics, both with regard to the main distribution and within each of the two major groups. Of all the genetic systems, the MNSs blood groups show the most striking picture. As compared with south east Asia, with its high frequency of gene *M*, the whole of New Guinea and the surrounding islands show rather low *M* and high *N*. The peoples of the interior of New Guinea have the highest *N* frequencies known anywhere in the world, nearly all above 90%. These peoples however (in contrast with the Australian Aborigines who also are high in *N*) possess the *S* gene, but almost exclusively as the complex *NS* which is rare in all other known peoples except the Ainu and some Lapps. Throughout New Guinea the *CDe* (R_1) complex reaches frequencies around 90%, again the highest known. The *CDE* (R_z) complex is almost totally absent, in contrast with Australia.

Certain other marker genes, including the Gerbich-negative (Ge^b) and a fast-migrating variant of malate dehydrogenase, are almost confined to New Guinea, and help in distinguishing between populations in the island.

The ancient populations of the interior are probably of comparable antiquity to the Australian Aborigines to whom they show a certain, albeit remote, genetic resemblance.

Simmons & Booth (1971) have prepared a most valuable stencilled typewritten Compendium of Melanesian genetic data, consisting of gene frequencies derived from all available genetic observations up to the earlier part of 1971 on the populations of Papua, New Guinea and the other islands of Melanesia, but not West Irian. Full bibliographies are given, and brief discussions of the data. The Compendium, in four parts, is available from the Commonwealth Serum Laboratories, Parkville, Victoria, Australia 3052.

Pacific islands

Dr J. M. Staveley and his colleagues in New Zealand have continued the work they have for many years been doing on Pacific populations, especially those of New Zealand itself, Fiji and Samoa (NZ/HA/7). No new results referring specifically to IBP work are, however, available.

The Caroline Islands

Hereditary blood factors have been studied in considerable detail by Professor N. Morton and his collaborators in the eastern islands of the Caroline group (USA/HA/25). The populations are classified as Eastern Micronesians. There are considerable differences between the islands, which is not surprising with such small and relatively isolated units, but almost everywhere frequencies of N, of $CDe(R_1)$ and of Hp^1 genes are remarkably high. Similarly high frequencies are, however, found for all these in the rather scanty data for the rest of Micronesia and (apart from the exceptional 75 % of Hp^1 in the Pingelapese) more extreme values are in all cases to be found in the Melanesians. Comparative data from adjacent groups of islands are needed for most of the other factors studied (immunoglobulins and red-cell enzymes).

Circumpolar populations

The peoples of the northern circumpolar region have formed the subject of an international multi-disciplinary study, the results of which have been brought together in *The Biology of Circumpolar Peoples*, a volume in the final IBP/HA series. The results of the genetic marker studies form the subject of a chapter in it, by Eriksson (see Eriksson *et al.*, 1977) setting them out in much greater detail than can be done here. In revising the present section, I have, however, had the benefit of reading a draft of their chapter. The latter, in addition to a description of the strictly circumpolar peoples, includes an account of the Ainu, who in the present chapter are considered together with the other inhabitants of Japan.

Since the time when man first reached the Polar regions he has spread eastwards through Siberia and America to Greenland and westward to Iceland (and thence, in very small numbers until recent centuries, to Greenland). In the course of IBP, circumpolar populations have been sampled mainly in Finland, Siberia, Alaska, Canada and Greenland.

The genetic markers of the blood

The hereditary blood factors of the Lapps have been studied by Finnish teams (HA/2) who have also compared them with Finnish populations and with other Finno-Ugric speaking populations, the Hungarians and the Mari of the Soviet Union (the latter in collaboration with Soviet workers). The results of extensive work done in Finland on the Lapps of that country have been summarised by Eriksson (1973) who has also tabulated a large body of data on the Lapps of Norway and Sweden; the Norwegian and Swedish investigations do not form an official part of the respective national HA programmes but are highly relevant to the Circumpolar project as a whole.

These investigations have confirmed the singular genetic status of the Lapps, as being something other than Caucasoid–Mongoloid hybrids, but having some features of both races, such as pigmented skin yet often blue eyes, uniquely high A_2 frequencies with high Fy^a (also uniquely high frequencies of C^w and P^c). They are certainly a very ancient ethnic group, and it may be that, as hinted by Eriksson (1977), they originated before the fully differentiated Caucasoids and Mongoloids emerged, or even that they survived the last Ice Age isolated in an unglaciated part of the Arctic. Whatever may, however, have been their ultimate origins, they have almost certainly been much modified as a whole by genetic drift, and it is clear that the small isolates which still exist, especially in Finland, have each been further modified by separate drift. Their Finno-Ugric language was probably acquired from Finnish immigrants about 500 B.C.

The Finns also are genetically unique among Europeans, but still unmistakably Caucasoid. In addition to the characteristic frequencies of the major marker genes, they have a set of individually rare congenital diseases, almost totally different from those of other Scandinavian peoples. They also have a remarkable set of low-frequency blood factor genes which appear to be almost totally confined to themselves.

The Finns appear to have separated from the Mari, and other Finno-Ugric peoples living in the Volga bend, only about 3000 years ago, but the Finns have probably mixed somewhat with Lapps while the Mari have acquired Asiatic genes such as Di^a.

A curious and not fully explained link between circumpolar populations is shown by the raised frequencies of the otherwise extremely rare superoxide dismutase gene, SOD^2. It has been found only in populations of Scandinavian origin, and mainly in those who are partly or wholly of Finnish descent. Frequencies above 1 % have been found in the northern Finns, the Swedish-speaking Finns of Perna, the Skolt Lapps of Nellim, the Swedes of Tornedalen on the Finnish border, and the population of the island of Westray (Orkney) who are largely of Scandinavian origin. No examples of the allele have been found among the Icelanders, or the Finno-Ugric Mari of the USSR, and it has not been found in any of the non-European populations tested (Greenland Eskimos, Ethiopians, Chinese, Japanese, Polynesians, Filipinos, Papuans).

27

A joint IBP study of Iceland populations by Icelandic and Finnish investigators began too late for results to be available, but a recent extensive survey (Bjarnason *et al.*, 1973) has done much to support the previously suspected close relationship of the Icelanders to the peoples of Ireland and northern Scotland.

Much work has been done on the peoples of Siberia including the far north. This has thrown some light on the origin of the indigenous peoples of America, who show very high frequencies of M and of cDE, and an almost total absence of B, characteristics which are not shown, for instance, by the Chinese or Japanese who, for lack of blood group data on other populations, have previously tended to be regarded as the typical eastern Asiatics. Though no single one of the peoples studied in Siberia shows a combination of all these characteristics, high frequencies of cDE are almost universal, and many of the peoples have rather high M and low B frequencies. It is thus, not unexpectedly, to Siberia rather than to the traditional 'Far East' that we must look for the origin of the indigenous Americans.

Eriksson (1973) has drawn attention to the peculiar distribution of the transferrin allele $Tf^{B}0\text{-}1$, which is found with frequencies of several per cent in the Finns and the Finno-Ugric speaking Komi of Russia, in the Nenets, Touvinians, and Evenks of Siberia (according to Spitsyn, 1973), and in Amerindians of middle America. It is present, but with lower frequencies, in the Hungarians, Estonians and Mari, all Finno-Ugric speakers, but is absent in the Finnish Lapps and in all other populations known to have been tested. The $Tf^{D_{CHI}}$ allele has a similar but less restricted distribution. These may represent the relics of an ancient circumpolar distribution.

For the Eskimos and the related Aleuts (and the probably related Chukchi) new (IBP) genetic data come from Siberia, Alaska, Canada and Greenland. For the conventional blood groups, new data come mainly from Siberia and Greenland, and some from Alaska, but considerable bodies of older data were already aiavlable for Alsaka and Canada, and the new information mostly agrees well with these. One surprising finding is a relatively high frequency of the Diego (Di^{a}) gene in the Siberian Eskimos and Aleuts, for the gene is almost completely absent from the Eskimos of Alaska and Canada (though present in many Amerindian populations).

One of the few genetic systems for which tests were done on nearly every population examined was that of the haptoglobins. Frequencies of Hp^{1} are moderate, 30–50%, in nearly all Eskimo populations tested, but they are 50 and 54%, respectively, in the Siberian and Alaskan Aleuts, figures which approach the high values found in most Amerindian populations. Gc^{2} frequencies are relatively high in Eskimo populations, being nearly all above 30%. This range includes the Siberian Eskimos with 30% and the Alaskan Aleuts with 32%.

Some red-cell enzyme tests were done on Alaskan Eskimos, a few on

Greenland Eskimos, and an exceptionally wide range of them on the Eskimos of Igloolik, Canada. In the acid phosphatase system the frequency of P^a is high in Alaskan Aleuts and Alaskan and Canadian Eskimos, averaging just over 50%, while P^c is everywhere rare or absent. The frequency of the phosphoglucomutase gene $PGM_1{}^2$ is highly variable, from 14% in Alaskan Aleuts to 35% in northwest Greenland Eskimos.

The frequency of the phenylthiocarbamide taster gene is the same in Siberian and Alaskan Eskimos, 49%, while in the Siberian (Komandorskiye Islands) Aleuts it is only 31%. Amerindian populations mostly show higher frequencies than these. Gm tests were done on Canadian and Greenland Eskimos, and lymphocyte histocompatibility antigen tests on those of Canada.

The Greenlanders, or Eskimos of Greenland, have been very fully studied by Gessain, by Jørgensen, by Persson, and by Fernet et al., who have produced extensive new data on the conventional blood groups and on serum groups. Eriksson et al. have studied red-cell enzymes of Greenlanders from Augpilaktok, northwest Greenland.

It will be seen that, in all, very numerous tests have been carried out on Eskimo populations, covering an exceptionally wide range of genetic systems. It is therefore unfortunate that, in general, the choice of systems differed greatly between the different parts of the Eskimo territory. It is to be hoped that the tests so far reported can be supplemented in the near future, in such a way as to make possible a full comparison of all the populations concerned, by such means as multivariate analysis. This is particularly to be desired as we possess such full data, largely obtained through IBP, about other characteristics of Eskimo populations throughout their range, as well as about their history as revealed by archaeology.

Some valuable conclusions can, however, be drawn. In particular it may be noted that new data on the Siberian Aleuts, taken in conjunction with IBP and earlier studies of the Alaskan Aleuts, are consistent with Laughlin's claim that the Aleuts and the Eskimos represent two separate lines of descent from the ancient fisher folk of the Bering Straits area. There are, however, indications that the Aleuts, in addition to their undoubted relationship to the Eskimos, may have a connection, not shared with Eskimos, with Alaskan Amerindians.

A hint at the completion of the last link in the circumpolar chain is provided by Persson (1969, 1970) who, on the basis of ABO and serum groups, suggests that the present Eskimos of the southern part of the west coast of Greenland incorporate genes derived from the early mediaeval Scandinavian colonists.

A. E. Mourant

Acknowledgments

The author of this chapter wishes to thank Mrs K. Domaniewska-Sobozak for the preparation of the bibliography, and Dr A. C. Kopeć for assistance in various ways, also the Royal Society for a grant which enabled the data to be prepared and the chapter written.

IBP Bibliography

Allison, A. C. (1954). Protection afforded by sickle-cell trait against subtertian malarial infection. *British Medical Journal*, 2, 290–4.

Araujo, H. M. M. de, Salzano, F. M. & Wolff, H. (1972). New data on the association between PTC and thyroid diseases. *Humangenetik*, 15, 136–44. (Brazil/ HA/23)

Arends, T., Brewer, G., Chagnon, N., Gallango, M. L., Gershowitz, H., Layrisse, M., Neel, J., Shreffler, D., Tashian, R. & Weitkamp, L. (1967). Intratribal genetic differentiation among the Yanomama Indians of Southern Venezuela. *Proceedings of the National Academy of Sciences, USA*, 57, 1252–9. (Venezuela/ HA/2 and USA/HA/26).

Arends, T., Weitkamp, L. R., Gallango, M. L., Neel, J. V. & Schultz, J. (1970). Gene frequencies and microdifferentiation among the Makiritare Indians. II. Seven serum protein systems. *American Journal of Human Genetics*, 22, 526–32. (Venezuela/HA/4).

Avčin, M. (1969). Gypsy isolates in Slovenia. *Journal of Biosocial Science*, 1, 221–33. (Yugoslavia/HA/10).

Beaven, G. H. (1973). Biological studies of Yemenite and Kurdish Jews in Israel and other groups in south-west Asia. X. Haemoglobin studies of Yemenite and Kurdish Jews in Israel. *Philosophical Transactions of the Royal Society*, B, 266, 185–93. (Israel/HA/7 and UK/HA/8)

Beaven, G. H., Fox, R. H. & Hornabrook, R. W. (1974). The occurrence of haemoglobin-J (Tongariki) and of thalassaemia on Karkar Island and the Papua New Guinea mainland. *Philosophical Transactions of the Royal Society*, B, 268, 269–77. (Australia/HA/2 and UK/HA/16/New Guinea)

Beckman, G. (1973). Population studies in northern Sweden. VI. Polymorphism of superoxide dismutase. *Hereditas*, 73, 305–10. (Sweden)

Beckman, G., Beckman, L. & Cedergren, B. (1971). Population studies in Northern Sweden. II. Red cell enzyme polymorphism in the Swedish Lapps. *Hereditas*, 69, 243–8. (Sweden)

Beckman, L. & Cedergren, B. (1971). Population studies in Northern Sweden. I. Variations of matrimonial migration distances in time and space. *Hereditas*, 68, 137–42. (Sweden)

Beckman, L., Cedergren, B., Collinder, E. & Rasmuson, M. (1972). Population studies in northern Sweden. III. Variations of ABO and Rh blood group gene frequencies in time and space. *Hereditas*, 72, 183–200. (Sweden)

Beckman, L., Cedergren, B. & Rasmuson, M. (1973). Population studies in northern Sweden. V. Regional heterogeneity of the A_2 blood group frequency. *Hereditas*, 73, 253–8. (Sweden)

Berg, K. & Eriksson, A. W. (1973). Genetic marker systems in Arctic populations, VII. Genetic variation in serum lipoproteins in Icelanders. *Human Heredity*, 23, 251–6. (Finland/HA/1/Iceland)

30

The genetic markers of the blood

Beroniade, S., Drăghicescu, T. & Aloman, S. (1973). Studiul repartiţiei şi al trans-
miterii ereditare a factorilor M–N în România. *Studii cercetări de Antropologie*,
10, 197–205. (Romania/HA/2)
Bhalla, V. (1972). Variations in taste threshold for PTC in populations of Tibet and
Ladakh. *Human Heredity*, 22, 453–8. (India/HA/3)
Bhatia, H. M. *et al.* (1974). Tribes of Maharashtra and Gujarat. Typescript.
(India/HA/4)
Bjarnason, O., Bjarnason, V., Edwards, J. H., Fridriksson, S., Magnusson, M.,
Mourant, A. E. & Tills, D. (1973). The blood groups of Icelanders. *Annals of
Human Genetics*, 36, 425–58. (Iceland)
Blackburn, C. R. B. & Hornabrook, R. W. (1969). Haptoglobin gene frequencies
in the people of the New Guinea Highlands. *Archaeology & Physical
Anthropology in Oceania*, 4, 56–63. (Australia/HA/4/New Guinea)
Boia, M., Vasiliu, L. Crainic, K., Danielescu, M. & Rîmneanţu, P. (1967). Cercetări
preliminare privind factorul Gm(a). (Preliminary studies concerning the Gm(a)
factor.) *Documentar de hematologie*, no. 3, 41–8. (Romania/HA)
Booth, P. B. (1971). A review of the Gerbich blood group system in Papua New
Guinea. *Papua New Guinea Medical Journal*, 14, 74–6. (Australia/HA/2/New
Guinea)
Booth, P. B. (1972). The occurrence of weak I^T red cell antigen among Melanesians.
Vox sanguinis, 22, 64–72 (Australia/HA/2/New Guinea).
Booth, P. B. (1974). Genetic distances between certain New Guinea populations
studied under the International Biological Programme. *Philosophical Trans-
actions of the Royal Society*, B, 268, 257–67. (Australia/HA/2/New Guinea)
Booth, P. B., Albrey, J. A., Whittaker, J. & Sanger, R. (1970). Gerbich blood group
system: a useful genetic marker in certain Melanesians of Papua and New
Guinea. *Nature, London,* 228, 462. (Australia/HA/2/New Guinea)
Booth, P. B. & Hornabrook, R. W. (1973). The MN^{A}SZ groups of some New Guinea
populations. *Human Biology in Oceania*, 11, 27–32. (Australia/HA/2/New
Guinea)
Booth, P. B., Hornabrook, R. W. & Malcolm, L. A. (1972). The red cell antigen N^A
in Melanesians: family and population studies. *Human Biology in Oceania*, 1,
223–8. (Australia/HA/2/New Guinea)
Booth, P. B. & McLoughlin, K. (1972) The Gerbich blood group system, especially
in Melanesians. *Vox sanguinis*, 22, 73–84. (Australia/HA/2/New Guinea)
Booth, P. B. & Simmons, R. T. (1972). Some thoughts on blood group genetic
work in Melanesia. *Papua New Guinea Medical Journal*, 15, 10–14. (Australia/
HA/2/New Guinea)
Booth, P. B. & Vines, A. P. (1967). Blood groups and other genetic data from
Bougainville, New Guinea. *Archaeology Physical Anthropology in Oceania*, 2,
227–35. (Australia/HA/2/New Guinea)
Booth, P. B., Wark, L., McLoughlin, K. & Spark, R. (1972). The Gerbich blood
group system in New Guinea. I. The Sepik district. *Human Biology in Oceania*,
1, 215–22. (Australia/HA/2/New Guinea)
Bouloux, C. J. (1968). Contribution à l'étude biologique des phénomènes puber-
taires en très haute altitude (La Paz). Enquête portant sur 80 enfants. Toulouse,
Centre Régional de Transfusion Sanguine et d'Hématologie, Thèse, Université
de Toulouse. (France/HA/8.3/Bolivia)
Bouloux, C., Gomila, J. & Langaney, A. (1972). Hemotypology of the Bedik.
Human Biology, 44, 289–302. (France/HA/5)
Brain, P. (1953). The sickle-cell trait: a possible mode of introduction into Africa.
Man, 53, 154. (Africa)

31

A. E. Mourant

Brönnestam, R., Beckman, L. & Cedergren, B. (1971). Genetic polymorphism of the complement component C3 in Swedish Lapps. *Human Heredity*, 21, 267–71. (Sweden)

Cabannes, R., Beurrier, A. & Monnet, B. (1965). Études des protéines, des haptoglobulines et des transferrines chez les Indiens de Guyane française. *Nouvelle Revue française d'Hématologie*, 5, 247–60. (France/HA/8.3/French Guiana)

Cabannes, R. & Schmidt-Beurrier, A. (1966). Recherches sur les hémoglobines des populations indiennes de l'Amérique du sud. *Anthropologie, Paris*, 70, 331–41. (France/HA/8.3/South America)

Cabannes, R., Schmidt-Beurrier, A. & Monnet, B. (1966). Étude des protéines, des haptoglobines, des transferrines et des hémoglobines d'une population noire de Guyane française (Boni). *Bulletin de la Société de Pathologie exotique*, 59, 908–16. (France/HA/8.3/French Guiana)

Carles-Trochain, E. (1968). *Étude hémotypologique des pêcheurs du lac Titicaca*. Paris: Hermann. (France/HA/8.3/Bolivia)

Cavalli-Sforza, L. L. & Edwards, A. W. F. (1967). Phylogenetic analysis. Models and estimation procedures. *American Journal of Human Genetics*, 19, 233–57.

Chapman, A. M. & Jacquard, A. (1971). Un isolat d'Amérique Centrale – Les Indiens Jicaques du Honduras. In *Génétique et Populations*, Paris: PUF, Institut Nationale d'Études démographiques. (France/HA/8.3/Honduras)

Chapman, J. A., Grant, I. S., Taylor, G., Mahmud, K., Sardar-ul-Mulk & Shahid, M. A. (1972). Endemic goitre in the Gilgit Agency, West Pakistan. With an appendix on dermatoglyphics and taste-testing. *Philosophical Transactions of the Royal Society*, B, 263, 459–90. (UK/HA/10/Pakistan)

Chilling-Rathford [*sic*], Larrouy, G., Mille, R. & Ruffié, J. (1966). Étude hémotypologique de la population indienne de l'île Dominique. *Anthropologie, Paris*, 70, 319–30. (France/HA/8.3/Dominica)

Constans, J. & Lefèvre-Wittier, Ph. (n.d.) Nouvelles données hémotypologiques en pays Touareg. I. Étude des haptoglobines à Idélès (Ahaggar-Sahara algérien) et chez les Kel Kummer des Touareg Iwellemeden (Mali). Typescript. (France/HA/8.1/Sahara)

Constans, J. & Quilici, J. C. (1973). Hémotypologie des Indiens Chipaya de Bolivie. *Cahiers d'Anthropologie et d'Ecologie humaine*, no 1, 147–59. (France /HA/8.3/Bolivia)

Curtain, C. C., Gajdusek, D. C., Kidson, C., Gorman, J. G., Champness, L. & Rodrigue, R. (1965). Haptoglobins and transferrins in Melanesia: relation to hemoglobin, serum haptoglobin and serum iron levels in population groups in Papua – New Guinea. *American Journal of Physical Anthropology*, 23, 363–79. (USA/HA/3/New Guinea)

Curtain, C. C., Gajdusek, D. C., Kidson, C., Gorman, J., Champness, L. & Rodrigue, R. (1965). Serum pseudocholinesterase levels and variations in the peoples of Papua and New Guinea. *American Journal of Tropical Medicine and Hygiene.* 14, 671–7. (USA/HA/3/New Guinea)

Curtain, C. C., Kidson, C., Gajdusek, D. C. & Gorman, J. G. (1962). Distribution pattern, population genetics and anthropological significance of thalassemia and abnormal hemoglobins in Melanesia. *American Journal of Physical Anthropology*, 20, 475–83. (USA/HA/3/New Guinea)

Dausset, J. & Colombani, J. (ed.) (1973). *Histocompatibility Testing* 1972. Copenhagen: Munksgaard.

Dhaliwal, S. S. & Chan, K. L. (1973). Population genetics (with particular reference to isoenzyme polymorphisms and blood groups). Typescript. (Malaysia/HA/5)

The genetic markers of the blood

Dumitru, M., Beroniade, S., Drăghicescu, T. & Aloman, S. (1966). Les hapto-globines en Roumanie. *Annuaire roumain d'Anthropologie* **3**, 71–7. (Romania/HA)

Duncan, I. W, Scott., E. M. & Wright, R. C. (1974). Gene frequencies of erythro-cytic enzymes of Alaskan Eskimos and Athabaskan Indians. *American Journal of Human Genetics*, **26**, 244–6. (USA/HA/13)

Edholm, O. G., Samueloff, S., Mourant, A. E., Fox, R. H., Lourie, J. A., Lehmann, H., Bavly, S., Beaven, G. & Even-Paz, Z. (1973). Biological studies of Yemenite and Kurdish Jews in Israel and other population groups in south-west Asia. XIII. Conclusions and summary. *Philosophical Transactions of the Royal Society*, B, **226**, 221–4. (Israel/HA/7 and UK/HA/8)

Ehnholm, C. & Eriksson, A. W. (1969). Haptoglobin subtypes among Finnish Skolt Lapps. *Annales medicinae experimentalis et biologiae Fenniae*, **47**, 52–4. (Finland/HA/1)

Eickstedt, E. von (1944). *Rassendynamik von Ostasien: China und Japan, Tai und Kmer von der Urzeit bis heute*, Berlin: Walter de Gruyter. (East Asia)

Eriksson, A. W. (1973). Genetic polymorphisms in Finno-Ugrian populations: Finns, Lapps and Maris. *Israel Journal of Medical Science*, **9**, 1156–70. (Finland/HA/2/Circumpolar)

Eriksson, A. W., Lehmann, W. & Simpson, N. E. (1977). Genetic studies of circum-polar populations of Europe and North America. In *The Biology of Circum-polar Peoples*, ed. F. A. Milan, London: Cambridge University Press (in press).

Eriksson, A. W., Fellman, J., Kirjarinta, M., Eskola, M.-R., Singh, S., Benkmann, H.-G., Goedde, H. W., Mourant, A. E., Tills, D. & Lehmann, W. (1971). Adenylate kinase polymorphism in populations in Finland (Swedes, Finns, Lapps), in Maris, and in Greenland Eskimos. *Humangenetik*, **12**, 123–30. (Finland/HA/2/Circumpolar)

Eriksson, A. W., Fellman, J. O., Workman, P. L. & Lalouel, J. M. (1973). Popula-tion studies on the Åland Islands. 1. Prediction of kinship from migration and isolation by distance. *Human Heredity*, **23**, 422–33. (Finland/HA/2)

Eriksson, A. W., Kirjarinta, M., Fellman, J., Eskola, M.-R. & Lehmann, W. (1971). Adenosine deaminase polymorphism in Finland (Swedes, Finns, and Lapps), the Mari Republic (Cheremisses), and Greenland (Eskimos). *American Journal of Human Genetics*, **23**, 568–77. (Finland/HA/2/Circumpolar)

Eriksson, A. W., Kirjarinta, M., Lehtosalo, T., Kajanoja, P., Lehmann, W., Mourant, A. E., Tills, D., Singh, S., Benkmann, H.-G., Hirth, L. & Goedde, H. W. (1971). Red cell phosphoglucomutase polymorphism in Finland-Swedes, Finns, Finnish Lapps, Maris (Cheremisses) and Greenland Eskimos, and segregation studies of PGM_1 types in Lapp families. *Human Heredity*, **21**, 140–53. (Finland/HA/2/Circumpolar)

Eriksson, A. W., Zolotareva, I., Kozentsev, A., Shevchenko, A., Escola, M-R., Kirjarinta, M., Partanen, K. & Fellman, J. (1971). Genetic studies on the Maris (Cheremis). Typescript, partly after a paper read by A. W. Eriksson at the Symposium on the Mari People, Moscow. (Finland/HA)

Fernet, P., Larrouy, G. & Ruffié, J. (1964). Étude hémotypologique des populations indiennes de Guyane française. II. Les groupes sériques du système Gm. *Bulletin de la Société d'Anthropologie*, série 11, **7**, 119–23. (France/HA/8.3/French Guiana)

Fernet, P., Mortensen, W. S., Langaney, A. & Robert, J. (1971). Hémotypologie du Scoresbysund (Est Groenland). *Bulletin de la Société d'Anthropologie*, series 12, **8**, 177–85. (France/HA/9/Greenland)

Fitch, W. M. & Margoliash, E. (1967). Construction of phylogenetic trees. *Science, Washington*, **155**, 279–84.

Fitch, W. M. & Neel, J. V. (1969). The phylogenic relationships of some Indian tribes of Central and South America. *American Journal of Human Genetics*, **21**, 384–97. (USA/HA/26/Central and South America)

Fix, A. G. (1971). Semai Senoi population structure and genetic microdifferentiation. Ph.D. thesis, Ann Arbor, University of Michigan. (Malaysia/HA/1)

Fraser, G. R., Steinberg, A. G., Defaranas, B., Mayo, O., Stamatoyannopoulos, G. & Motulsky, A. G. (1969). Gene frequencies at loci determining blood-group and serum-protein polymorphisms in two villages of northwestern Greece. *American Journal of Human Genetics*, **21**, 46–60. (USA/HA/25/Greece)

Freire-Maia, N., Karam, E. & Mehl, H. (1967). Phenylthiocarbamide and schizophrenia. *Lancet*, **i**, 576. (Brazil/HA/7)

Friedlaender, J. S., Sgaramella-Zonta, L. A., Kidd, K. K., Lai, L. Y. C., Clark, P. & Walsh, R. J. (1971). Biological divergences in south-central Bougainville: an analysis of blood polymorphism gene frequencies and anthropometric measurements utilizing tree models, and a comparison of these variables with linguistic, geographic, and migrational 'distances'. *American Journal of Human Genetics*, **23**, 253–70. (Melanesia)

Friedlaender, J. S. & Steinberg, A. G. (1970). Anthropological significance of gamma globulin (Gm and Inv) antigens in Bougainville Island, Melanesia. *Nature, London*, **228**, 59–61. (USA/HA/2/Melanesia)

Fujiki, N., Yamamoto, M., Takenaka, S., Ishimaru, T., Takanashi, T., Sugimoto, N., Nakajima, K. & Masuda, M. (1968). A study of inbreeding in some isolated populations. *Japanese Journal of Human Genetics*, **12**, 205–25. (Japan/HA/3)

Gershowitz, H., Junqueira, P. C., Salzano, F. M. & Neel, J. V. (1967). Further studies on the Xavante Indians. III. Blood groups and ABH-Le^a secretor types in the Simões Lopes and São Marcos Xavantes. *American Journal of Human Genetics*, **19**, 502–13. (Brazil/HA/3 and USA/HA/26)

Gershowitz, H., Layrisse, M., Layrisse, Z., Neel, J. V., Brewer, C., Chagnon, N. & Ayres, M. (1970). Gene frequencies and microdifferentiation among the Makiritare Indians. I. Eleven blood group systems and the ABH-Le secretor traits. A note on Rh gene frequency determinations. *American Journal of Human Genetics*, **22**, 515–25. (Venezuela/HA/2 and USA/HA/26)

Gershowitz, H., Layrisse, M., Layrisse, Z., Neel, J. V., Chagnon, N. & Ayres, M. (1972). The genetic structure of a tribal population, the Yanomama Indians. II. Eleven blood-group systems and the ABH-Le secretor traits. *Annals of Human Genetics*, **35**, 261–9. (Venezuela/HA/2 and USA/HA/26)

Gessain, R., Ruffié, J., Kane, Y., Kane, O., Cabannes, R. & Gomila, J. (1965). Note sur la séro-anthropologie de trois populations de Guinée et du Sénégal: Coniagui Bassari et Bédik (groupes ABO, MN, Rh, P, Kell, Gm et hémoglobines). *Bulletin de la Société d'Anthropologie*, série 11, **8**, 5–18. (France/HA/5)

Glasgow, B. G., Goodwin, M. J., Jackson, F., Kopeć, A. C., Lehmann, H., Mourant, A. E., Tills, D., Turner, R. W. D. & Ward, M. P. (1968). The blood groups, serum groups and haemoglobins of the inhabitants of Lunana and Thimbu, Bhutan. *Vox sanguinis*, **14**, 31–42. (UK/HA/2/Bhutan)

Godber, M. J., Kopeć, A. C., Mourant, A. E., Tills, D. & Lehmann, E. E. (1973). Biological studies of Yemenite and Kurdish Jews in Israel and other population groups in south-west Asia. IX. The hereditary blood factors of the Yemenite and Kurdish Jews. *Philosophical Transactions of the Royal Society*, B, **266**, 169–84. (Israel/HA/7 and UK/HA/8)

The genetic markers of the blood

Glindeman, V. P., Koreshkova, L. P. & Khorvat, G. N. (1971). Raspredeleniye grupp krovi sistem *ABO*, *MN* i *Rhesus* (*D* i *C*) sredi korennogo naseleniya Uzbekistana. (The distribution of the blood group systems ABO, MN and Rhesus (D and C) among native population of Uzbekistan.) *Voprosy antropologii*, **37**, 140–42. (USSR/HA/6)

Grimm, H. (1973). Neues Material zur beschreibenden Populationsgenetik und Genogeographie der Bevölkerung der Deutschen Demokratischen Republik. *Biologische Rundschau*, **11**, 95–106. (DDR/HA/20)

Gürtler, H. (1971). Personal communication, quoted in *The Distribution of the Human Blood Groups and Other Biochemical Polymorphisms*, by A. E. Mourant, A. C. Kopeć & K. Domaniewska-Sobczak (1976), London: Oxford University Press. (Denmark/HA/2/Greenland)

Harada, S., Akaishi, S., Kudo, T. & Omoto, K. (1971). Distribution of phenotypes and gene frequencies of six red cell enzymes in the district of Tohoku, northern part of Japan. *Journal of the Anthropological Society of Japan*, **79**, 356–66. (Japan/HA/3)

Harada, S., Misawa, S. & Omoto, K. (1971). Detection of the red cell acid phosphatase variant allele in Japanese. *Japanese Journal of Human Genetics*, **16**, 22–9. (Japan/HA/3)

Harrison, G. A., Küchemann, C. F., Moore, M. A. S., Boyce, A. J., Baju, T., Mourant, A. E., Godber, M. J., Glasgow, B. G., Kopeć, A. C., Tills, D. & Clegg, E. J. (1969). The effects of altitudinal variation in Ethiopian populations. *Philosophical Transactions of the Royal Society*, B, **256**, 147–82. (UK/HA/3/Ethiopia)

Harvey, R. G., Godber, M. J., Kopeć, A. C., Mourant, A. E. & Tills, D. (1969). Frequency of genetic traits in the Caribs of Dominica. *Human Biology*, **41**, 342–64. (UK/HA/20/Dominica)

Hautvast, J. (1971). Finger print patterns and blood group frequencies in Kisi schoolchildren (Tanzania). *Zeitschrift für Morphologie und Anthropologie*, **63**, 215–19. (Netherlands/HA/Tanzania)

Hiernaux, J. (1964). La mesure de la différence morphologique entre populations pour un ensemble de variables. *Anthropologie, Paris*, **68**, 559–68.

Ikin, E. W., Lehmann, H., Mourant, A. E. & Thein, H. (1969). The blood groups and haemoglobins of the Burmese. *Man*, n.s. **4**, 118–22. (Burma)

Jenkins, T. (1973). Genetic polymorphisms of man in Southern Africa, M.D. Thesis, University of London. (South Africa/HA/2)

Jenkins, T. (1974). Sero-genetic studies on the Pedi. Typescript. (South Africa/HA/2)

Jenkins, T. & Corfield, V. (1972). The red cell acid phosphatase polymorphism in Southern Africa: population data and studies on the R, RA and RB phenotypes. *Annals of Human Genetics*, **35**, 379–91. (South Africa/HA/2)

Jenkins, T., Harpending, H. C., Gordon, H., Keraan, M. M. & Johnston, S. (1971). Red-cell-enzyme polymorphisms in the Khoisan peoples of Southern Africa. *American Journal of Human Genetics*, **23**, 513–32. (South Africa/HA/2)

Kang, Y. S., Paik, S. G. & Lee, C. C. (n.d.). The researches of the Korean population genetics (xvi). Studies on the incidences of erythrocyte glucose-6-phosphate dehydrogenase deficiency and acetylator phenotypes. Typescript. (Korea/HA/1)

Karam, E. & Freire-Maia, N. (1961). Phenylthiocarbamide and mental immaturity. *Lancet*, i, 622–3. (Brazil/HA/7)

Khazanova, A. B. & Shamlyan, N. P. (1970). K antropologii i populyatsionnoy genetike loparey kol'skovo poluostrova. (Anthropology and population genetics of Lapps of the Kola peninsula.) *Voprosy̌ Antropologii*, **34**, 71–8. (USSR/HA/2)

Kirk, R. L., Blake, N. M., Lai, L. Y. C. & Cooke, D. R. (1969). Population genetic studies in Australian aborigines of the Northern Territory. The distribution of some serum protein and enzyme groups among the Malag of Elcho Island. *Archaeology and Physical Anthropology in Oceania*, **4**, 238–51. (Australia/HA/1)

Kornstad, L. (1972). Distribution of the blood groups of the Norwegian Lapps. *American Journal of Physical Anthropology*, **36**, 257–66. (Norway)

Larrouy, G., Marty, Y. & Ruffié, J. (1964). Étude hémotypologique des populations indiennes de la Guyane Française. I. Les groupes érythrocytaires. *Bulletin de la Société d'Anthropologie*, série 11, **7**, 107–17. (France/HA/8.3/French Guiana)

Lefévre-Wittier, P. & Vergnes, H. Situation et structure génétiques de la communauté d'Idélès. (Centre de culture de l'Ahaggar – Sahara algérien.) Typescript. (France/HA/B.1/Sahara)

Lehmann, H., Ala, F., Hedeyat, S., Montazemi, K., Nejad, H. K., Lightman, S., Kopeć, A. C., Mourant, A. E., Teesdale, P. & Tills, D. (1973). Biological studies of Yemenite and Kurdish Jews in Israel and other population groups in south-west Asia. XI. The hereditary blood factors of the Kurds of Iran. *Philosophical Transactions of the Royal Society*, B, **266**, 195–205. (UK/HA/22/Iran)

Lie-Injo Luan Eng (n.d.) Blood genetic studies in different racial groups in Malaysia and Indonesia. Typescript. (Malaysia/HA/4)

Lie-Injo Luan Eng, Bolton, J. M. & Fudenberg, H. H. (1967). Haptoglobins, transferrins and serum gamma globulin types in Malayan aborigines. *Acta criminologiae et medicinae legalis japonica*, **33**, 155–6. (Malaysia/HA/4)

Lucarelli, P., Agostino, R., Palmarino, R., & Bottini, E. (1971). Adenosine deaminase polymorphism in Sardinia. *Humangenetik*, **14**, 1–5. (Italy/HA/5)

Mabuchi, Y. (1971). Studies on the placental polymorphic systems in Japanese. *Bulletin of the Osaka Medical School*, **17**, 51–9. (Japan)

McAlpine, P. J., Chen, S.-H., Cox, D. W., Dossetor, J. B., Giblett, E., Steinberg, A. G. & Simpson, N. E. (1974). Genetic markers in blood in a Canadian Eskimo population with a comparison of allele frequencies in circumpolar populations. *Human Heredity*, **24**, 114–42. (Canada/HA/1)

Macgregor, A. & Booth, P. B. (1973). A second example of anti-Ge[1], and some observations on Gerbich subgroups. *Vox sanguinis*, **25**, 474–8. (Australia/HA/2/New Guinea)

Malcolm, L. A., Booth, P. B. & Cavalli-Sforza, L. L. (1971). Intermarriage patterns and blood group gene frequencies of the Bundi people of the New Guinea highlands. *Human Biology*, **43**, 187–99. (Australia/HA/2/New Guinea)

Marengo-Rowe, A. J., Aviet, K., Godber, M. J., Kopeć, A. C., Mourant, A. E., Tills, D. & Woodhead, B. J. (1974). The inherited blood factors of the inhabi-, tants of Southern Arabia. *Annals of Human Biology*, **1**, 311–26. (Southern Arabia)

Matoušek, V. & Seemanova, E. (n.d.) A contribution to the estimate of genetic load in human populations. Summary. Typescript (Czechoslovakia/HA/1)

Matsumoto, H. & Miyazaki, T. (1972). Gm and Inv allotypes of the Ainu in Hidaka area, Hokkaido. *Japanese Journal of Human Genetics*, **17**, 20–6. (Japan)

Matsumoto, H., Miyazaki, T. & Fong, J. M. (1973). Further data on the Gm and Am allotypes of the Takasago in Taiwan. *Japanese Journal of Legal Medicine*, **27**. (Japan/HA/?/Taiwan)

Matsumoto, H., Miyazaki, T., Fong, J. M. & Mabuchi, Y. (1972). Gm and Inv allotypes of the Takasago tribes in Taiwan. *Japanese Journal of Human Genetics* **17**, 27–37. (Japan/HA/?/Taiwan)

The genetic markers of the blood

Matsumoto, H., Miyazaki, T. & Lin, J. Y. (n.d.) Gm and Inv allotypes of the Taiwanese. Typescript submitted to the *Japanese Journal of Human Genetics*. (Japan/HA/?/Taiwan)

Matsumoto, H., Miyazaki, T., Lin, J. Y. & Hotta, S. (n.d.) Gamma-globulin allotypes of the Indonesians from Java-Madura in Indonesia. Typescript submitted to the *Japanese Journal of Human Genetics*. (Japan/HA/?/Indonesia)

Matsumoto, H., Miyazaki, T. & Ohkura, K. (n.d.) Gammaglobulin allotypes of the inhabitants on the Caspian Sea coast in Iran. Typescript submitted to the *Japanese Journal of Human Genetics*. (Japan/HA/?/Iran)

Matsumoto, H. & Takatsuki, K. (1968). Gm factors in Japan: population and family studies. *Japanese Journal of Human Genetics* 13, 10–19. (Japan)

Matsunaga, E., Shinoda, T. & Handa, Y. (1965). A genetic study on the quantitative variation in erythrocyte glucose-6-phosphate dehydrogenase activity of apparently healthy Japanese. *Japanese Journal of Human Genetics*, 10, 1–12. (Japan/HA/3)

Mesa, M. S. (1973). Estudio del polimorfismo genético de los sistemas Rh y Kell en la comarca natural de La Mancha. *Trabajos de Antropologia* 16, 245–51. (Spain/HA/3)

Modiano, G., Bernini, L., Carter, N. D., Santachiara Benerecetti, S. A., Detter, J. C., Baur, E. W., Paolucci, A. M., Gigliani, F. Morpurgo, G., Santolamazza, C., Scozzari, R., Terrenato, L., Khan, P. M., Nijenhuis, L. E., & Kanashiro, V. K. (1972). A survey of several red cell and serum genetic markers in a Peruvian population. *American Journal of Human Genetics*, 24, 111–23. (Italy/HA/4/Peru)

Monn, E., Berg, K., Reinskou, T. & Teisberg, P. (1971). Serum protein polymorphisms among Norwegian Lapps. Studies on the Lp, Ag, Gc and transferrin systems. *Human Heredity*, 21, 134–9. (Norway)

Montalenti, G. (n.d.). A genetic survey of the Sardinian population. Typescript. (Italy/HA/5)

Morton, N. E. & Lalouel, J. M. (1973). Bioassay of kinship in Micronesia. *American Journal of Physical Anthropology*, 38, 709–19. (USA/HA/12/Caroline Islands)

Morton, N. E. & Lalouel, J. M. (1973). Topology of kinship in Micronesia. *American Journal of Human Genetics*, 25, 422–32. (USA/HA/12/Caroline Islands)

Morton, N. E. & Yamamoto, M. (1973). Blood groups and haptoglobins in the Eastern Carolines. *American Journal of Physical Anthropology*, 38, 695–8. (USA/HA/12/Caroline Islands)

Mourant, A. E. (1974). The hereditary blood factors of the peoples of New Guinea and the surrounding regions. *Physiological Transactions of the Royal Society*, B, 268, 251–5. (UK/HA/16/New Guinea)

Mourant, A. E., Kopeć, A. C. & Domaniewska-Sobczak, K. (1976). *The Distribution of the Human Blood Groups and Other Polymorphisms*, 2nd edn. London: Oxford University Press.

Mya-Tu, M. & Ma Than Saw (1970). A note on the ABH secretion among the Burmese. *Union of Burma Journal of Life Sciences*, 3, 101. (Burma/HA/1)

Mya-Tu, M., May-May-Yi & Thin-Thin-Hlaing. (1968). Blood groups of the Inthas and Tavoyans of Burma. *Union of Burma Journal of Life Sciences*, 1, 353–7. (Burma/HA/1)

Mya-Tu, M., May-May-Yi & Thin-Thin-Hlaing. (1971). Blood groups of the Burmese population. *Human Heredity*, 21, 420–30. (Burma/HA/1)

Nakajima, H. (1971). The Rh, MNSs, Duffy and Xg blood group frequencies in Japanese – further tests on unrelated people. *Journal of the Anthropological Society of Japan*, 79, 178–81. (Japan)

Nakajima, H., Ohkura, K., Inafuku, S., Ogura, Y., Koyama, T., Hori, F. & Takahara, S. (1967). The distribution of several serological and biochemical traits in East Asia. II. The distribution of ABO, MNSs, Q, Lewis, Rh, Kell, Duffy and Kidd blood groups in Ryukyu. *Japanese Journal of Human Genetics*, 12, 29–37. (Japan)

Necrasov, O. (1974). Recherches hématotypiques en Roumanie. Typescript. (Romania/HA/2)

Necrasov, O., Botezatu, D. & Iacob, M. (1967). Considérations sur la répartition des groupes sanguins du système OAB et de leurs facteurs héréditaires en Roumanie. *Annuaire roumain d'Anthropologie*, 4, 17–31. (Romania/HA)

Necrasov, O., Iacob, M. & Botezatu, D. (1968). Sur la répartition des facteurs Rh (D) en Roumanie. *Annuaire roumaine d'Anthropologie*, 5, 37–42. (Romania/HA)

Neel, J. V. (1968). The American Indian in the International Biological Program. In *Biomedical Challenges Presented by the American Indian*. Pan American Health Organization, pp. 47–54. (USA/HA/26)

Neel, J. V. (1972). The genetic structure of a tribal population, the Yanomama Indians. I. Introduction. *Annals of Human Genetics*, 35, 255–9. (USA/HA/26/Venezuela and Brazil)

Neel, J. V. (1973). Diversity within and between South American Indian tribes. *Israel Journal of Medical Science*, 9, 1216–24. (USA/HA/26/South America)

Neel, J. V. (1973). Inferences concerning evolutionary forces from genetic data. *Israel Journal of Medical Science*, 9, 1519–32. (USA/HA/26/South America)

Neel, J. V. (1973). 'Private' genetic variants and the frequency of mutation among South American Indians. *Proceedings of the National Academy of Sciences, USA*, 70, 3311–15. (USA/HA/26/South America)

Neel, J., Arends, T., Brewer, C., Chagnon, N., Gershowitz, H., Layrisse, M., Layrisse, Z., MacCluer, J., Migliazza, E., Oliver, W., Salzano, F., Spielman, R., Ward, R. & Weitkamp, L. (1971). Studies on the Yanomama Indians. In *Human Genetics*, Proceedings of the 4th International Congress on Human Genetics, Paris, 1971, pp. 96–111. (USA/HA/26/Venezuela and Brazil)

Neel, J. V., Rothhammer, F. & Lingoes, J. C. (1974). The genetic structure of a tribal population, the Yanomama Indians. X. Agreement between representations of village distances based on different sets of characteristics. *American Journal of Human Genetics*, 26, 281–303. (USA/HA/26/Brazil)

Neel, J. V. & Salzano, F. M. (1966). A prospectus for genetic studies on the American Indians. In *The Biology of Human Adaptability*, ed. P. T. Baker & J. S. Weiner, pp. 245–74. Oxford: Clarendon Press. (Brazil/HA/3 and USA/HA/26)

Neel, J. V. & Salzano, F. M. (1967). Further studies on the Xavante Indians. X. Some hypotheses – generalizations resulting from these studies. *American Journal of Human Genetics*, 19, 554–74. (Brazil/HA/3 and USA/HA/26)

Neel, J. V., Salzano, F. M., Junqueira, P. C., Keiter, F. & Maybury-Lewis, D. (1964). Studies on the Xavante Indians of the Brazilian Mato Grosso. *American Journal of Human Genetics*, 16, 52–140. (Brazil/HA/3 and USA/HA/26)

Neel, J. V. & Ward, R. H. (1970). Village and tribal genetic distances among American Indians, and the possible implications for human evolution. *Proceedings of the National Academy of Sciences, USA*, 65, 323–30. (USA/HA/26/Amerinds)

Neel, J. V. & Ward, R. H. (1972). The genetic structure of a tribal population, the Yanomama Indians. VI. Analysis by F-statistics (including a comparison with the Makiritare and Xavante). *Genetics*, 72, 639–66. (USA/HA/26/Venezuela and Brazil)

The genetic markers of the blood

Nijenhuis, L. E. & Gemser-Runia, Jeltje. (1965). Hereditary and acquired blood factors in the Negroid population of Surinam. III. Blood group studies. *Tropical Geographical Medicine*, 17, 69–79. (Netherlands/HA/3/Surinam)

Omoto, K. (1973) Polymorphic traits in peoples of Eastern Asia and the Pacific. *Israel Journal of Medical Science*, 9, 1195–215. (Japan)

Omoto, K. & Harada, S. (1969). Polymorphism of red cell acid phosphatase in several population groups in Japan. *Japanese Journal of Human Genetics*, 14, 17–27. (Japan/HA/?3)

Omoto, K. & Harada, S. (1970). Frequencies of polymorphic types of four red cell enzymes in a Central Japanese population. *Japanese Journal of Human Genetics*, 14, 298–305. (Japan/HA/?3)

Omoto, K., Ishizaki, K., Harada, S., Akaishi, S., Kudo, T. & Takahashi, K. (1973). The distribution of serum protein and red cell enzyme types among blood donors of Okinawa Is., the Ryukyus. *Journal of the Anthropological Society of Japan*, 81, 159–73. (Japan/HA/?3)

Palatnik, M. (1975). Genetica de la población Toba del Chaco Argentino. *Acta VI Congreso argentino de Biología*, Tucuman 1973. Tucuman: Fundación Miguel Lillo. (Argentina/HA)

Paolucci, A. M., Ferro-Luzzi, A., Modiano, G., Morpurgo, G. & Kanashiro, V. K. (1971). Taste sensitivity to phenylthiocarbamide (PTC) and endemic goiter in the Indian natives of Peruvian Highlands. *American Journal of Physical Anthropology*, 34, 427–30. (Italy/HA/4/Peru)

Pauls, F. P. & Allen, F. H. (1969). Blood groups of Wainwright Eskimos. Typescript. (USA/HA/9/Alaska)

Persson, I. (1968). The distribution of serum types in West Greenland Eskimos. *Acta genetica et statistica medica*, 18, 261–70. (Denmark/HA/2/Greenland)

Persson, I. (1969). The fate of the Icelandic Vikings in Greenland. *Man*, n.s. 4, 620–28. (Denmark/HA/2/Greenland)

Persson, I. (1970). Anthropological investigations of the population of Greenland. *Meddelanden från Grønland*, 180, 1–78. (Denmark/HA/2/Greenland)

Persson, I., Melartin, L. & Gilberg, A. (1971). Alloalbuminemia. A search for albumin variants in Greenland Eskimos. *Human Heredity*, 21, 57–9. (Denmark/HA/2/Greenland)

Persson, I., Rivat, L., Rousseau, P. Y. & Ropartz, C. (1972). Ten Gm factors and the Inv system in Eskimos in Greenland. *Human Heredity*, 22, 519–28. (Denmark/HA/2/Greenland)

Persson, I. & Swan, T. (1971). Serum β-lipoprotein polymorphism in Greenlanders. Frequency of the Ag(x) factor. *Human Heredity*, 21, 384–7. (Denmark/HA/2/Greenland)

Persson, I. & Tingsgård, P. (1968). Serum protein types in East Greenland Eskimos. *Acta genetica et statistica medica*, 18, 61–9. (Denmark/HA/2/Greenland)

Pik, C., Loos, J. A., Jonxis, J. H. P. & Prins, H. K. (1965). Hereditary and aquired blood factors in the negroid population of Surinam. II. The incidence of haemoglobin anomalies and the deficiency of glucose-6-phosphate dehydrogenase. *Tropical Geographical Medicine*, 17, 61–8. (Netherlands/HA/3/Surinam)

Planas, J., Fusté, M., Diaz, J. M. & Pons, J. (1969). Blood groups (Rh, ABO) in the population of Gran Canaria (Canary Islands, Spain). *Human Heredity*, 19, 185–9. (Spain/HA/3)

Planas, J., Pons, J. & Triginer, J. (1968). Haptoglobin types in the population of Asturias (Spain). *Acta genetica et statistica medica*, 18, 155–8. (Spain/HA/3)

Polunin, I. & Sneath, P. H. A. (1953). Studies of blood groups in South-East Asia. *Journal of the Royal Anthropological Institute*, 83, 215–51. (South east Asia)

Pons, J. (1970). Anthropological and genetical studies in Spain. Typescript. (Spain/HA/3)

Pons, J., Fusté, M. Diaz, J. M. & Planas, J. (1968). Haptoglobin types in the population of the Gran Canaria. *Acta genetica et statistica medica*, **18**, 579–83. (Spain/HA/3)

Pons, J., Triginer, J. & Planas, J. (1967). Distribución de los tipos de haptoglobinas en la población asturiana. *Revista de la Facultad de Ciencias, Oviedo* **8** (1), 13–16. (Spain/HA/3)

Post, R. H., Neel, J. V. & Schull, W. J. (1968). Tabulations of phenotype and gene frequencies for 11 different genetic systems studied in the American Indian. In *Biomedical Challenges Presented by the American Indians*, pp. 141–85. Pan American Health Organization (USA/HA/26/Amerinds)

Quilici, J. C. (1968). *Les altiplanides du corridor interandin. Étude hémotypologique.* Toulouse: Centre Régional de Transfusion Sanguine et d'Hématologie. (France/HA/8.3/Bolivia)

Quilici, J. C., Ruffié, J. & Marty, Y. (1970). Hémotypologie d'un groupe paléoamérindien des Andes: les Chipaya. *Nouvelle revue française d'Hématologie*, **10**, 727–38. (France/HA/8.3/Bolivia)

Rothhammer, F. & Spielman, R. S. (1972). Anthropometric variation in the Aymará: genetic, geographic, and topographic contributions. *American Journal of Human Genetics*, **24**, 371–80. (USA/HA/26/Peru)

Ruffié, J., Benabadji, M. & Larrouy, G. (1966). Étude hémotypologique des populations sédentaires de la Saoura (Sahara occidental). I. Les groupes sanguins érythrocytaires. *Bulletin de la Société d'Anthropologie*, séries 11, **9**, 45–53. (France/HA/8.1/Algeria)

Ruffié, J., Carles-Trochain, E., Quilici, J. C. & Bouloux, C. (1969). Étude hémotypologique et épidémiologique des Maya de la région de Peto (Yucatan mexicain). *Bulletin de la Société d'Anthropologie*, séries 12, **4**, 281–94. (France/HA/8.3/Mexico)

Ruffié, J., Larrouy, G. & Vergnes, H. (1966). Hématologie comparée des populations amérindiennes de Bolivie et phénomènes adaptatifs. *Nouvelle Revue française d'Hématologie*, **6**, 544–52. (France/HA/8.3/Bolivia)

Ruffié, J., Quilici, J. C., Bouloux, C. & Carles-Trochain, E. (1968). Répartition des caractères sanguins chez les populations du Corridor Interandin. *8th International Congress of Anthropological and Ethnological Science*, Tokyo and Kyoto, 1968, **1**, 210–12. (France/HA/8.3/Bolivia)

Russell, D. A., Wigley, S. C., Vincin, D. R., Scott, G. C., Booth, P. B. & Simmons, R. T. (1971). Blood groups and salivary ABH secretion of inhabitants of the Karimui Plateau and adjoining areas of the New Guinea Highlands. *Human Biology in Oceania*, **1**, 79–89. (Australia/HA/2/New Guinea)

Rychkov, Yu. G. & Borodina, S. R. (1969). Gepersensitivnost' k feniltiomochevine v odnom iz izolyatov Sibiri. Vozmozhnaya gipoteza nasledovaniya. (Hypersensitivity to phenylthiorea, PTC, in an isolated population of Eastern Siberia. A possible hypothesis of its inheritance.) *Genetika*, **5**, 116–23 (USSR/HA/6)

Rychkov, Yu. G., Perevozchikov, I. V., Sheremet'yeva, V. A., Volkova, T. V. & Bashlay, A. G. (1969). K populyatsionnoy genetike korennogo naseleniya Sibiri. Vostochnyye Sayany. (To the population genetics of the native peoples of Siberia, Eastern Sayans.) *Voprosÿ antropologii*, **31**, 3–32. (USSR/HA/6)

Rychkov, Yu. G. & Sheremet'yeva, V. A. (1972). Populyatsionnaya genetika aleutov Komandorskikh Ostrovov (v svyazi s problemami istorii narodov i adaptatsii naseleniya drevney Beringii). (Population genetics of Aleuts in

The genetic markers of the blood

Commander Islands, (in connection with peoples history and population of ancient Beringean land).) *Voprosÿ antropologii*, **40**, 45–70; **41**, 3–18. (USSR/ HA/6)

Rychkov, Yu. G. & Sheremet'yeva, V. A. (1972). Populyatsionnaya genetika narodov Severa Tikhookeanskogo basseyna v svyazi s problemami istorii i adaptatsii naseleniya. iii. Populyatsii aziatskikh eskimosov i chukchey poberezh'ya Beringova morya. (Population genetics of peoples in the North of Pacific Ocean basin; the problems of their history and adaptation. iii. Populations of Asian Eskimos and Chukchi of Bering Sea coast.) *Voprosÿ antropologii*, **42**, 3–30. (USSR/HA/6)

Salzano, F. M., Franco, M. H. L. P. & Ayres, M. (1974). Alloalbuminemia in two Brazilian populations: a possible new variant. *American Journal of Human Genetics*, **26**, 54–8. (Brazil/HA/3)

Salzano, F. M., Gershowitz, H., Junqueira, P. C., Woodall, J. P., Black, F. L. & Hierholzer, W. (1972). Blood groups and H-Le^a salivary secretion of Brazilian Cayapo Indians. *American Journal of Physical Anthropology*, **36**, 417–25. (Brazil/HA/3 and USA/HA/26)

Salzano, F. M., Moreno, R., Palatnik, M. & Gershowitz, H. (1970). Demography and H-Le^a salivary secretion of the Macá Indians of Paraguay. *American Journal of Physical Anthropology*, **33**, 383–8. (Brazil/HA/3/Paraguay)

Salzano, F. M. & Neel, J. V. (1967). Fertility patterns and genetic structure of Xavante Indian populations. *Ciência e Cultura*, **19**, 64–6. (Brazil/HA/3 and USA/HA/26)

Salzano, F. M., Neel, J. V. & Maybury-Lewis, D. (1967). Further studies on the Xavante Indians. i. Demographic data on two additional villages: genetic structure of the tribe. *American Journal of Human Genetics*, **19**, 463–89. (Brazil/HA/3 and USA/HA/26)

Salzano, F. M., Neel, J. V., Weitkamp, L. R. & Woodall, J. P. (1972). Serum proteins, hemoglobins and erythrocyte enzymes of Brazilian Cayapo Indians. *Human Biology*, **44**, 443–58. (Brazil/HA/3 and USA/HA/26)

Salzano, F. M., Steinberg, A. G. & Tepfenhart, M. A. (1972). Gm and Inv allotypes of Brazilian Cayapo Indians. *American Journal of Human Genetics*, **25**, 167–77. (Brazil/HA/3)

Salzano, F. M., Woodall, J. P., Black, F. L., Weitkamp, L. R. & Franco, M. H.L.P. (1974). Blood groups, serum proteins and hemoglobins of Brazilian Tiriyo Indians. *Human Biology*, **46**, 81–7. (Brazil/HA/3 and USA/HA/26)

Sanghvi, L. D., Balakrishnan, V., Bhatia, H. M., Sukumaran, P. K. & Undevia, J. V. (ed.) (1974). *Human Population Genetics in India*. Proceedings of the 1st Conference of the Indian Society of Human Genetics, Bombay 1974. New Delhi: Orient Longman, Ltd.

Santachiara-Benerecetti, A. S., Beretta, M., Ulizzi, L. & Terrenato, L. The frequency of the red cell NADH diaphorase Dia^2 allele in Sardinia. *Human Heredity*, **22**, 290–3. (Italy/HA/5)

Santachiara-Benerecetti, S. A., Brunelli, F., Gigliani, F., Latte, B., Modiano, G., Negri, M., Santolamazza, C., Scozzari, R. & Terrenato, L. (1969). Further population data on red cell acid phosphatase, phosphoglucomutase and adenylate-kinase polymorphisms in Sardinia. *Rendiconti della Accademia dei Lincei*, series 8, **47**, 122–5. (Italy/HA/5)

Santachiara-Benerecetti, S. A. & Modiano, G. (1969). Studies on African Pygmies. ii. Red cell phosphoglucomutase studies in Babinga Pygmies: a common PGM$_2$ variant allele. *American Journal of Human Genetics*, **21**, 315–21. (Italy/HA/1/ Central Africa)

Schott, L. (1965). Untersuchungen zur PTC-Schmeckfähigkeit an Potsdamer Studenten. *Anthropologischer Anzeiger*, **29**, 213–27. (DDR/HA/15)

Sendrail, A. M. & Lefèvre-Wittier, Ph. (n.d.) Nouvelles données hémotypologiques en pays Touareg. II. Étude des hémoglobines dans le village d'Idélès (Ahaggar, Sahara algérien) et dans la fraction Kel Kummer (Touareg Iwellemeden Kel Attaram, Mali). Typescript. (France/HA/8.1/Sahara)

Sendrail, A. & Quilici, J. C. (1970). Étude des hémoglobines des habitants du Corridor Interandin. *Anthropologie, Paris*, **74**, 269–74. (France/HA/8.3/ Bolivia and Peru)

Série, C., Vergnes, H. & Samson, M. (1968). Les enzymes érythrocytaires dans les populations de la Guyane française. *Bulletin de la Société d'Anthropologie*, séries 12, **3**, 283–8. (France/HA/8.3/French Guiana)

Shinoda, T. (1969). A note on the frequency of red cell acid phosphatase types in Japan. *Japanese Journal of Human Genetics*, **13**, 249–55. (Japan/HA/?3)

Shinoda, T. (1970). Inherited variation in tetrazolium oxidase in human red cells. *Japanese Journal of Human Genetics*, **15**, 144–52. (Japan/HA/?3)

Shinoda, T. (1970). Inherited variations in red cell phosphoglucos eisomerase among Japanese. *Japanese Journal of Human Genetics*, **15**, 159–65. (Japan/HA/?3)

Shinoda, T. (1970). Polymorphism of red cell adenosine deaminase in the Japanese population. *Japanese Journal of Genetics*, **45**, 147–52. (Japan/HA/?3)

Shinoda, T. & Matsunaga, E. (1967). Serum alkaline phosphate variants and their association with ABO blood groups in a Japanese sample. *Japanese Journal of Human Genetics*, **12**, 170–6. (Japan/HA/?3)

Shinoda, T. & Matsunaga, E. (1970). Polymorphism of red cell phosphoglucomutase among Japanese. *Japanese Journal of Human Genetics*, **14**, 316–23. (Japan/HA/ ?3)

Shinoda, T. & Matsunaga, E. (1970). Studies on polymorphic types of several red cell enzymes in a Japanese population. *Japanese Journal of Human Genetics*, **15**, 133–43. (Japan/HA/?3)

Shreffler, D. C. & Steinberg, A. G. (1967). Further studies on the Xavante Indians. IV. Serum protein groups and the SC_1 trait of saliva in the Simões Lopes and São Marcos Xavantes. *American Journal of Human Genetics*, **19**, 514–23. (USA/HA/26/Brazil)

Simmons, R. T. (1969). The Diego (Di) blood group: some recent observations. *Anthropologist*, special vol., 45–51. (Australia)

Simmons, R. T. (1970). The apparent absence of the Diego (Di[a]) and the Wright (Wr) blood group antigens in Australian aborigines and in New Guineans. *Vox sanguinis*, **19**, 533–6. (Australia/HA/1)

Simmons, R. T. (1970). X-linked blood groups, Xg, in Australian aborigines and New Guineans. *Nature, London*, **227**, 363. (Australia)

Simmons, R. T. & Booth, P. B. (1971). A compendium of Melanesian genetic data. I. ABO, MNSs and Rh blood groups. Typescript. (Melanesia)

Simmons, R. T. & Cooke, D. R. (1969). Population genetic studies in Australian aborigines of the Northern Territory. Blood group genetic studies in the Malag of Elcho Island. *Archaeology & Physical Anthropology in Oceania*, **4**, 252–9. (Australia/HA/1)

Simmons, R. T. & Graydon, J. J. (1971). Population genetic studies in Australian aborigines of the Northern Territory. Blood group genetic studies on populations sampled at 16 localities including Arnhem Land and Groote Eylandt. *Human Biology in Oceania*, **1**, 23–53. (Australia/HA/1)

Simmons, R. T., Graydon, J. J., Gajdusek, D. C., Alpers, M. P. & Hornabrook, R. W. (1972). Genetic studies in relation to kuru. II. Blood-group genetic patterns in kuru patients and populations of the Eastern Highlands of New Guinea. *American Journal of Human Genetics*, **24**, suppl., S39–S71. (Australia/HA/4/New Guinea)

Singer, R. (1960). Some biological aspects of the Bushman. *Zeitschrift für Morphologie und Anthropologie*, **51**, 1–6. (USA/HA/4/Africa)

Singer, R. (1961). Serum haptoglobins in Africa. *South African Medical Journal*, **35**, 520–3. (USA/HA/4/Africa)

Singer, R. (1962). The significance of the sickle cell in Africa. *Leech*, **32** (5), 152–61. (USA/HA/4/Africa)

Singer, R. (1965). The biology of the Bushman. In *The Bushman*, ed. P. V. Tobias. Johannesburg: Institute for the Study of Man in Africa. (USA/HA/4/South Africa)

Singer, R. (n.d.) Evolution and genetics of South African races of man. Typescript. (USA/HA/4/South Africa)

Singer, R. (1970). Investigations on the biology of Hottentot and Bushman populations in Southern Africa. Proceedings of the IBP/HA symposium Man in Africa. *Materiały i Prace antropologiczne*, **78**, 37–48. (USA/HA/4/South Africa)

Singer, R. & Weiner, J. S. (1963). Biological aspects of some indigenous African populations. *Southwest Journal of Anthropology*, **19**, 169–76. (USA/HA/4/Africa)

Singer, R., Weiner, J. S. & Zoutendyk, A. (1963). The blood groups of the Hottentots. *2nd International Congress of Human Genetics*, Rome 1961, pp. 884–7 Rome: Istituto Gregorio Mendel. (USA/HA/4/Africa)

Sinnett, P., Blake, N. M. Kirk, R. L., Lai, L. Y. C. & Walsh, R. J. (1970) Blood, serum protein and enzyme groups among Enga-speaking people of the Western Highlands, New Guinea, with an estimate of genetic distance between clans. *Archaeology & Physical Anthropology in Oceania*, **5**, 236–52.

Spielman, R. S. (1973). Differences among Yanomama Indian villages: do the patterns of allele frequencies, anthropometrics and map locations correspond? *American Journal of Physical Anthropology*, **39**, 461–79. (USA/HA/26/Venezuela and Brazil)

Spitsyn, V. A. (1973). Interpretation of population-genetic investigations of blood enzymes with special reference to the genesis of the Finno-Ugric peoples. In 'Studies in the Anthropology of the Finno-Ugrian Peoples'. Helsinki: Archaeological Institute of the University of Helsinki. Stencil no. 7.

Stamatoyannopoulos, G., Panayotopoulos, A. & Motulsky, A. G. (1966). The distribution of glucose-6-phosphate dehydrogenase deficiency in Greece. *American Journal of Human Genetics*, **18**, 296–308. (USA/HA/25/Greece)

Stamatoyannopoulos, G., Voigtlander, V., Kotsakis, P. & Akrivakis, A. (1971). Genetic diversity of the 'Mediterranean' glucose-6-phosphate dehydrogenase deficiency phenotype. *Journal of Clinical Investigation*, **50**, 1253–61. (USA/HA/25/Greece)

Steinberg, A. G., Damon, A. & Bloom, J. (1972). Gammaglobulin allotypes of Melanesians from Malaita and Bougainville, Solomon Islands. *American Journal of Physical Anthropology*, **36**, 77–84. (USA/HA/2/Melanesia)

Steinberg, A. G. & Kirk, R. L. (1970). Gm and Inv types of aborigines in the Northern Territory of Australia. *Archaeology & Physical Anthropology in Oceania*, **5**, 163–72. (Australia/HA/1)

Steinberg, A. G. & Morton, N. E. (1973). Immunoglobulins in the Eastern Carolines. *American Journal of Physical Anthropology*, **38**, 699–702. (USA/HA/12)

Steinberg, A. G., Tilikainen, A., Eskola, M.-R. & Eriksson, A. W. (1974). Gammaglobulin allotypes in Finnish Lapps, Finns, Åland Islanders, Maris (Cheremis), and Greenland Eskimos. *American Journal of Human Genetics*, **26**, 223–43. (Finland/HA/2)

Sunderland, E. & Coope, E. (1973). Biological studies of Yemenite and Kurdish Jews in Israel and other groups in south-west Asia. XII. Genetic studies in Jordan. *Philosophical Transactions of the Royal Society*, B, **266**, 207–20. (UK/HA/8/Jordan)

Sunderland, E., Tills, D., Bouloux, C. & Doyle, J. (1973). Genetic studies in Ireland. In *Genetic Variation in Britain*, ed. D. F. Roberts & E. Sunderland, Symposium of the Society for the Study of Human Biology, Vol. **12**, pp. 141–59. London: Taylor & Francis. (UK/HA/19/Ireland)

Szewko-Szwaykowska, I. (1966). Badania ludności osady kaszubskiej Kuźnica na Helu. (Investigations of a Kashubian settlement in Kuźnica at the peninsula of Hel.) *Materiały i prace antropologiczne*, **73**, 129–60. (Poland/HA/10)

Tanis, R. J., Neel, J. V., Dovey, H. & Morrow, M. (1973). The genetic structure of a tribal population, the Yanomama Indians. IX. Gene frequencies for 18 serum protein and erythrocyte enzyme systems in the Yanomama and five neighboring tribes: nine new variants. *American Journal of Human Genetics*, **25**, 655–76. (USA/HA/26/South America)

Tashian, R. E., Brewer, G. J., Lehmann, H., Davies, D. A. & Rucknagel, D. L. (1967). Further studies on the Xavante Indians. V. Genetic variability in some serum and erythrocyte enzymes, hemoglobin, and the urinary excretion of β-aminoisobutyric acid. *American Journal of Human Genetics*, **19**, 524–31. (USA/HA/26/Brazil)

Terrenato, L., Loghem, E. van, Bernini, L., Santachiara-Benerecetti, S. A., Modiano, G., Santolamazza, C., Scozzari, R., Ulizzi, L. & Beretta, M. (1971). Preliminary data on the genetic heterogeneity among the Sardinian isolates. *Rendiconti della Accademia dei Lincei*, series 8, **51**, 249–53. (Italy/HA/5)

Tills, D. (1969). Protein and enzyme polymorphisms in human populations with special reference to Middle Eastern populations including their blood groups. M.Phil. thesis, Council for National Academic Awards, London.

Tills, D. (1975). *Population genetics of the Irish*. Ph.D. thesis, Council for National Academic Awards, London.

Tills, D. (1976). Red cell and serum proteins of the Irish. *Annals of Human Biology* (in press).

Tills, D., Teesdale, P. & Mourant, A. E. (1976). Blood groups of the Irish. *Annals of Human Biology* (in press).

Tjong Tjin Joe, J., Prins, H. K. & Nijenhuis, L. E. (1965). Hereditary and acquired blood factors in the Negroid population of Surinam. I. Origin, collection, and transport of the blood samples; characteristic blood groups. *Tropical Geographical Medicine* **17**, 56–60. (Netherlands/HA/3/Surinam)

Tobias, P. V. (1966). The peoples of Africa south of the Sahara. In *The Biology of Human Adaptability*, ed. P. T. Baker & J. S. Weiner, pp. 111–200. Oxford: Clarendon Press. (South Africa/HA/2)

Tobias, P. V. (1972). Recent human biological studies in southern Africa, with special reference to Negroes and Khoisans. *Transactions of the Royal Society of South Africa*, **40** (iii), 109–33. (South Africa/HA/2)

Tobias, P. V. (1974). The biology of the Southern African Negro. In *The Bantu-speaking Peoples of Southern Africa*, 2nd edn, ed. W. D. Hammond-Tooke, pp. 3–45. London & Boston: Routledge & Kegan Paul.

Triginer, J. & Pons, J. (1967). Contribución al estudio de los grupos sanguíneos MN y Rh en los asturianos. *Revista de la Facultad de Ciencias, Oviedo*, **8**, (2), 3–7. (Spain/HA/3)

The genetic markers of the blood

Valaoras, V. G. (1968). The A.B.O. blood groups in Greece [In Greek]. *Praktika tes Akademias Athenon*, **48**, 479–97. (Greece/HA/6)

Valaoras, V. G. (1970). Biometric studies of army conscripts in Greece. Mean stature and ABO blood-group distribution. *Human Biology*, **42**, 184–201. (Greece/HA/6)

Valaoras, V. G. (1971). Corporal development and other characteristics in school children and university students in Greece. *Social Biology*, **18**, 397–405. (Greece/HA/5)

Vergnes, H. & Larrouy, G. (1967). Les déficits en G6PD dans les populations des Andes boliviennes. *Nouvelle Revue française d'Hématologie*, **7**, 124–8. (France/HA/8.3/Bolivia)

Vergnes, H. & Quilici, J. C. (1967). Le gène E_1^a de la pseudo-cholinestérase sérique (A.C.A.H.) chez les Amérindiens. *Annales de Génétique*, **13**, 96–9. (France/HA/8.3/South America)

Villiers, H. de (1976). A study of morphological variables and genetic markers in urban and rural South African Negro male populations. *9th International Congress of Anthropological and Ethnological Sciences*, Chicago 1973 (in press). (South Africa/HA/7)

Vogel, F. & Chakravartti, M. R. (1966). ABO blood groups and smallpox in a rural population of West Bengal and Bihar (India). *Humangenetik*, **3**, 166–80. (BRD/HA/2/India)

Vogel, F. & Chakravartti, M. R. (1966). ABO blood groups and the type of leprosy in an Indian population. *Humangenetik*, **3**, 186–8. (BRD/HA/2/India)

Vogel, F., Krüger, J., Chakravartti, M. R., Flatz, G. & Ritter, H. (1971). Inv phenotypes and quantitative gamma globulin determinations in leprosy patients and control populations from India and Thailand. *Humangenetik*, **12**, 35–41. (BRD/HA/2/India)

Vogel, F., Krüger, J., Chakravartti, M. R., Ritter, H. & Flatz, G. (1971). ABO blood groups, Inv serum groups, and serum proteins in leprosy patients from West Bengal (India). *Humangenetik*, **12**, 284–301. (BRD/HA/2/India)

Walter, H., Kellermann, G., Bajatzadeh, M., Krüger, J. & Chakravartti, M. R. (1972). Hp, Gc, Cp, Tf, Bg and Pi phenotypes in leprosy patients and healthy controls from West Bengal (India). *Humangenetik*, **14**, 314–25. (BRD/HA/2/India)

Walter, H. & Nemeskéri, J. (1967). Demographical and sero-genetical studies on the population of Bodrogköz (NE Hungary). *Human Biology*, **39**, 224–40. (Hungary/HA/2)

Walter, H. & Nemeskéri, J. (1969). Blut- und Serumgruppendaten aus zwei Hegyköz-Orten: Kovácsvágás und Végardó. *Anthropologiai Közlemények* **13**, 69–78. (In Hung.; German summary.) (Hungary/HA/2)

Walter, H., Neumann, S. & Nemeskéri, J. (1968). Investigations on the occurrence of glucose-6-phosphate-dehydrogenase deficiency in Hungary. *Acta genetica et statistica medica*, **18**, 1–11. (Hungary/HA/2)

Ward, R. H. (1972). The genetic structure of a tribal population, the Yanomama Indians. v. Comparisons of a series of genetic networks. *Annals of Human Genetics*, **36**, 21–43. (USA/HA/3/Venezuela and Brazil)

Ward, R. H. & Neel, J. V. (1970). Gene frequencies and microdifferentiation among the Makiritare Indians. iv. A comparison of a genetic network with ethno-history and migration matrices; a new index of genetic isolation. *American Journal of Human Genetics*, **22**, 538–61. (USA/HA/26/Venezuela)

Weiner, J. S. & Huizinga, J. (ed.) (1972). *The Assessment of Population Affinities in Man*. Oxford: Clarendon Press.

Weitkamp,, L. R., Arends, T., Gallango, Maria L., Neel, J. V., Schultz, J. & Shreffler, D. C. (1972). The genetic structure of a tribal population, the Yanomama Indians. III. Seven serum protein systems. *Annals of Human Genetics*, 35, 271–9. (USA/HA/26/Venezuela and Brazil)

Weitkamp, L. & Neel, J. V. (1970). Gene frequencies and microdifferentiation among the Makiritare Indians. III. Nine erythrocyte enzyme systems. *American Journal of Human Genetics*, 22, 533–7. (USA/HA/26/Venezuela and Brazil)

Weitkamp, L. R. & Neel, J. V. (1972). The genetic structure of a tribal population, the Yanomama Indians. IV. Eleven erythrocyte enzymes and summary of protein variants. *Annals of Human Genetics*, 35, 433–44. (USA/HA/26/ Venezuela)

Welch, S. G., Barry, J. V., Dodd, B. E., Griffiths, P. D., Huntsman, R. G., Jenkins, G. C., Lincoln, P. J., McCathie, M., Mears, G. W. & Parr, C. W. (1973). A survey of blood group, serum protein and red cell enzyme polymorphisms in the Orkney Islands. *Human Heredity*, 23, 230–40. (UK/Circumpolar)

Welch, Q. B. & Lie-Injo Luan Eng. (1972). Serum albumin variants in three Malaysian racial groups. *Human Heredity*, 22, 503–7. (Malaysia/HA/4)

Welch, Q. B., Lie-Injo Luan Eng & Bolton, J. M. (1971). Adenylate kinase and malate dehydrogenase in four Malaysian racial groups. *Humangenetik*, 14, 61–3. (Malaysia/HA/4)

Yamamoto, M. & Fu, L. (1973). Red cell isozymes in the Eastern Carolines. *American Journal of Physical Anthropology*, 38, 703–7. (USA/HA/12/Caroline Is)

Yamamoto, M., Wada, T., Watanabe, T., Kanazawa, H., Saito, R., Kondo, M., Hosokawa, K., Masuda, M., Nakai, T. & Fujiki, N. (1972). Genetic polymorphisms in four isolated communities in Kinki district. *Japanese Journal of Human Genetics*, 17, 273–85. (Japan/HA/3).

Yanase, T., Fujiki, N., Handa, Y. Yamaguchi, M., Kishimoto, K., Furusho, T., Tsuji, Y. & Tanaka, K. (1973). Genetic studies on inbreeding in some Japanese populations. XII. Studies of isolated populations. *Japanese Journal of Human Genetics*, 17, 332–66. (Japan/HA/1)

3. The genetic process in the system of ancient human isolates in North Asia

YU. G. RYCHKOV & V. A. SHEREMETYEVA

Isolated fringe tribes and peoples make up part of that *terra incognita* that had always whetted man's thirst for knowledge and kindled his imagination. As the modern concept of evolution matured and took shape in the nineteenth century, it drew the attention of those who, in the kaleidoscopic diversity of their cultures, customs and biological features, sought both proof of boundless evolutionary transformations and evidence of evolution on the wane, of stagnation and degeneration.

By the mid twentieth century, with the evolution doctrine firmly entrenched and the type of civilisation that engendered it spreading to every continent, interest in the scant surviving fragments of these all-but-lost worlds, far from flickering out, had strengthened.

Ideas, as a Russian philosopher aptly noted, are indeed the shadows of events to come. With man disrupting the balance of our planet's natural forces, at last everyone is beginning to hear the underground rumblings of the gathering catastrophe thus generated. Warnings, declarations and laws adopted by both international organisations and governments, meant to protect and stabilise man's position in the environment, suggest an understanding of what is essentially a self-regulatory process.

Awareness of how limited is this knowledge nourishes the interest in isolates – those ancient segregated human populations that were able, throughout a long sequel of generations, to attain and maintain an equilibrium with the environment.

The problem of the stability of isolates merits the attention even of those people who keenly follow mankind's prospects in this space age, with its different time and spatial dimensions against which our planet is itself but an isolate of life and reason.

Let us, then, examine the isolate, not with the preconception that it is some symbol of degeneration and evolution on the wane but in the hope of discovering precious clues to the mechanisms and paths leading to the stability and equilibrium which are so urgently needed today.

Every isolate is a population. Organisation of life at the population level needs no explanation. However alluring the problems of the molecular basis of genetic information and the principles of heredity underlying the development and functioning of organisms, the fate of genes is determined by the principles of population structure and existence. Compared to the life span of a single organism, a population is an immortal super-organism.

Of all the diverse approaches used to investigate this level of human society, we chose the genetic one. The theory and methods of population genetics permit the fate of genetic information in a population to be described in the most general form and expressed quantitatively. It is for this reason that the basic concepts of population genetics assume the role of a framework in today's complicated concept of evolution.

Below, we shall just briefly touch on those key points in the population genetics theory which the reader might find helpful in grasping the ideas, methods and structure of this work and also in critically comprehending its conclusions.

Problems of the evolution and stability of a population in the light of population genetics

It is assumed that evolution in populations proceeds by genetic reorganisation from generation to generation. A population is an elementary evolving unit and evolution is unthinkable without genetic reorganisation. The content of genetic information is determined by the concentration of genes in a population: a change of gene concentration represents an elementary evolutionary action and such a change in a population is, therefore, the central theme of population genetics. The genetic differences between populations, established both by individual genes and by a set of gene loci, are measured by the criteria of kinship and generalised genetic distances and are interpreted as a function of genetic dynamics (of the evolution of populations). These dynamics are based on the processes of mutation, natural selection, gene migration and random gene drift. The effect of each of these factors was established in human populations, and became the subject of special investigations, now regarded as classical, among which we would like to mention those by Allison (1955) on natural selection, Glass (1956) on random gene drift and Glass & Li (1953) on gene migration. A detailed review of data on the mutation process can be found in the book by Cavalli-Sforza & Bodmer (1971).

In nature, a population is subjected to the simultaneous action of all the factors of evolution dynamics. The change of gene concentration in each generation is so small that one can regard a population's evolutionary transformation as a continuous, uniform probability process. Considerable deviations in gene concentration can accumulate only with a great number of generations (something, of course, out of the experimenter's reach), and will give rise to essentially new biological characters. However, the methods of population genetics make it possible to trace even those minute changes that occur from one generation to another. Yet, possibilities for studying changes in gene concentration are not restricted simply to observations over varying periods of time. Let us imagine something like a bee swarm: every new generation leaves the parent population to form a new population, the number of which corresponds to the number of generations. All of them, having dispersed and simultaneously existing in space, will differ one from another, and from the parent population, in gene concentration. Thus, spatial differences between populations can be likened to the changes occurring in time within the ancestral population itself, reproducing itself in new points in space, and, consequently, in new environmental conditions.

Thus, the principle of ergodicity, extended from the realm of physics to that of population genetics, allows one to transform a static picture of spatial gene distribution into a process unfolding in time. Man's genogeography of the world, presented, for example, in Mourant's compendium (1954), and the book by Mourant, Kopeć & Domaniewska-Sobczak (1958), indicates two basic types of spatial gene distribution: regular and irregular. These distributions usually overlap: within a distribution that is on the whole clinal, one encounters gene frequencies deviating more or less significantly from the expected gradient. More often than not these deviations can be detected on a (general enough) regional or global scale, i.e. the large is better viewed from afar. 'From afar', however, one sees not simply 'the large' (the main) but what is distinguished for greater stability over historical, prehistorical, and, sometimes, geological time. Therefore, in order to approximate spatial gene distribution by means of dynamic models of population genetics, it is necessary at the same time to turn to models of a genetically stable population. Had not a population, subjected to systematic and random environmental pressures, the possibility of remaining genetically stable (i.e. of preserving its genetic structure constant), it would be impossible to find any regularities in a spatial projection of the process of genetic evolution which were manifested further back than a few immediate generations.

However, the practical problems of the stability of genetic information in populations, also elaborated theoretically, have so far been much less extensively investigated. Models of population genetics presuppose the existence of so-called equilibrium gene concentration values. Apart from causing changes in gene frequencies, the above-mentioned factors disrupting the stability of a population are capable, either in partial or general interaction with one another and after a certain number of steps from the beginning of operation, of returning the population to a state of genetic stability. The new state of equilibrium at new gene concentrations is determined by a balance of opposing forces, engendered by the pressure of these factors.

Should one expect other than rapid attainment of stable gene concentration equilibrium, given the simultaneous effect of all the so-called evolution-

Yu. G. Rychkov and V. A. Sheremetyeva

ary forces on a population? It could be expected that the genetic reorganisation of populations in generations or in space, recorded in a series of observations, would be represented by fluctuations around some level of gene frequencies corresponding to equilibrium values, balanced by the counteraction of diverse environmental influences on the population. Such genetic reorganisation would evolve into a stationary process. The types of distribution of gene concentrations possible in this case are envisaged by the stationary distribution models suggested by Wright (1931), at the very beginning of the development of population genetics, which, nevertheless, have failed to attract attention. Later, we shall turn to the search for stationary states of populations in nature. For the moment, though, we shall discuss another essential point – the choice of the subject to be investigated.

Justifying the choice of subject

The study of population genetics is as much an experimental as it is a theoretical discipline. Many of its models of selection and evolution were subjected to experimental verification in the laboratory, where the investigator himself is both the creator and controller of a population's history.

Yet nothing other than field work can furnish information on the actual processes of genetic reorganisation occurring in natural populations. The task here is to learn not only the genetic structure, but all the components that together make up the natural history of a population In this respect, it is possible to carry out examinations of human populations which involve all the information required by theoretical models but seldom available when studying other populations. It goes without saying that the planning of such field work is as essential as the laboratory research. The available genetic information has been obtained mainly by methods of serology and biochemistry. These, however, are merely the techniques. More important is the selection of an isolated subject. The effects of isolation among religious sectarian communities, e.g. the Dunkers, Mormons and others in the USA, are widely known. One may question, though, whether they reflect the actual genetic history of isolated human populations. The emergence of isolates of religious sects is in essence equal to sympatric formation, which doesn't occur naturally very often. They only became segregated as late as the early eighteenth century from the total mass of central Europe's population, which was dense and already saturated with mutations. Clearly, isolated communities in this case may serve to crystallise such mutations and to accumulate corresponding anomalous features; a theory which is now being borne out by population genetics studies.

On the whole it is doubtful that such an isolated subject will fit both the particular state of population isolation with which we are concerned and the

50

general separation that dominated humanity right up to the age of the great geographical discoveries that culminated in the Russian discoveries of the eighteenth century in North Asia and North America.

So, let us turn to North Asia, an area where the continuous development of populations goes back far earlier than the eighteenth century. A good deal of time 'reserve' is essential in our search, for only then shall we be able to ascertain the actual stability of populations, in the hope that time will have left its imprint on their genetic structure. The affinities of Siberian and American populations and the peopling of America via North Asia is generally recognised: this event is radiocarbon dated at *c.* 30000 to 25000 years B.P. The Upper Palaeolithic human sites discovered in Siberia and in Kamchatka are radiocarbon dated at *c.* 35000 to 15000 years B.P. The Neolithic sites in southern Siberia, around Lake Baikal and in eastern Siberia belong to a period of up to 7000 years B.P. Plenty of even later sites are known.

But records of population existence over a long period of time are not enough to justify selection for study. There must be other criteria: chiefly demographic and cultural stability in the region selected. For instance, the region should incorporate at least one stable population. During the same time spans development of populations went on rapidly in adjacent regions: the growth of multimillion peoples in Europe and the Far East, and the formation of the empires of the Huns, Turks, Tungus-Manchus and the Mongols on the southern borders of North Asia. Against this background the region we are concerned with did incorporate populations which were stable in regard to size and culture dynamics and remained so practically up to recent times.

We can trace the genealogy of populations by analysis of linguistic data (Fig. 3.1). The basic branches of the genealogical tree coincide with the classification of anthropological types, confirming that we are indeed looking at the result of the differentiation in time of an initial group of early invaders of Siberia (see Rychkov, 1969). As we see, the dramatic events in the history of the peoples of adjacent countries did not essentially affect the region we selected for investigation. The entire region looks like a giant isolate on the extreme fringe of the Old World.

We can overlook the seeming contradiction, that an isolate is associated with a small exclusive group whereas here we are dealing with a mammoth territory, because such sources as the Yasak books reveal that the total Siberian population in the seventeenth century was only 200000 to 250000 (Dolgikh, 1960; Rychkov, 1965*b*). For population genetics it is only the effective reproducing part of the population that is important, and this, after the exclusion of elderly people, children and juveniles, came to no more than 100000 at that time. Today this number has grown by only four times, which has happened to only three out of the 30 North Asian peoples. These numbers are comparable with individual local populations in other species of animals. As it happened, this restricted group was

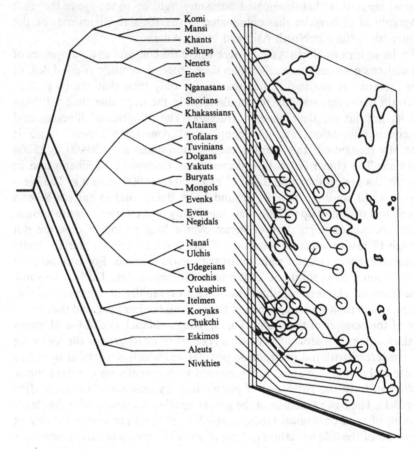

Fig. 3.1. Approximate ethnolinguistic tree of the indigenous populations of North Asia, projected onto their geographical positions.

able to disperse freely over a vast subcontinent and formed an aggregate of populations separated in different degrees. What should be the approach to this situation?

Planning research

Irrespective of how versed the reader is in population genetics, he will hardly find in its theory any indications of the principles on which an investigation should be organised. However, on the basis of the prominence given by theory to the fate of genes in an individual elementary population. choosing some typical group and examining it in detail would seem the best place to start.

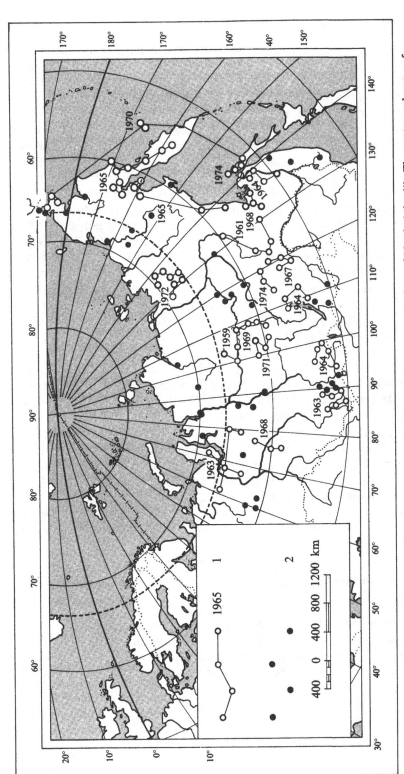

Fig. 3.2. The routes and stations of anthropological genetic studies of the indigenous populations of North Asia. (1) The routes and years of Moscow University Expeditions conducted by the authors and their collaborators; (2) working stations of other anthropological and medical expeditions.

Yu. G. Rychkov and V. A. Sheremetyeva

But how is one to know that the group chosen is the most typical? How much more characteristic are the Tofalars (Karagas) of the Sayan mountains than the Tungus-speaking Evenks on the Yenisei, and the latter more than the Tungus of the Sea of Okhotsk, the Koryak of Kamchatka, the Eskimo of Chukotka, or the Nentsi of Yamal? Where is the certainty that the results of the most detailed investigation of such a population unit will be true and representative of all other units? Siberia and its indigenous peoples have entered a stage of rapid socio-economic and environmental transformations which, accordingly, leave very little time to the investigators. In view of this, the authors conducted a genetic study of a randomised representative sample from all the indigenous Siberian populations, within the boundaries of the North Asian subcontinent, in the period from 1963 to 1975. The work was greatly assisted by students of Moscow University (Rychkov, 1965a, b, 1968, 1969, 1973, 1975; Rychkov et al., 1969; Rychkov & Borodina, 1969, 1973; Rychkov & Sheremetyeva, 1972a, b, c, 1974, 1976a, b, c; Rychkov, Rusakova, Rappoport & Sheremetyeva, 1973; Rychkov et al., 1974a, b; Sheremetyeva, 1971, 1973, 1975; Zhukova, Bronnikova, Rychkov & Sheremetyeva, 1975). In addition, it was predominantly during this period that a large number of other investigators were working in Siberia, and their published works have also been used in this review (Zhinkin, 1927; Shluger, 1940; Debetz & Trofimova, 1941; Yarkho, 1947; Debetz, 1951; Levin, 1958a, b, 1959; Galperin & Zhitomirski, 1967; Spitzin, 1967, 1976; Zolotaryova, 1968; Alexeyev et al., 1968; Alexeyeva et al., 1970; Gorokhov, 1971; Alexeyeva et al., 1972; Spitzin & Irissova, 1973; Seleznyova, 1974; Voronina, 1975).

Before turning to a description of samples and techniques let us note that work in North Asia required an adaptation of serological methods to field conditions and a laboratory that would not only carry out standard serological, cytogenetical, physiological and other techniques but be small in size and transportable by both modern and also traditional Siberian means, such as boats and draught animals, including reindeer. It was with such a laboratory that during 1963–74 the authors travelled the routes shown in Fig. 3.2.

Samples and techniques

Samples

The main purpose of the present review is a genetic characterisation of an isolate at its different structural levels, and a summary of the population samples at three levels (total, ethnic, and elementary populations) is given in Table 3.1. Thus, the total Siberian population is characterised on the basis of a 2 to 0.06% sampling of individuals. The population at the ethnic level is characterised on the basis of an 88 to 19% sample of the total number of ethnic groups in Siberia, and a 20 to 2% sample of the mean size of an ethnic group.

Table 3.1. *A summary characterisation of data on the genetics of the indigenous population of North Asia*

Locus	Number of ethnic groups	Number of populations	Number of individuals
ABO	28 (25)	212 (137)	15929 (5883)
Se (ABH)	13 (10)	48 (43)	972 (745)
Lewis	18 (17)	78 (67)	3278 (1913)
MNSs:			
MN	28 (25)	180 (136)	10005 (5362)
Ss	21 (21)	53 (53)	1362 (1362)
Rhesus:			
Dd	28 (25)	155 (129)	6824 (4084)
Cc	22 (21)	77 (74)	2131 (1685)
Ee	22 (21)	77 (74)	2131 (1685)
P	28 (23)	115 (100)	4372 (2886)
Kell	13 (10)	45 (38)	1663 (1022)
Diego	6 (6)	24 (24)	464 (464)
Duffy	9 (9)	38 (38)	944 (944)
Hp	20 (10)	59 (38)	2698 (1371)
Tf	17 (10)	53 (29)	2071 (979)
Gc	9 (7)	29 (13)	899 (143)
PTC	17 (10)	60 (40)	2823 (1303)

Total scope of data – including those of the authors, which are given in parentheses.

An elementary population is characterised on the basis of a 20 to 2.5 % sample from the total number of populations and a 20 % sample based on the mean population size.

Techniques

Blood groups

Bearing in mind that the summary consists in considerable measure of our own data, we present a brief description of the techniques used. All blood group determinations were carried out on the spot, in the field laboratory on the same day, or the day after, samples were drawn. For the preservation of samples, such natural cold storages were used as the frost in winter time and permafrost in summer. All agglutination reactions (with the exception of the ABO and MN blood groups) were checked microscopically against a five-point scale. The following techniques were used: saline, Fisk & MacGuy, Bashlai, and Coomb's indirect test. The iso-agglutinins of the ABO system and the incomplete anti-D antibodies produced by the Moscow City Blood Transfusion Station, antisera of the MN, P, and Lewis systems produced by the Moscow Institute of Forensic Medicine; specificity sera anti-C,

55

anti-c, anti-E, anti-e, anti-(Fy^a + Kell), and anti-Human-globulin from Dr A. E. Mourant at the Blood Group Reference Laboratory, England, anti-Fy^a, anti-Kell and anti-E from the National Reference Laboratory, Canadian Red Cross Blood Transfusion Service, Toronto; anti-Fy^a, anti-Kell, anti-Cellano, and anti-Di^a from the American National Red Cross; anti-Fy^a and anti-Kell from King County Centre Blood Bank, Seattle, Washington; anti-Di^a from Dr M. Layrisse, Venezuela; anti-C, anti-c, anti-E, anti-e, and anti-s from the firm of Schwab–Reagenzien Vienna, Austria.

In some cases when working on the northern sea coasts, sea water was used as a saline medium.

Serum proteins

The determination of the electrophoretic variants of haptoglobin, transferrin and the group-specific component were conducted at the end of each expedition in a laboratory in Moscow. Horizontal and vertical electrophoresis in a starch gel was used.

PTC taste reaction

The determination of the minimum sensitivity threshold to a solution of phenylthiocarbamide (PTC) was conducted according to the modified method of Harris & Kalmus (Rychkov & Borodina, 1973).

Colour perception

(Rabkin's (1965) polychromatic tables were used. Not one of the populations studied revealed significant colour blindness so this character was not used further.

General genetic-anthropological characteristics of the indigenous population of North Asia

Owing to a number of specific difficulties facing the systematist in this part of the globe, the anthropological classification of the physical types of Siberian peoples is still far from complete. These difficulties are connected with the highly fractional ethnic composition of the population, with small individual ethnic groups and extremely large and frequently overlapping ethnic habitats. Populatious from the same ethnic group, in different parts of the area, reveal differences frequently commensurate with the total value of differences between ethnic groups in Siberia.

The stability of combinations of characters that have singled out one or other physical type is not very great, while the taxonomic validity of any

Table 3.2. *Mean population values for some anthropological characters of the indigenous populations of North Asia*

Characters	N	$\bar{M}_x$	σ_x	As_x	Es_x	F	CV_x	$\bar{\sigma}_x$
Stature	158	160.89	3.55	−0.4391***	+1.3219***	3.51	2.21	5.62
Head length	185	190.23	2.33	−0.0303	+0.2157	7.59	1.22	5.98
Head breadth	185	157.23	3.67	−0.0224	−0.9825***	2.98	2.33	5.17
Head height	20	127.36	2.41	+0.0039	−0.2348	11.39	1.89	4.41
Minimum frontal diameter	103	104.52	2.64	+0.1825	−0.2591	3.79	2.25	7.77
Bizygomatic diameter	182	146.72	3.36	−0.2944***	−0.3619	3.26	2.29	5.05
Jaw bigonial diameter	153	115.09	3.15	−0.2040	+0.0038	4.52	2.74	5.91
Physiognomic height of face	128	193.64	4.33	−0.4699**	−0.7106	4.97	2.24	8.64
Morphological height of face	141	134.20	3.63	−0.4345***	+0.3495	4.59	2.70	6.88
Height of nose from nasion	124	62.65	2.96	+0.3604**	−0.6959	3.38	4.72	4.56
Width of nose	142	37.25	1.13	+0.0730	−0.3353	5.46	3.03	2.74
Width of mouth	92	53.61	1.75	+0.4020	−0.2207	5.82	3.26	3.84
Height of upper lip	95	17.81	1.45	−0.0612	+0.7303	4.36	8.14	2.65
Lip thickness	85	16.59	1.65	−0.3589	−0.0128	6.65	9.95	3.92
Cephalic index	180	81.00	2.27	+0.0196	−0.7229	2.94	2.80	3.17
Facial morphological index	98	90.71	2.25	−0.5688***	+0.9660***	5.50	2.48	4.77
Nose index	132	59.98	3.13	−0.1836	+0.3969	4.30	5.22	5.69
Eye colour, percentage of light and mixed	110	24.40	15.87	—	—	—	65.04	—
Hair colour, percentage of No. 27, Fisher's scale	87	34.02	20.58	—	—	—	60.48	—
Skin colour, percentage of swarthy hues (Nos. 9–14, Lushan's scale)	54	39.57	21.40	—	—	—	54.08	—
Epicanthus, % incidence	134	47.31	17.81	—	—	—	37.64	—

N = number of populations, M_x = populational mean, σ_x = standard interpopulation deviation, As_x = coefficient of skewness, Es_x = coefficient of kurtosis, F = total variance to interpopulation variance ratio, CV_x = $100\sigma_x/M_x$ = interpopulation coefficient of variation, $\bar{\sigma}_x$ = mean standard intrapopulation deviation.

individual character varies broadly throughout the territory. A particular physical type may combine simultaneously with both the most and the least Mongoloid characters. For example, in the Baikal type of the North Asian race (classification by Roginski & Levin, 1955) or the Baikal group of types (classification by Debetz, 1951) a low bridge of the nose, a flat and broad face and high incidence of the epicanthus is combined with a low face, very soft hair and considerable depigmentation of eyes, skin and hair.

Whereas the classification of Roginski & Levin (1955) still preserves the hierarchical principle of singling out the large race, the small race, and types within the small race, Debetz's classification, which takes into account to a greater degree the difficulties of the hierarchical systematics of Siberian peoples, is actually built on singling out and describing geographical groups of physical types without attempting to bring them together in a systematic hierarchy. The difficulties and contradictions which arise result from the fact that in considering human variability under Siberian conditions population laws take first place and, therefore, a study of variability requires in this case a population approach. Table 3.2 gives a composite anthropological description of the total Siberian indigenous population, based on individual population mean values of anthropometric and anthroscopic characters. Beside the number of individual populations (N), there are also the following parameters of interpopulational distribution of a character: mean ($M_{\bar{x}}$), standard deviation ($\sigma_{\bar{x}}$), coefficient of skewness ($As_{\bar{x}}$), coefficient of kurtosis ($Es_{\bar{x}}$), coefficient of variation ($CV_{\bar{x}}$) of mean populational values ($\bar{x}$), and also mean value of the standard intrapopulational deviation (σ). The ratio (F) of the total to the inter-population variance, given in Table 3.2, does not reach the critical level of heterogeneity in any of the characters. The same is indicated by the distribution of interpopulation mean values given in Figs. 3.3 and 3.4. One may conclude that the differences between Siberian populations are variations relative to one type of North Asian Mongoloids, whose features are fully conveyed by the population characters' mean values. Combinations of these characters may of course be singled out, just as is done in the above mentioned classifications. However, the systematic value of these combinations may be determined only by considering knowledge of the specific features of the evolution of genes in populations in the territory of North Asia.

The distribution of some blood groups, serum proteins and other genetic markers in the total sample from the total aggregate of Siberia's indigenous populations is shown in Table 3.3. It is worthwhile looking at the reasons behind our organisation of the genetic data.

A reader of this review, unfamiliar either with the peoples or the local ethnic groups of Siberia, might find it helpful to think in general terms of the distribution of genetic markers throughout this whole group of the world's population, irrespective of the degree in which it is differentiated internally. For instance, were Siberian blood and serum samples to be sent for gene

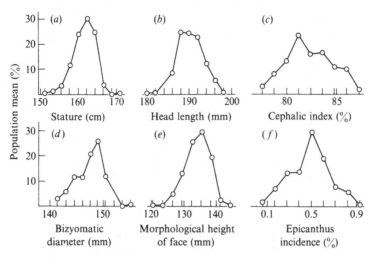

Fig. 3.3. Distribution of population means of some anthropological characters of the indigenous populations of North Asia.

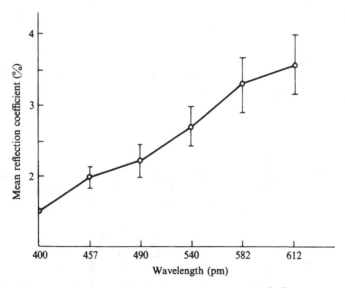

Fig. 3.4. The mean reflectometric hair colour curve among the indigenous populations of North Asia. Mean value in 95 % confidence interval.

Table 3.3. *Distribution of blood groups and other genetic markers in total sampling from North Asian indigenous populations*

Marker	Number Tested[a]	Phenotypes (%)		χ^2 (d.f.)	Genes (%) q	$\bar{q} \pm s_{\bar{q}}$
		Observed	Expected	(d.f. = 1)		
ABO groups	15929	O	37.60	37.48	0.06	$O = 61.22$	62.14+0.85
	(212)	A	27.40	27.53	0.09	$A = 19.41$	19.47±0.60
		B	27.33	27.47	0.11	$B = 19.37$	18.40±0.75
		AB	7.67	7.52	0.44		
	Total		100.00	100.00	0.70		
				(d.f. = 2)		
A_1A_2BO groups	821	O	47.26	47.65	0.02	$O = 69.03$	
	(20)	A_1	28.14	28.37	0.02	$A_1 = 17.75$	
		A_2	2.80	2.83	0.002	$A_2 = 2.02$	
		B	16.57	16.72	0.001	$B = 11.20$	
		A_1B	4.99	3.98	1.94		
		A_2B	0.24	0.45	0.69		
	Total		100.00	100.00	2.673		
ABH secretor system	972 (48)	Secretors	87.24			$se = 35.72$	21.20±3.73
		Non-secretors	12.67				
Lewis groups (samples tested with anti-Le^a and anti-Le^b)	3278 (78)	Le a− b+	61.04			$le = 57.46$	56.86±2.19
		Le a+ b−	5.95				
		Le a− b−	33.10				
					(d.f. = 1)		
MN groups	10005	MM	37.34	33.22	51.16	$M = 57.64$	65.50±1.41
	(180)	MN	40.59	48.83	139.13	$N = 42.36$	34.50±1.41
		NN	22.07	17.95	94.56		
	Total		100.00	100.00	284.85		
					(d.f. = 1)		
Ss groups	1362	SS	30.11	21.40	49.66	$S = 46.26$	40.68±2.33
	(76)	Ss	32.30	49.72	82.97		
		ss	37.59	28.88	36.03		
	Total		100.00	100.00	167.66		

[*For note refs. see p. 62*]

Table 3.3. *cont.*

Marker	Number tested[a]	Phenotypes (%)		χ^2 (d.f.)	Genes (%)	
					q	$\bar{q} \pm s_{\bar{q}}$
		Observed	Expected			
MNSs groups (samples tested with anti-M, anti-N, anti-S and anti-s)	1362 (76)	MSMS	11.89	7.41	36.84	
		MSMs	11.67	16.28	17.88	
		MsMs	13.00	8.94	24.89	
		MSNS	11.16	10.36	0.86	$MS = 27.23$ — 25.26 ± 5.62
		MNSs	14.61	24.36	52.48	$Ms = 29.89$ — 32.93 ± 5.29
		MsNs	15.34	14.26	1.16	$NS = 19.03$ — 15.45 ± 1.31
		NSNS	7.06	3.62	4.51	$Ns = 23.85$ — 26.34 ± 1.61
		NSNs	6.02	9.08	13.56	
		NsNs	9.25	5.69	31.18	
	Total		100.00	100.00	183.36	
Rhesus groups (samples tested with anti-D only)	6824 (155)	D+	98.92			$d = 10.41$ — 4.07 ± 0.75
		D−	1.08			
					(d.f. = 1)	
Cc groups	2131 (79)	CC	18.25	19.68	2.21	$C = 44.36$ — 42.16 ± 1.44
		Cc	52.23	49.36	3.52	
		cc	29.52	30.96	1.42	
	Total		100.00	100.00	7.15	
					(d.f. = 1)	
Ee groups	2131 (79)	EE	17.45	16.72	0.68	$E = 40.89$ — 43.59 ± 1.89
		Ee	46.88	48.34	0.94	
		ee	35.67	34.94	0.32	
	Total		100.00	100.00	1.94	
					(d.f. = 8)	
Rhesus groups[b] (samples tested with anti-D, anti-C, anti-C^w, anti-E, anti-c and anti-e)	2131 (79)	CCDEE	1.17	0.48	21.75	
		CCDEe	4.60	5.17	1.34	
		CCDee	12.48	14.02	3.60	
		CCddee	0.00	0.003	0.07	$CDE(R^z) = 6.90$ — 12.10 ± 1.00
		CcDEE	5.63	4.69	4.03	$CDe (R^1) = 33.78$ — 29.92 ± 1.65
		CcDEe	30.36	28.44	2.77	$C^w De (R_1^w) = 3.09$ — 2.83 ± 0.54
		CcDee	16.14	16.15	0.00	$Cde (R') = 0.58$ — 0.10 ± 0.06
		CcddEe	0.00	0.003	0.08	$cDE (R^2) = 33.67$ — 27.82 ± 1.34
		Ccddee	0.09	0.083	0.07	$cdE (R'') = 0.31$ — 0.19 ± 0.14
		ccDEE	10.65	11.54	1.47	$cDe (R^o) = 14.99$ — 26.28 ± 1.73
		ccDEe	11.87	14.67	11.43	$cde(r) = 6.67$ — 1.20 ± 0.48
		ccDee	6.29	4.26	20.91	
		ccddEE	0.00	0.001	0.02	
		ccddEe	0.05	0.04	0.02	
		ccddee	0.67	0.44	2.15	
	Total		100.00	100.00	69.73	

61

Table 3.3. *cont.*

Marker	Number tested[a]	Phenotypes (%)		χ^2 (d.f.)	Genes (%)	
					q	$\bar{q} \pm s_{\bar{q}}$
		Observed	Expected			
P groups	4372 (115)	P+ 52.29			$P^2 = 69.07$	64.90 ± 2.26
		P− 47.71				
Kell groups	1663 (45)	K+ 8.36			$K = 4.27$	6.57 ± 0.80
		K− 91.64				
Diego groups	464 (24)	Di^a+ 18.10			$Di^a = 9.50$	8.82 ± 1.25
		Di^a− 81.90				
Duffy groups	944 (38)	Fy a+ 41.00			$Fy^a = 23.19$	27.85 ± 8.17
		Fy a− 59.00		(d.f. = 2)		
Transferrins	2071 (53)	Tf CC 96.43	96.46	0.0002	$Tf^C = 98.22$	97.10 ± 0.71
		Tf CD$_{chi}$ 1.88	1.86	0.0060	$Tf^D = 0.94$	1.85 ± 0.61
		Tf B$_{0-1}$C 1.69	1.66	0.0108	$Tf^B = 0.84$	1.03 ± 0.32
		The rest 0.00	0.02	0.4142		
	Total	100.00	100.00	0.4312		
				(d.f. = 1)		
Hapto-globins	2698 (59)	Hp 1-1 10.45	10.34	0.03	$Hp^1 = 32.16$	33.15 ± 1.62
		Hp 2-1 43.41	43.63	0.03		
		Hp 2-2 46.14	46.03	0.01		
	Total	100.00	100.00	0.07		
				(d.f. = 1)		
Group-specific component	899 (29)	Gc 1-1 39.27	37.70	0.57	$Gc^1 = 61.40$	50.54 ± 6.38
		Gc 2-1 44.27	47.40	1.84		
		Gc 2-2 16.46	14.90	1.46		
	Total	100.00	100.00	3.87		
PTC taster test	2823 (60)	Sensitive 79.81			$t = 44.93$	26.94 ± 2.55
		Non-sensitive 20.19		(d.f. = 1)		
PTC taster test	844 (37)					
Four-model distribution of sensitivity thresholds and corresponding genotypes	1	tt 22.51	22.30	0.02	$t = 47.22$	42.65 ± 3.18
	2	tT$_1$+T$_1$T$_1$ 51.42	51.15	0.01	$T_1 = 38.48$	40.74 ± 3.08
	3	T$_1$T$_2$ 10.43	11.00	0.25	$T_2 = 14.30$	16.61 ± 2.30
	4	tT$_2$+T$_2$T$_2$ 15.64	15.55	0.005		
	Total	100.00	100.00	0.285		

[a] Number of populations is given in parentheses. q = weighted mean, $\bar{q} + s_{\bar{q}}$ = unweighted mean with standard error.

[b] Gene frequency R^w is determined in sample of 1247 persons from 56 populalations.

typing to some laboratory in another part of the world, the workers of that laboratory would obviously be concerned with the part of the globe where the samples had been collected rather than with the specific population. One could consider the results given in Table 3.3 as having been obtained in such an imaginary laboratory.

Knowing beforehand how vast is the subcontinent of North Asia and the spatial isolation of its individual populations, a considerable disruption of genetic equilibrium could be expected when evaluating it by the Hardy–Weinberg theorem, which is obviously inapplicable in this case. Genetic equilibrium is indeed absent in many loci, but by no means from all. Equilibrium is ideally fulfilled in ABO, Ee, Hp, Tf and PTC loci. However, in some other loci, equilibrium is disrupted because of a deficit of heterozygotes, which is to be expected with such spatial organisation of the material, and equilibrium is absent in yet other loci (the Cc locus, for example) because of an excess of heterozygotes.

These varying results in evaluating genetic equilibrium may be interpreted as preliminary evidence of the non-similarity of environmental factors determining the total population's genetic dynamics for each of the independent loci. The nature of these factors will be elucidated further on. Later, we shall be using the unweighted mean gene frequencies ($\bar{q}$), shown in Table 3.3, in parallel with the weighted gene frequencies (q) assessed on the basis of a total sample of the North Asian population.

Now we have to become acquainted with the real populations from which the above-described material was derived.

Size of population

The size of the population as described in this section neglects the fact that the bulk of the Siberian population today consists of Russian, Ukrainian and other peoples of non-Siberian origin, whose numbers greatly exceed the size of the indigenous populations. In the same way ignoring the fact of Siberia's present-day industrialisation and urbanisation, this section describes the population at the pre-socialist revolution stage of economic development, when Siberia's few traditional economic and cultural patterns have easily traceable links with the Palaeolithic and Neolithic types of economy (Levin, 1958 *b*). This conventional approach is dictated by the fact that the size of a population is important as one of the fundamental parameters of the process of genetic evolution, whereas the genetic characteristics of the populations are determined with elimination of the effects of the intensive mixing of the indigenous population of North Asia with the peoples of non-Siberian origin (individuals of mixed origin in traceable generations were excluded from the sample).

Table 3.4 gives the size of 28 Siberian ethnic groups according to the USSR

Table 3.4. *Sizes of ethnic groups in North Asia (data of USSR Census, 1959)*

	N			N
1 Evenks	24 583	15 Koryaks		6 168
2 Evens	9 023	16 Ulchis		2 049
3 Nenets	22 845	17 Nanai		7 919
4 Selkups	3 704	18 Nivkhies		3 690
5 Khants	19 246	19 Buryats		251 504
6 Mansi	6 318	20 Tofalars		600
7 Enets	439	21 Tuvinians		99 864
8 Nganasans	721	22 Altaians		44 654
9 Dolgans	4 083	23 Shorians		14 938
10 Yakuts	236 125	24 Khakassians		56 032
11 Yukaghirs	440	25 Kets		1 017
12 Chukchi	11 680	26 Itelmen		1 096
13 Eskimos	1 111	27 Udegeians		1 395
14 Aleuts	399	28 Orochis		779

$\bar{x} = 29\,729 \pm 12\,144$; $\sigma_x = 64\,269$; $H = 1639$.

x = mean population size; σ_x = standard deviation, H = harmonic mean.

census of 1959 (published 1963). An individual ethnic group is characterised by a mean size of about 30 000 with a harmonic mean of 1640 persons. The distribution of population size is asymmetric to the same degree.

The contemporary indigenous population of North Asia is distributed over an area exceeding 13 million km^2 at an average density of 7.5 persons per 100 km^2. This makes it clear that the spatial isolation of the populations is extremely great and the borders between them are distinctly expressed. Nevertheless, determining the size of a population is a very difficult task: first, because of the considerable variability of size and, secondly, because of man's high mobility in the conditions of extremely low population density and the traditional type of economy. The structure of the population at the lowest level is extremely tenuous and an individual family or familial group can be easily confused with a population proper, unless the huge distances over which marriage contracts occur between individual families are taken into account.

The direct determinations of the total (N_t) and effective (N_e) size of a population after Wright (1931), i.e. with account taken of the mean number of gametes (k) a parent leaves to the next generation and of the variance σ_k^2 of this number, have been carried out in only a few Siberian populations (Rychkov & Sheremetyeva, 1972a, c; Rychkov et al., 1974 b). Twenty assessments of the absolute N_e value obtained in these works constitute an obviously non-representative sampling from the total aggregate of Siberian populations. More important in these works is another result: a very low variability of the ratio N_e/N_t, i.e. of the effective to total size, has been detected. This ratio proved close to the mean value of 0.28 ± 0.01 of $\bar{N}_e/\bar{N}_t$ in most diverse ethnic

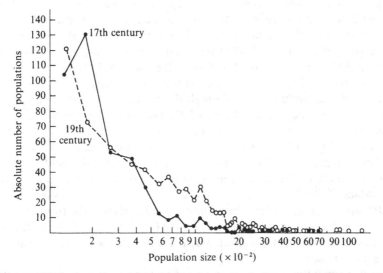

Fig. 3.5. Distribution of total population size in the indigenous population groups of North Asia in the seventeenth and nineteenth centuries. Semi-logarithmic scale.

and geographic groups of the population of Northern Asia: the Tungus-speaking Evenks of Central Siberia (0.27), the Yakuts of the Yana river basin (0.27), the Koryaks of Kamchatka (0.28), the Chukchi of the Chukotka peninsula (0.32), the Eskimo of the Chukotka peninsula (0.28) and the Aleuts of the Komandorski Islands (0.29) (Rychkov & Sheremetyeva, 1976c).

Bearing in mind this stability of the effective share in the total population size, we can turn to the distribution of the total size of population in North Asia. The data used here are those of the 1897 census, elaborated by Patkanov (1911, 1912) with regard to the sizes of local groups of the indigenous Siberian population. Taking into account what has been said above that such a local group often proves to be not a population but an individual family, the size of the smallest administrative structural unit existing at that time was taken as a 'unit of observation'. The distribution of the total size of each such units is given in Fig. 3.5. It is characterised by the mean value of $\bar{N}_t = 925.5 \pm 51$, $\sigma_{N_t} = 1333$, harmonic mean $H_{N_t} = 213$. By multiplying the mean value of the total size by the above cited size of the effective share 0.28, we obtain $\bar{N}_t (N_e/N_t) = 259 \pm 14$, which is close to the harmonic mean value of 213.

The second source of data on the distribution of population sizes came from seventeenth-century archival documents on the localisation of individual population groups and the number of 'yasak souls' per group. The number of 'yasak souls' is the taxable number of able-bodied men; the yasak being a tax in pelts and other products of hunting. These archival data were processed by Dolgikh (1960) to determine the size of Siberia's total indigenous population in the seventeenth century. When assessing the size of an individual population group Dolgikh introduced the coefficient of 0.25 for the proportion of able-bodied men in the group and the coefficient of 0.50 for the proportion of able-bodied men and women in the entire group. It might be that by using the 0.50 coefficient the value N_t is somewhat under-estimated, but this is not a critical matter for our task since it does not affect the possibility of evaluating N_e through the recorded number of physically mature men in the population, $N_e = 0.5\,N_t$. The distribution of N_t is charac-terised by the parameters of $\bar{N}_t = 436 \pm 36$, $\sigma_{N_t} = 722$, from which $N_e = 0.5$ $\bar{N}_t = 218 \pm 18$ (Rychkov, 1965b). In Fig. 3.5 the distribution of the total population in the seventeenth century is combined with the distribution at the end of the nineteenth century. For these two distributions we obtain three close values of N_e (213, 218 and 259) of which we take $N_e = 218$ for further use. Its advantage lies in that, being intermediate between the two others, it represents at the same time a size value for the historical period when Siberian populations had not yet been affected by the socio-economic transformations taking place today.

However small the absolute mean size of any one Siberian population, the actual sizes of populations are in some cases an order of magnitude below the $\bar{N}_e$. In such extremely small populations the effects of isolation become manifest with particular clarity. We can cite, by way of example, the data on one of the Siberian peoples which occupies a geographically central position. The Evenks, a Tungus-speaking people, number less than 30000. According to their traditional cultural and economic structure they are taiga hunters and reindeer breeders, using the reindeer as a draught animal only. With such a small total population size they are distributed along a taiga belt from the left bank of the Yenisei to the coast of the Sea of Okhotsk. Less than 5000 Evenks dwell in the central part of Siberia known as the Central Siberian Plateau, with an area of 796000 km². Elementary Evenk population groups are, on average, 400 km apart. Some characteristics of such groups – N_t and N_e, as well as the inbreeding coefficients (F) – are shown in Table 3.5 (Rychkov *et al.*, 1974a, b). The inbreeding coefficient is established genealogically on the basis of studies of the pedigrees of 1222 persons (26% of the total Evenk population), coming from most diverse parts of the Central Siberian Plateau. Their levels of kinship with one another are on the average those of siblings thrice removed. This degree of kinship in a population scattered over a terri-tory greater than that of a number of European countries is evidence that the total Evenk population in Central Siberia constitutes no less a real population group than the elementary populations which it comprises. Marriage contracts among the Evenks occur throughout the entire Central Siberian Plateau and are regulated by a system of gentile exogamy. A statis-tical analysis of gentile marriage relationships showed the certain existence of an additional system whereby marriage contacts are avoided by individual

Table 3.5. *Total (N_t) and effective (N_e) sizes and inbreeding coefficients (F) of Evenk groups of Central Siberia*

Populations	N_t	N_e	$F \pm$ S.E.$_F$
1	95	26	0.0137 ± 0.0041
2	103	20	0.0081 ± 0.0011
3	130	30.5	0.0106 ± 0.0046
4	67	31	0.0053 ± 0.0027
5	185	40.5	0.0018 ± 0.0009
6	95	21.5	0.0032 ± 0.0015

Table 3.6. *A comparison of some indices in the inbred and non-inbred Evenk families of Central Siberia*

Marriages	Indices	Number of marriages	Number of living children	Number of deceased children	Number of mis-carriages	Mean number of living children per family ($\pm$ S.E.)	Mean total of children per family
Evenk marriages	Inbred	25	57	38	10	2.28 ± 0.29	3.80
	Non-inbred	106	337	104	1	3.18 ± 0.028	4.16
Significance of difference, P			> 0.99	> 0.99	> 0.999	> 0.95	

gentile groups. This system, which is not apparently traditional, may be either a relic of deep antiquity or a new superstructure. However, the effect of this superstructure is clearly shown in the data in Table 3.6: the mean inbreeding level under this system is lower than in the rest of the population, and some demographic indices are more favourable. Having thus briefly glimpsed at a vivid example of isolated populations, we shall proceed with an analysis of the central theme of this review – the specific features of the process of genetic evolution within the system of isolated populations of Siberia.

Gene migrations and their structure in Siberia

We would like to avoid giving the impression that we are treating the whole of the spatially and historically isolated North Asian populations as a system without adequate substantiation. However large the territory of North Asia, it is, after all, the territory of a single subcontinent. However deep the roots of the numerous peoples inhabiting this subcontinent, anthropologically, archaeologically and culturally all the evidence points to their coming from

Yu. G. Rychkov and V. A. Sheremetyeva

Table 3.7. *Gene migration matrix at the ethnic level of population structure in North Asia*

n_i^u	Ethnic group	1	2	3	4	5	6	7	8	9	10	11	12
1	Evenks, Evens	1295	.	.	.	.	.	.	.	.	95.25	.	.
2	Komi	.	176	26	.	1	2	.	.	.	.	.	.
3	Nenets	.	21	387	3	17	2	11	1	2	.	.	.
4	Selkups	17	.	11	147	20	.	.	.	.	.	.	.
5	Khants	2	2	16	2	200	9	.	.	.	.	.	.
6	Mansi	.	.	.	.	.	—	.	.	.	.	.	.
7	Enets	.	.	.	.	.	.	—	.	2	.	.	.
8	Nganasans	.	.	2	.	.	.	5	219	1	.	.	.
9	Dolgans	.	.	.	.	.	.	.	.	—	.	.	.
10	Yakuts	25	.	.	.	.	.	.	.	.	394	.	.
11	Yukaghirs	15	.	.	.	.	.	.	.	.	9	30	24
12	Chukchi	21	.	.	.	.	.	.	.	.	4	12	265
13	Eskimos	.	.	.	.	.	.	.	.	.	.	.	25
14	Aleuts	.	.	.	.	.	.	.	.	.	.	.	.
15	Koryaks	23	.	.	.	.	.	.	.	.	.	.	24
16	Negidals	7	.	.	.	.	.	.	.	.	.	.	.
17	Ulchis	7	.	.	.	.	.	.	.	.	0.25	.	.
18	Nanai	.	.	.	.	.	.	.	.	.	.	.	.
19	Nivkhies	.	.	.	.	.	.	.	.	.	.	.	.
20	Buryats	7	.	.	.	.	.	.	.	.	.	.	.
21	Tofalars	1	.	.	.	.	.	.	.	.	.	.	.
22	Todzhinians	.	.	.	.	.	.	.	.	.	.	.	.
23	Tuvinians	.	.	.	.	.	.	.	.	.	.	.	.
24	Tubalars	.	.	.	.	.	.	.	.	.	.	.	.
25	Telengets	.	.	.	.	.	.	.	.	.	.	.	.
26	Altai-Kijis	.	.	.	.	.	.	.	.	.	.	.	.
27	Lebedians	.	.	.	.	.	.	.	.	.	.	.	.
28	Kumandinians	.	.	.	.	.	.	.	.	.	.	.	.
29	Shorians	.	.	.	.	.	.	.	.	.	.	.	.

n_i^u = direct migration to population of ith ethnic group from others (across);

a single origin in antiquity. However great the diversity of economic and cultural patterns as types of social adaptation by Siberian peoples to the environment, all of them are the most ancient types of natural economy and, and this is the main thing, have developed and are functioning on the principle of an equilibrium between the population and the environment.

The central argument purporting that Siberian populations are organised into a system is their interaction throughout the vast expanses of North Asia. From the population genetics aspect this interaction is conceived of as an exchange of genes between populations, through which every individual population receives genetically significant information on the state of the system as a whole, and on the state of the environment not only in the nearby but also in the extreme peripheral parts of the area.

68

n_i^u

13	14	15	16	17	18	19	20	21	22	23	24	25	26	27	28	29
.	.	16	.	1	.	.	3	.	.	.	.	.	.	.	.	.
.	.	.	.	.	.	.	.	.	.	.	.	.	.	.	.	.
.	.	.	.	.	.	.	.	.	.	.	.	.	.	.	.	.
.	.	.	.	.	.	.	.	.	.	.	.	.	.	.	.	.
.	.	.	.	.	.	.	.	.	.	.	.	.	.	.	.	.
.	.	.	.	.	.	.	.	.	.	.	.	.	.	.	.	.
.	.	.	.	.	.	.	.	.	.	.	.	.	.	.	.	.
.	.	.	.	.	.	.	.	.	.	.	.	.	.	.	.	.
.	.	.	.	.	.	.	.	.	.	.	.	.	.	.	.	.
.	.	.	.	.	.	.	.	.	.	.	.	.	.	.	.	.
42	.	.	.	.	1	.	.	.	.	.	.	.	.	.	.	.
176	.	.	.	.	.	.	.	.	.	.	.	.	.	.	.	.
.	52	.	.	.	.	.	.	.	.	.	.	.	.	.	.	.
.	.	771	.	.	.	.	.	.	.	.	.	.	.	.	.	.
.	.	.	54	7	.	7	.	.	.	.	.	.	.	.	.	.
.	.	.	11	214	18	15	.	.	.	.	.	.	.	.	.	.
.	.	.	.	.	323	.	.	.	.	.	.	.	.	.	.	.
.	.	.	3	12	1	252	1	.	.	.	.	.	.	.	.	.
.	.	.	.	.	.	.	200	.	.	.	.	.	.	.	.	.
.	.	.	.	.	.	.	.	165	.	0.5	.	.	.	.	.	.
.	.	.	.	.	.	.	.	.	162	.	7	.	.	.	.	.
.	.	.	.	.	.	.	.	.	7	201	.	.	.	.	.	.
.	.	.	.	.	.	.	.	.	.	.	8.5	3	.	2	.	4
.	.	.	.	.	.	.	.	.	.	.	.	157	2	.	.	.
.	.	.	.	.	.	.	.	.	.	.	.	2	244	1	1	6
.	.	.	.	.	.	.	.	.	.	.	2	.	.	5	.	.
.	.	.	.	.	.	.	.	.	.	.	.	.	.	.	5	144
.	.	.	.	.	.	.	.	.	.	.	.	.	3	1	124	1

n_i^v = reverse migration from population of ith ethnic group to others (down).

Because of the great ethnic differentiation of the Siberian population, gene exchange between these populations can be marked by ethnonyms, a method we used for analysing gene migrations all over North Asia. The matrix of migrations between populations at the ethnic level is given in Table 3.7. The asymmetry of this matrix reflects two components of migration: cases of direct migration (n_i^u) into the population of the ith ethnic group from the other ones and cases of reverse migration of (n_i^v) from the ith ethnic group to others. Diagonally, cases of N_i are shown, indicating absence of migrations, i.e. cases of endogamy. All the values of N_i, n_i^u, n_i^v represent the numbers of direct descendants from marriages of these types. A spatial structure of gene migration corresponding to this matrix is given in Fig. 3.6; from this it

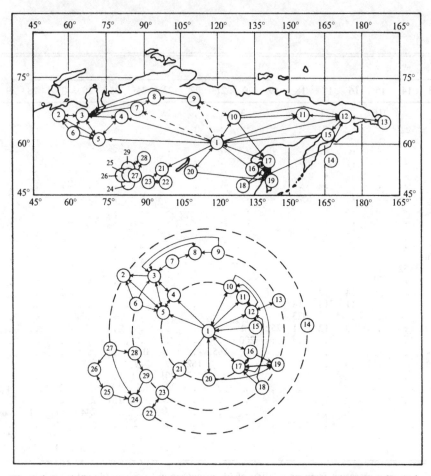

Fig. 3.6. Spatial structure of gene migrations in indigenous populations of North Asia established from sampling data. Dashed line = migrations not represented in sampling material, but known from other sources. The numbers correspond to those in Table 3.7, column 1.

follows clearly that Siberian ethnic groups, however small and segregated, are actually linked into a single system through intergroup gene exchanges among them.

A quantitative evaluation of migration intensity (*m*) can be made on the basis of the migration matrix, which, however, requires additional assumptions about the type of relations between gene migrations and the spatial structure of the population system.

We have made from two such assumptions which are equally realistic under the conditions of the North Asian subcontinent and therefore supplement and control one another.

70

Table 3.8. *Gene migration coefficients at the ethnic level of populations and within the population system of North Asia as a whole*[a]

Island model

	N_i	n_i^u	k_i^u	n_i^v	k_i^v	m_{ei}^u	m_{ei}^v
1 Evenks and Evens	1295	115.25	4	125	10	0.0136	0.0366
2 Komi	176	29	3	23	2	0.0176	0.0096
3 Nenets	387	57	7	55	4	0.0374	0.0207
4 Selkups	147	48	3	5	2	0.0308	0.0027
5 Khants	200	31	5	38	3	0.0280	0.0200
6 Mansı	—	—	—	—	—	—	—
7 Enets	—	—	—	—	—	—	—
8 Nganasans	219	8	3	1	1	0.0044	0.0002
9 Dolgans	—	—	—	—	—	—	—
10 Yakuts	394	25	1	108.5	4	0.0025	0.0360
11 Yukaghirs	30	48	3	12	1	0.0769	0.0119
12 Chukchi	365	80	5	73	3	0.0375	0.0208
13 Eskimos	176	25	1	42	1	0.0052	0.0080
14 Aleuts	50	0	0	0	0	0	0
15 Koryaks	771	47	2	16	1	0.0203	0.0009
16 Negidals	54	21	3	14	2	0.0365	0.0179
17 Ulchis	214	51.25	5	20	3	0.0403	0.0107
18 Nanai	323	0	0	20	3	0	0.0076
19 Nivkhies	252	17	4	22	2	0.0110	0.0070
20 Buryats	200	7	1	4	2	0.0196	0.0017
21 Tofalars	165	1.5	2	0	0	0.0008	0
22 Todzhinians	162	7	1	7	1	0.0018	0.0018
23 Tuvinians	201	7	1	7, 5	2	0.0015	0.0031
24 Tubalars	85	9	3	2	1	0.0125	0.0010
25 Telengets	157	2	1	5	2	0.0006	0.0027
26 Altai-Kijis	244	10	4	5	2	0.0201	0.0018
27 Lebedians	5	2	1	4	3	0.0124	0.0580
28 Kumandinians	144	5	1	11	3	0.0015	0.0092
29 Shorians	124	5	3	6	2	0.0051	0.0040

$$m_e = 0.0141 \pm 0.0023$$

[a] For symbols see text.

Assumption 1: taking into account that geographical distances between populations are measured in many hundreds of kilometres, an individual ethnic group may be represented as an island to which genes come in from outside – both from the central nucleus of the system, clearly visible in Fig. 3.6, and from other peripheral and very distant populations. Such an assumption corresponds to one of the possible versions of Wright's (1943) 'island' model. In this case the content of genetic information transmitted through migration is determined not by the distance over which migration occurred, but by the number of populations that served as the source of migration. When, along with the notion of migration intensity (m), we introduce the notion of migration efficiency (m_e), the latter value will be directly propor-

Yu. G. Rychkov and V. A. Sheremetyeva

Table 3.9. *Total intensity of gene migration to a population over different distances and* (2) *migration coefficient* (*migration variance*) *within the population system of North Asia*

Stepping-stone model

	Migration distance		
	1 step	2 steps	3 steps
1 Evenks and Evens	0.040 862	0	0
2 Komi	0.070 732	0	0
3 Nenets	0.058 559	0.004 505	0.001 126
4 Selkups	0.123 077	0	0
5 Khants	0.062 771	0.004 329	0
6 Mansi	—	—	—
7 Enets	—	—	—
8 Nganasans	0.002 203	0.011 015	0.004 406
9 Dolgans	—	—	—
10 Yakuts	0.029 833	0	0
11 Yukaghirs	0.307 692	0	0
12 Chukchi	0.084 270	0.004 494	0.001 124
13 Eskimos	0.062 189	0	0
14 Aleuts	0	0	0
15 Koryaks	0.028 728	0	0
16 Negidals	0.140 000	0	0
17 Ulchis	0.096 152	0.000 471	0
18 Nanai	0	0	0
19 Nivkhies	0.027 881	0.001 859	0.001 859
20 Buryats	0.016 908	0	0
21 Tofalars	0.003 030	0	0.001 501
22 Todzhinians	0	0.020 710	0
23 Tuvinians	0.016 826	0	0
24 Tubalars	0.047 872	0	0
25 Telengets	0.006 289	0	0
26 Altai-Kijis	0.015 748	0.003 937	0
27 Lebedians	0.142 857	0	0
28 Kumandinians	0.016 778	0	0
29 Shorians	0.003 875	0.015 504	0
Average	0.013 510	0.000 321	0.000 032
Migration variance	$V_m = 0.015\,082$		

tional to the number of migration sources (k_i), which is expressed by the following formula (Rychkov & Sheremetyeva, 1974):

$$m_e = (2K)^{-1} \sum_{u,v} m_{ei}, \tag{3.1}$$

where $m_{ei} = n_i k_i (n_i + N_i)^{-1} (K-1)^{-1}$, and K = the total number of populations, i.e. potential sources of migration.

Assumption 2: however great the geographical distances over which migrations take place in Siberia, it is still obvious that the most extreme populations of the southwestern and northeastern or northwestern and southeastern borders are denied any chance of direct gene exchange with one another and, consequently, migrations, reduced by space, become step-like in character. This assumption agrees with the well-known 'stepping-stone' model of population structure of Kimura & Weiss (1964). The migration matrix in Table 3.7 indicates that the actual structure of the Siberian ethnic group system differs from the ideal version of the stepping-stone model in that migration takes place over more than a single step. For this case Kimura & Weiss recommend the use of the gene migration variance (V_m) as a measure of intensity.

The two-dimensional version of the stepping-stone model of population structure was used for our study. The results of gene migration evaluations on the basis of the island and the stepping-stone models are given in Tables 3.8 and 3.9. Both sets of values are close, though the ones for the stepping-stone model ($V_m = 0.0151$) do not appear to be effective. In anticipation, let us point out that $m_e = 0.0106$ for the stepping-stone model proves lower than $m_e = 0.0141$ for the island model. We must keep in mind, however, that both structure types are merely assumptions, and their verification and selection constitute the next stage of our task.

Predicting the gene frequency variance

The most direct method of checking the reliability of such parameters of population structure as N_e and m_e consists of comparing an empirical gene frequency variance in a population with a variance calculated on the basis of an assumed population structure model. The variance, standardised by the mean gene frequency $f = V_q[\bar{q}(1-\bar{q})]^{-1}$ in the case of Wright's island model, is:

$$f = (1 + 4N_e m_e)^{-1}, \tag{3.2}$$

and in the two dimensional stepping-stone model of Kimura & Weiss is

$$f = (1 + 2N_e C_0)^{-1}, \tag{3.3}$$

where

$$C_0^{-1} = \left\{ \frac{1}{M_1} K\left(\frac{1}{M_1}\right) + \frac{1}{M_2} K\left(\frac{1}{M_2}\right) \right\} \Big/ 2\pi\sqrt{(\sigma^2_{mA}\sigma^2_{mB})},$$

and

$$M_1 = \sqrt{[(1 + m_\infty/2\sigma^2_{mA})(1 + m_\infty/2\sigma^2_{mB})]},$$

$$M_2 = \sqrt{\left[\left(1 - \frac{2 - m_\infty}{2\sigma^2_{mA}}\right)\left(1 - \frac{2 - m_\infty}{2\sigma^2_{mB}}\right)\right]}$$

and $K(\ )$ stands for the complete elliptic integral of the first order. For further details it is advisable to turn to the primary source (Kimura & Weiss, 1964), while here we shall just explain the value m_∞, characterising the systematic pressure experienced by the population system as a whole. We are unwilling to introduce into this value the quite probable but unpredictable pressure of

Table 3.10. *Theoretical f_e values of standardised gene frequency variance calculated on the basis of an effective size of $N_e = 218$, and the population group migration matrix of Table 3.7*

Assumed population structure	f_e	Variances ratio F
Island model	0.0753 ⎫	
Two-dimensional	0.0974 ⎬	1.29
stepping-stone model		

natural selection, but we assume it to be equal to 4×10^{-5}, i.e. somewhat higher than the mutation pressure level. Such an m_∞ value was used in assessing the gene frequency variances both in the island and stepping-stone models. It should also be noted that the stepping-stone model variance determined by equation 3.3 is additionally corrected (see Kimura & Weiss, 1964) for both the restricted number of populations in the system and spatial limitations, however strange this might sound in regard to the subcontinent of North Asia. The gene frequency variances predicted in this way are given in Table 3.10. Processing the observed variance values in this theoretical manner, we must compare them, imparting to each a number of degrees of freedom equal to infinity. In this case they differ significantly from each other, since the stepping-stone to island variances ratio is 1.29. However, if we discard such a formal method and recall that the prediction was based on empirical demographic figures of population size and gene migrations, the proximity of the two theoretical values becomes obvious and one can say beforehand that the structure of the system of isolates in North Asia must combine within it the properties of both the island and the stepping-stone model types.

The theoretically predicted interval for the gene frequency variance is so narrow it affords a very precise prediction against which the empirical data on gene frequency variances at the specific loci can be tested,

Gene frequency variance as a measure of genetic differentiation in the system of isolated populations of North Asia

In the last few decades, very wide attention has been drawn to studying gene frequency variances as indices of differentiation and the degree of inbreeding and kinship. The most striking example of such an analysis is the work of Neel & Ward (1972) for determining Wright's F statistics in populations of a number of Indian tribes of South America. Both in this work and in all the others, the determination of the genetic variance (F_{ST} statistic of Wright), as formally required, is made by using the mean weighted gene frequency in the totality of populations studied.

In this review, however, we estimate gene frequency variances relative to the non-weighted mean gene frequency. Arguments in favour of this approach are as follows.

(1) The formal statistical procedure for analysing variances is elaborated for the case of randomised samplings from an unlimited homogeneous general total, and it is precisely in such a case that the size of the sample has the significance of a statistical weight.

(2) Any isolated population *per se* represents a general total and, therefore, an evaluation of the gene frequency variance is done at the level of the aggregate of general totals. In this case the actual weight of gene frequency q_i of the ith population can only be its effective size N_{ei}, whereas the size of the sample from the ith population can serve as a weight only on the condition that the investigation is planned in such a manner that the sizes of all the samples are proportional to the effective sizes of the population. We have not seen any work in which this condition has been observed; more especially it was not fulfilled when populations in Siberia were studied previously. This being so, the use of the sample size as a weight and the utilisation of a weighted mean gene frequency introduces a misleading element into the analysis, since the size of the sample is determined by extraneous factors in the organisation and practice of field investigations that have nothing to do with the biological (in general) and genetic (in particular) characteristic of a population.

(3) The more elementary a population's structure – and in the case of isolated populations the researcher strives to work as near as possible to the elementary level – the greater the degree in which an isolated population is susceptible to 'accidents' during its time span. Not only do such accidents produce a random gene drift at a limited, constant population size, but they also cause a random drift of the population size itself. In other words, the more elementary an isolated population, the less resistant it is, and has been in the past. To use their sizes at a given point in time as statistical weight when studying gene frequency variances in such populations means giving importance to accidents in their past. For example, to take Siberian conditions, several centuries ago the Yakuts were less numerous than the Tungus are today. In the subsequent time span this proportion was likely to change and indeed has changed, as one can see from Table 3.4. Moreover, precisely because of the small numbers of the Tungus, researchers in Siberia collected larger samples from them than they did from the Yakut population. Thus, neither the differences in population size as they exist today, nor the differences in the sample sizes from these populations conform to the concept of weight that is used when analysing gene frequency variances. The reverse is true, as will be shown further, for it is not the weighted, but the unweighted mean gene frequency in a system of isolated populations that carries information about the properties of a population in the long term, i.e. a time span greater than the duration of the individual isolate.

Table 3.11. *Assessments of observed standardised genetic variances in a gene loci sampling, studied in populations of North Asia and their comparison with theoretical f_e values*

Loci	Number of population K	Mean size of sampling N	Number of standardised variance values used n	$\overline{f_i}$	Comparison with theoretical values (P) % Island model $f_{e_{IS}}$ 0.0753	Stepping-stone model $f_{e_{ST}}$ 0.0974
ABO	205	85	6	0.0556	< 2.5	< 0.5
Se	38	17	1	0.2337	< 0.5	< 0.5
Le	76	35	1	0.1291	< 0.5	< 5
MNSs	76	25	12	0.0894	> 5	> 5
Rh	79	26	29	0.0865	> 25	> 25
P	111	34	1	0.2317	< 0.5	< 0.5
K	40	25	1	−0.1180	< 0.1	< 0.1
Di	24	19	1	−0.1185	< 2.5	< 2.5
Fy	37	25	1	0.0773	> 5	> 5
Tf	49	39	6	0.0012	≪ 0.5	≪ 0.5
Hp	49	51	1	0.0480	< 2.5	< 0.5
Gc	19	20	1	0.2843	< 0.5	< 0.5
PTC	36	23	6	0.1230	1	> 5

Total of f_i estimates = 67
$\overline{f_w} = 0.0810 \pm 0.0071$ $P_{IS} > 25\%$ $P_{ST} > 25\%$

An evaluation of a standardised gene frequency variance, with the above-mentioned qualification regarding an unweighted mean has been carried out according to the variances and covariances for the case of loci with multiple alleles (Nei, 1965), i.e.

$$f_g = V_g/\overline{q}_i(1-\overline{q}_i), \quad \text{and} \quad f_g = -Cov_g/\overline{q}_i\overline{q}_j \qquad (3.4)$$

where $\quad V_g = V_0 - V_s, \quad Cov_g = Cov_0 - Cov_s,$

V_0 = the variance observed, V_s = the sampling variance, Cov_0 = the covariance observed, Cov_s = the sampling covariance, and $\overline{q}_i$ = the mean (unweighted) frequency of the *i*th allele in the locus.

Genes were excluded from the analysis with a frequency of $\overline{q} < 0.01$, or $\overline{q} > 0.99$, on the grounds of not having reached the level of real polymorphism: i.e. genes R' and R'' in the rhesus blood group system and colour blindness genes mentioned above.

All in all, 67 values of a standardised genetic variance were obtained for

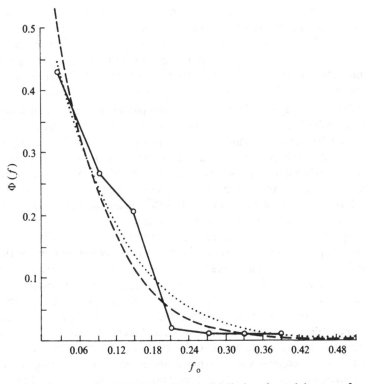

Fig. 3.7. Distribution of empirical f_0 values (solid line) estimated by gene frequency variances and covariances, and theoretical distribution curves f_e in the island and stepping-stone models of population structure. Dotted line = island model, dashed curve = stepping-stone model. The interval on the abscissa up to 0.06 gives f_e values estimated by the variances and covariances of the genes A; O; A, B; A, O; NS; MS, NS; Ms, Ns; NS, Ns; d; $C + C^w$; C; R^z; R_1^w; R^2; R^z, R_1^w; R^z, R^2; R^z, R^0; R^1, R^2; R_1^w, R^2; R^1, r; $R^1 + R_1^w, R^2$; $R^1 + R_1^w$, R^0; R^0, r; Hp^1; K; Di^a; Tf^B; Tf^C; Tf^B; Tf^B, Tf^D; in the interval 0.06–0.12, from the genes B; B, O; MS; Ms; Ns; MS, Ms; e; R^1; $R_1^w + R^1$; R^0; R^2, R^0; Fy^a; Tf^C; Tf^D; Tf^C, Tf^D; T^1; T^2; T^1, T^2; in the interval 0.12–0.18 of the genes N; S; MS, Ns; Ms, NS; r; R^z, R^1; $R^2, R_1^w + R_1$; R^1, R_1^w; R^1, R^0; R_1^w, R^0; le; t; t, T^1; t, T^2; in the interval 0.18–0.24 of the genes: P^2; se; in the interval 0.24–0.30 of the gene Gc^1; in the interval 0.30–0.36 of the genes R^z, r; in the interval 0.36–0.42 of the genes $R^1 + R_1^w, r$; in the interval 0.42–0.48 of the genes R^z, r. The criteria of agreement between empirical distribution with the stepping-stone model $\chi^2 = 4.17$, d.f. = 6, $P > 0.5$; for comparison with the island model $\chi^2 = 11.32$, d.f. = 4, $P < 0.025$, which indicates that the distribution of gene frequency variances differs from the structure of the island type and that the population approximates to the stepping-stone type structure, thereby supplementing the analysis of the mean values $\bar{f_i}$ and $\bar{f_w}$.

13 loci. The following values were examined: $\bar{f_i}$, the mean variance size in ith locus, and $\bar{f_w}$, the weighted mean standardised variance such that

$$\bar{f_w} = (\Sigma w_i)^{-1} \Sigma \bar{f_i} w_i, \tag{3.5}$$

where w_i is the number of variance values at the ith locus and $\Sigma w_i = 67$.

The results are shown in Table 3.11 and reflected in Fig. 3.7.

77

Let us now turn to a comparison of gene frequency variance values, in this Table with the values predicted from North-Asian empirical demographic data on the effective size and gene migrations in a theoretical model of a population structure (Table 3.10). As we see, the value $\bar{f}_w = 0.0810 \pm 0.0071$ lies almost strictly in the middle of the theoretically predicted interval between the structure's island ($f_{e_{IS}} = 0.0753$) and stepping-stone ($f_{e_{ST}} = 0.0974$) models. Thus we may conclude that the observed level of genetic differentiation in the Siberian population system indicates a type of structure that is intermediate between the island and stepping-stone types. It follows, that the interaction, through gene exchange, of populations in the system is only partially limited by distance and approaches a type of interaction independent of distance. Such a type of population structure seems to be an echo of the remote Palaeolithic past in which, with little dependence on distance, man penetrated and peopled the American continent.

Having satisfied ourselves as to the reliability of theoretically predicted standardised gene frequency variance values we can compare them with the observed standardised gene frequency variances in each of the loci studied in the population system of North Asia. Such an approach to a comparison of variances seems to us to be more effective than the usual method of evaluating the heterogeneity of variances from different loci. Such heterogeneity, if established, still says nothing about the causes that engendered it. In our case however, it is highly probable that the cause of any discrepancy between empirical and theoretical values arises from differences in the degree of the selective neutrality at specific gene loci. Having constructed the theoretical values of a standardised gene frequency variance without taking into account the factor of natural selection and being satisfied as to the reliability of these values, we are now able, from the nature of the deviation of the empirical standardised variance from the theoretical values, to judge whether one locus or another is subjected to natural selection. From the results of such a comparison (Table 3.11) we can conclude that ABO, Kell, Diego, Tf and Hp are under natural selection pressure of a stabilising type, while Se, P and Gc are subjected to a differentiating natural selection pressure. Matters are more complicated with the Rh locus, whose mean standardised variance value (see Table 3.11) corresponds to a selectively neutral heterogeneity, whereas the $29 f_i$ values used for calculating it by variances and covariances reveal high heterogeneity. Interallelic correlations also point to both stabilising and differentiating effects of selection at this locus. A more detailed description of this case is beyond the framework of reference of this review. It can be easily seen that, on the whole, the results of comparing standardised frequency variances of loci with theoretical values agree with the degree and type of deviation from the Hardy–Weinberg law in the Siberian population system, regarded as a whole (see Table 3.3 and text relating).

Thus, not only isolation and gene migration but also natural selection, i.e.

all the basic factors of a population's evolutionary dynamics, operate in the system of ancient human isolates in North Asia. But how did the pressure of these evolutionary forces affect the spatial distribution of genes in the population system?

The genogeography of North Asia

This section is predominantly of a descriptive character, since we are unable to examine, within the scope of this review, the spatial distribution of all the genes studied in Siberia. We may recall that before the IBP began, North Asia was represented as a large blank area on the majority of the world maps showing the distribution of human genes (Mourant, 1954). Now due to the work of Soviet investigators over the last decade the situation is very different as can be seen for the ABO system by comparing Figs. 3.10 and 3.11 with those published by Mourant *et al.* (1958). However, bearing in mind that each new discovery of human genes, turns not only Siberia but the world as a whole back into a blank area and that the work towards eliminating such areas is endless, we shall attempt in this section to trace through the geographical distribution of genes a reconstruction of the evolutionary forces operating in the populations of North Asia.

From the results of analysing gene frequency variances, we chose the loci of ABO blood groups (assumed stabilising selection pressure), MN blood groups (assumed selective neutrality) and P blood groups (assumed diversifying selection pressure). The scales of the maps (Figs. 3.8–3.12) correspond to the degree of knowledge of genetics in the territory, with account taken of the average area per population and the number of populations known in Siberia in the seventeenth century.

If the populations of North Asia were isolates in the full sense of the word, the plotting of a genogeographic map would be pointless because of the unpredictability of random gene drift in each of the individual isolates. A chaotic and totally patternless spatial differentiation is all that could be expected in this event. Such an effect of random gene drift is really partially manifested on the maps, but only partially, since the degree of random spatial differentiation in North Asia is certainly less than is possible from the number of population data used: 213 populations for the distribution of ABO genes, 177 for the *M* gene, and 111 for the P genes.

Considerable areas of the territory on all the maps are characterised by compact and uniform gene frequency distribution. But, as our preceding sections have shown, random gene drift is obviously not the only force operating within Siberian populations. It is counter-acted first of all by gene migrations. It is on this assumed equilibrium between random drift and gene migrations that the theoretical values of gene frequency variance were worked out. We shall turn, therefore, to the map of gene *M* distribution, the fre-

79

quency variance of which does not differ significantly from the theoretically expected one. Purely visually one may get two impressions from perusing this map: (1) compared to other maps, the space of North Asia is weakly differentiated, and (2) considerable areas of this space are occupied by a gene concentration that is close to the mean gene frequency value throughout the whole population system. But in this case the mean gene frequency within the system is an equilibrium frequency dependent on the structure of gene migrations. Let us recall that this structure possesses an internal nucleus located in the internal regions of the system from which and through which gene migration takes place (see Fig. 3.6 and text relating).

Now let us turn to the gene distribution maps for the ABO locus which substantially differ from the M gene map. The territory of North Asia on these maps is more differentiated, but the areas with mean gene frequencies (corresponding to frequencies in the population system as a whole) occur in many different places, including the periphery. In the central regions of the area, individual gene frequencies in a population group are by no means closer to the mean values and, consequently, gene migrations and their spatial structure cannot fully account for the picture observed. The impression given is that throughout North Asia the genes of the ABO locus are subjected to the additional effect of some factor, other than migration, and that the effect of this factor is such that gene frequencies close to mean values appear in the most diverse parts of the subcontinent. As one may gather from acquaintance with genogeography supported by analyses of variances, these frequencies appear as points of equilibrium in the stabilising type of selection, affecting the entire system of North Asian populations.

Lastly, let us turn to the gene P_1 distribution map. This map differs markedly from all the preceding ones in that regions with gene frequencies close to the system's mean, are virtually absent except for a very narrow strip bordering on the west, south and east of Siberia. Beyond this strip – on the western, southern and eastern fringes of Siberia, with the exception of the northern one, a steep rise in P_1 concentration can be observed. Thus, this gene's distribution map agrees least of all with the spatial structure of gene migrations, and indicates that a selection is operating on both sides of the narrow zone of the mean gene frequency. This occurs through all Siberia, and is directed in favour of the gene on the periphery and against it in the internal and northern regions. The geographical regularity observed, permits us to assume a connection between the distribution of alleles of P locus with the operation of some factor present in the natural environment. Noteworthy in this light is the rather good conformity of this genogeographical map to the geocryological map of Siberia: exactly at the point where continuous permafrost is replaced by irregular permafrost, we observe a steep rise in P_1 gene concentration. Differences in the natural environment of these territories are very great and involve not only the soil and vegetative

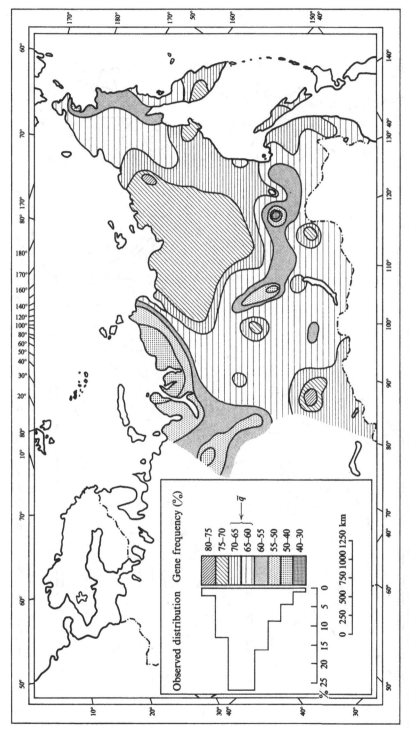

Fig. 3.8. Geographical distribution of the MN blood groups' gene *M* in the total indigenous population of North Asia.

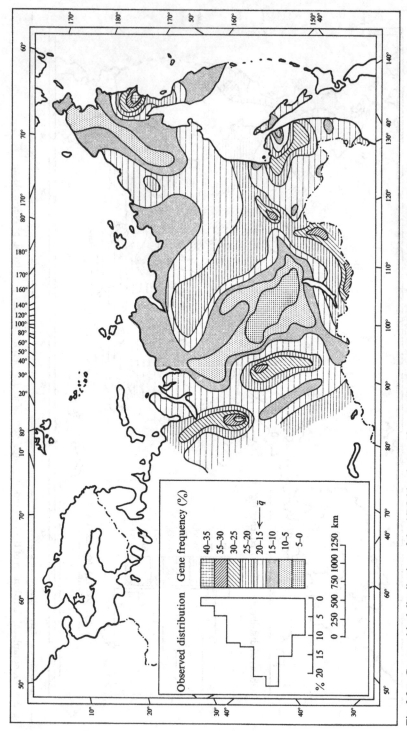

Fig. 3.9. Geographical distribution of the ABO blood groups' gene *A* in the total indigenous population of North Asia.

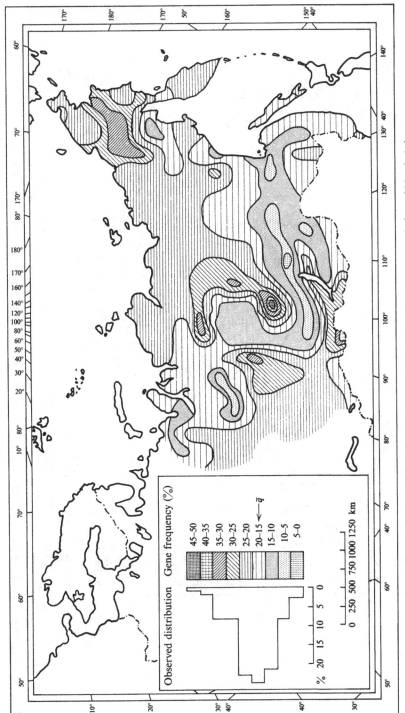

Fig. 3.10. Geographical distribution of the ABO blood groups' gene *B* in the total indigenous population of North Asia.

Observed distribution Gene frequency (%)

45–50
40–35
35–30
30–25
25–20
20–15 ←— $\bar{q}$
15–10
10–5
5–0

0 250 500 750 1000 1250 km

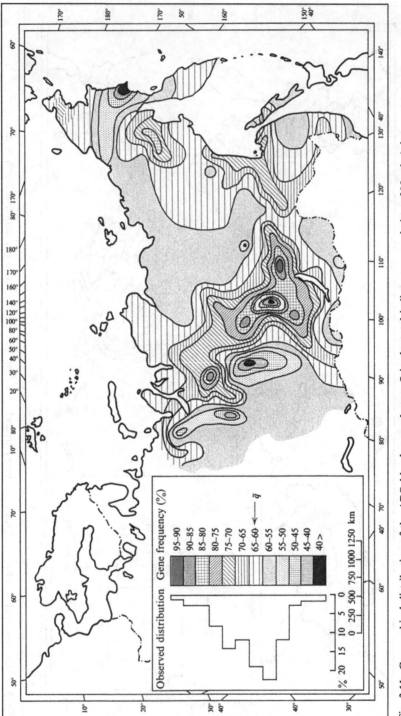

Fig. 3.11. Geographical distribution of the ABO blood groups' gene *O* in the total indigenous population of North Asia.

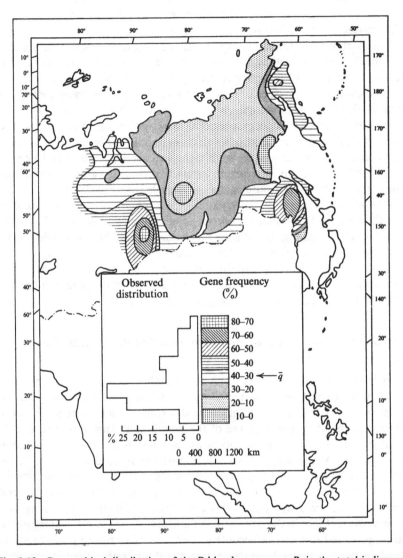

Fig. 3.12. Geographical distribution of the P blood group gene P_1 in the total indigenous population of North Asia.

cover, but many pheno-logical characteristics and their seasonal dynamics as well. We found it difficult to assess the degree to which this parallelism with the genogeography of P blood groups is causal, all the more so since outside North Asia it is apparently absent.

Returning to the genetic facts we note that the results of analysing the variances indicate a significant increase in genetic differentiation, as compared with that expected under the selectively neutral hypothesis and agree with genogeography. Moreover, genogeography in this case reinforces our hypothesis about the differentiating selection pressure and points to the fact that, the human population system in North Asia undergoes disruptive selection at the P blood group locus. The spatial localisation of the disruption zone indicates that the factor of selection in this case is some factor (or factors) still unknown in the natural environment of North Asia.

This discussion of genogeographic data was carried out in relation to the results of analysing gene frequency variances and an earlier analysis of gene migrations. The genogeography itself points to such a complicated genetic structure of the Siberian population system that, operating with gene distribution maps alone, it would seem impossible to consider the Siberian population system as a single entity at all. Let us recall, however, that underlying the treatment of isolated Siberian populations as a system is the established fact of the uniting role of gene migrations within the boundaries of the North Asian subcontinent. It follows therefore that genogeography reflects the process of genetic reorganisation, and the properties of this process are revealed as we go from examining populations within geographical co-ordinates to examining them within genetic co-ordinates.

Stationary distribution of gene frequencies in the Siberian population system

As we have remarked earlier, the pressure of evolutionary forces on a population, and their contribution to the formation of the genetic structure, can be the more easily traced by means of simple deterministic mathematical models, the more elementary the structure of the population. This approach is only possible, however, in experiments with a laboratory or computer simulation model. When we find such a population in nature, we can be no longer confident that the population, within its observed boundaries and its genetic structure established by observation, is a stable rather than ephemeral formation. We see a way out of this contradictory situation by taking into account the fact that in the course of its natural history any elementary population, provided it does not become extinct, may give rise to a multitude of populations. It is precisely in this entire multitude, rather than in any individually taken elementary daughter population, that the parental population reproduces itself with all its properties, including its genetic structure. Consequently, in the case of long-lasting isolates, the most appropriate unit for study is the historically formed population aggregate, i.e. a system of populations. Since the study of the process of genetic dynamics requires many generations, we shall replace this by an equivalent number of populations in a system.

This will require the fulfilment of a number of additional conditions. The populations, however diverse they appear at present, must have originated from a common ancestral population, must be aboriginal to the given region and have experienced the minimum of outside genetical influences. They must interact among themselves with some intensity within the boundaries of some spatially single region, in the framework of which the populations should have adapted to the environment both natural and social, through essentially the same patterns of culture and economy, thus ensuring equilibrium with the environment and stabilising their demography.

Given all these conditions, populations appear as a population system, each element of which is a population of a definite hierarchic level that may be regarded as a sample appearing (and disappearing) in the course of history, from the given natural-historical population system, and partially reproducing the system's specific features. A randomised representative sampling of such populations should adequately reflect the properties of the system and the factors which affected its genetic structure throughout the course of its natural history.

The system of human populations in the framework of the North Asian subcontinent apparently meets these requirements. Moreover, the choice of this system as the highest hierarchic unit of observation has the advantage that the greater the system's geographical area, the lesser the probability of the whole subcontinent's natural conditions systematically changing in one particular direction. Indeed, the natural environmental condition of North Asia remained, on the whole, stable, at least since the end of the glacial epoch, of which the above-mentioned permafrost is a reflection. This of course does not rule out local changes in natural conditions. But such local changes cannot be unidirectional and consequently cannot involve the population system as a whole in the process of directional genetic evolution.

Under these conditions, however complicated the gene distribution appears in the geographical co-ordinates of the population system, there is a high probability that the genetic reorganisation of the system occurring throughout the generations will turn, in the long run, into a stationary process. Such a process must be reflected in the type of gene frequency distribution in the individual ethnic groups.

The mathematical theory of stationary gene frequency distribution has been elaborated by Wright (1931, 1970). The types of such distributions are determined by those of the system parameters which not only are independent of its genetical structure but, on the contrary, appear to support the concept of the structure. The general formula of a multi-dimensional probability distribution for genes with multiple alleles, according to Wright (1970), is

$$\Phi(q_1, q_2, \ldots, q_n) = C\overline{W}^{2N} \prod_{i=1}^{n} [q_i^{4NmQ_i-1}] \qquad (3.6)$$

where N = the effective size; m = migration coefficient; Q_i = gene frequency at source of migration (and in our case, as earlier argued above, $Q_i = \overline{q}_i$); $\overline{W} = \Sigma f_i w_i$ = adaptiveness of population, w_i = adaptiveness of the ith genotype, assumed to be constant under constant environmental conditions; and C = the normalising multiple.

It is equation 3.6 that we shall use to analyse the properties of the process of genetic evolution in the indigenous populations of North Asia.

Though all forces of evolutionary significance affect the population simultaneously, it is advisable to analyse their effects in the hierarchical sequence (Rychkov *et al.*, 1973; Rychkov & Sheremetyeva, 1974; Rychkov, 1973, 1975). We propose a general null-hypothesis for migration pressure ($m = 0$) and natural selection ($W_i = 1$ at every i) as systematic pressures (neglecting for obvious reasons the effect of mutation). In this case the probability distribution (equation 3.6) is a distribution in the order of random gene drift and is U-shaped. We predict by such a distribution the empirical gene frequency distribution and, in the case of it failing to agree with the proposed general null-hypothesis, we suggest another hypothesis – that of gene migration counteraction to random gene drift: $m > 0$, $W_i = 1$. This new hypothesis remains as the null-hypothesis in regard to the effect of natural selection. In the case of the empirical data disagreeing with this hypothesis, we propose a third hypothesis in which $m > 0$, $W_i = 1 \pm s_i$ where s_i is the selection coefficient of the ith genotype. The task boils down to finding the W_i value by the pre-set empirical density at established values of N, m, and Q. When it proves difficult to determine the constant C in equation 3.6 we used the least-square method for W_i assessments. By dividing the empirical distribution density by the theoretical distribution density conforming to the second hypothesis ($m > 0$, $W_i = 1$), we obtain a system of equations in which the unknown parameters of W_i are incorporated linearly (Rychkov *et al.*, 1973). Hence we obtain by the least-square method the initial approximations of W_i which are then used for finding the final values of the parameters by regression analysis (Korostelyov & Malyutov, 1975). A BESM-3M computer was used for which an ALGOL programme was compiled (Zilberman, 1975).

Some of the results of studying gene frequency distributions in the Siberian population system are reflected in Figs. 3.13–3.16. As we see, the distribution of gene M truly corresponds to the expected one, given equilibrium between random gene drift and gene migrations and at N_e values established on the basis of empirical non-genetic data and the type of migration structure of the Siberian population system. Thus, not only does the analysis of gene frequency variance and genogeographic data but also that of gene frequency distribution indicate that under Siberian conditions the MN blood groups' locus is to a considerable degree selectively neutral.

On the other hand, studying the ABO gene frequency distributions

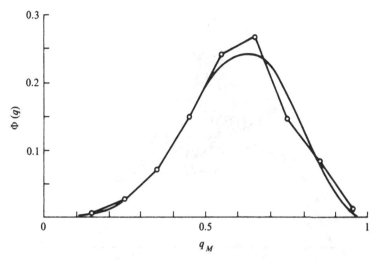

Fig. 3.13. Empirical gene M frequency distribution (q_M) in the total indigenous population of North Asia and its approximation to stationary distribution of species $\phi(q) = Cq^{4NmQ-1} (1-q)^{4Nm(1-Q)-1}$ $\chi^2 = 3.79$, d.f. $= 8$, $P = 85\%$.

(Fig. 3.14) not only confirms the results of analysing the variances and geno-geographic data on the effect on this locus of stabilising selection, but also demonstrates the mechanism of this selection. This is selection in favour of heterozygotes in which the genotype adaptiveness is:

$$W_{AA} \qquad W_{AB} \qquad W_{AO}$$
$$0.954 \pm 0.012 \quad 1.033 \pm 0.003 \quad 1.036 \pm 0.004$$

$$W_{BB} \qquad W_{BO} \qquad W_{OO}$$
$$0.983 \pm 0.009 \quad 1.024 \pm 0.003 \quad 1.008 \pm 0.001$$

The majority of North Asian populations can be located, in regard to the ABO locus, in the region of the adaptive peak and on its slopes (Fig. 3.14d), but the adaptive peak is not tied to any single region of the vast territory of North Asia (Fig. 3.15). It is the established ABO genotypic adaptive values that explain the occurrence of mean gene frequency values in the most diverse points of Siberian space, as mentioned in the preceding section. As Fig. 3.14d shows, the mean $\bar{q}$ values are very close to equilibrium values $\hat{q}$ pre-set by the genotype adaptive values, while the mean adaptive value of populations in $\bar{q}_A$, $\bar{q}_B$, $\bar{q}_O$, $\overline{W}_{\hat{q}} = 1.0172$ does not significantly differ from the maximum $\overline{W}_{\hat{q}} = 1.0174$ in equilibrium points $\hat{q}$ of genetic space. It follows, therefore, that not only the ABO locus but the gene migration structure itself of the Siberian population system may be regarded as adaptively significant. Such a structure of gene migrations when $Q = \bar{q} \rightarrow \hat{q}$ leads to the distribution of a population around the peak of the adaptive surface.

89

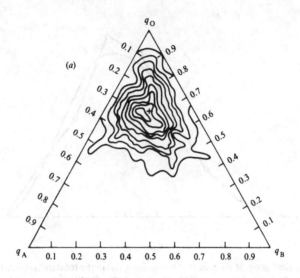

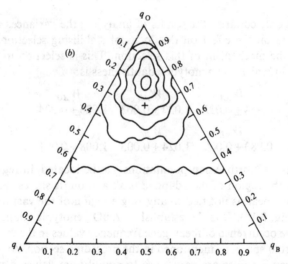

Fig. 3.14. Observed and theoretical distribution surfaces of ABO gene frequencies and adaptive landscape of North Asian populations within the co-ordinates of ABO genes.

(a) Empirical distribution surface.

(b) Expected distribution surface of species $\phi(q_1, q_2, q_3) = C \prod_{i=1}^{3} [q_i^{4NmQi-1}]$. Criterion of conformity to empirical surface $\chi^2 = 115.65$, d.f. $= 29$, $P < 0.5\%$. $+$ = location of $\bar{q}$ point.

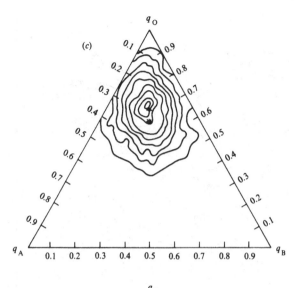

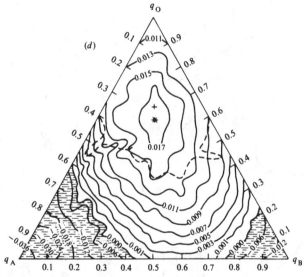

Fig. 3.14 (*cont.*).

(*c*) Expected distribution surface of species $\phi(q_1, q_2, q_3) = C\overline{W}^{2N} \prod_{i=1}^{3} [q_i^{4NmQ_i-1}]$. Criterion of conformity to empirical surface $\chi^2 = 23.38$, d.f. $= 23$, $P > 25\%$.

(*d*) Hypsometric map of the North Asian adaptive landscape within the co-ordinates of ABO gene frequencies. Section value in selection coefficient units $s = \overline{W} - 1$. For values $-0.036 \leqslant s < -0.001$, section size $= 0.006$; for values $s \geqslant -0.001$, section size $= 0.002$. Dashed line indicates empirical dissemination limit of Siberian populations.

    * $=$ location of $\hat{q}$ point.

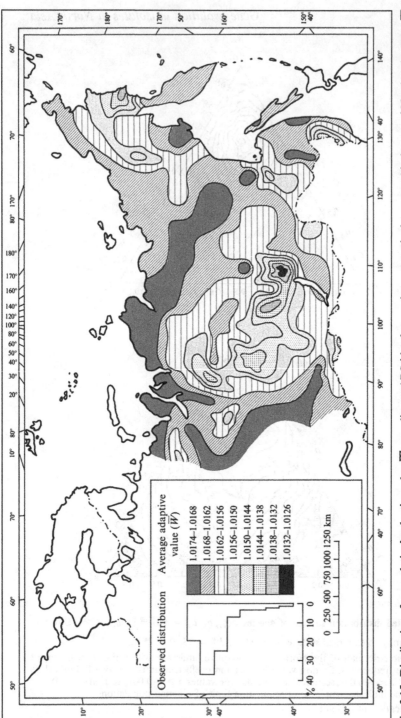

Fig. 3.15. Distribution of a population's adaptive value W according to ABO blood groups' genotypes in the geographical space of North Asia. Here W represents the mean selective value in Wright's interpretation (1970). A comparison of this Figure with Fig. 3.14d shows that the distribution surfaces W in the gene frequency space and in the geographical space relate one to another as a 'cone' to a 'funnel'; the high W values are distributed in rings along the periphery of the geographical area of the Siberian population system. By integrating the genogeographic data of ABO blood groups (Figs. 3.9–3.11), this map indicates that the combination of ABO gene frequencies at which $q_i \to q \approx \hat{q}$ at all is at this gene locus is more often realised on the periphery of the area.

Observed distribution

Average adaptive value ($\overline{W}$)

1.0174–1.0168
1.0168–1.0162
1.0162–1.0156
1.0156–1.0150
1.0150–1.0144
1.0144–1.0138
1.0138–1.0132
1.0132–1.0126

0   250   500   750 1000 1250 km

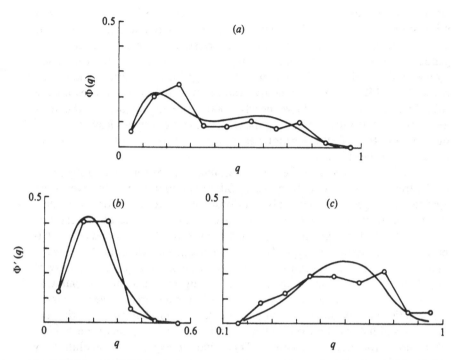

Fig. 3.16. Empirical distribution of P_1 gene frequenc yin North Asia as a whole (*a*), in the 'northern' internal (*b*) and in the 'southern' peripheral (*c*) parts of the subcontinent. Approximation by stationary distribution of species

$$\phi(q) = C\Sigma\phi'(q), \ \phi'(q) = Ce^{-4Nsq}q^{4NmQ-1}(1-q)^{4Nm(1-Q)-1}$$

Values of selection coefficents are determined from the condition of selection–migration equilibrium in each of the distributions (*b*) and (*c*) and $s_b = 0.0123$, $s_c = -0.0088$; $\chi_a^2 = 10.13$, d.f. $= 6$, $P > 10\%$.

The results of analysing gene frequency distribution in the P blood group locus fully conform to the genogeographic data. As Fig. 3.16 shows, the distribution of this gene in Siberia actually represents the sum of two distributions in two Siberian areas, conditionally designated as 'north' and 'south', and formed by gene migrations and selection, both against and in favour of, this gene, at the assumed intermediate fitness of heterozygotes. The integrative role of gene migrations in the case of this locus is strikingly apparent.

However, we see the most important result of studying gene frequency distribution in the fact that, in full conformity with the prerequisites of this analysis outlined above, all the diverse variants of empirical distributions are approximated as stationary type distributions. This is true also for the P alleles' distribution, since the sum of two stationary distributions is also

93

stationary. We have seen that the population system of Siberian isolates does not just maintain a state of equilibrium between isolation and migration, but is subjected in addition to the selective pressure of the environment, engendering the process of population adaptation. Suffice to say that, calculated by the total size of Siberia's contemporary indigenous population, and on the basis of the ABO genotypic adaptive values estimated, people possessing the *AO, BO, AB* genotypes leave nearly 7000 more decendants in the next generation, while those possessing the *AA, BB, OO* genotypes leave the same number less than if all the genotypes were selectively neutral in the natural conditions of North Asia.

Both the analysis of gene frequency variances and of genogeography point to the tremendous genetic dynamics of Siberian populations. However, only an analysis of gene frequency distributions which reveals stationary properties permits us to conclude that the intensive genetic dynamics of populations evolves into a stationary process, which is mirrored by the observed distributions of the stationary type. Hence the conclusion, which is essential for understanding the relationship between evolution and the genetic stability of populations. Under conditions ensuring a long existence of a population system, evolution proceeding at the system's lower hierarchic levels enhances the genetic stability of the system as a whole. The mechanisms ensuring the system's stability qualitatively differ from the Hardy–Weinberg mechanism of a non-structured population. This qualitatively higher evolutionary stability of a population system, ensured by the movement of its components, is a means of verifying the evolutionary significance of the genetic information contained in the system, through the population's much longer exposure, on a natural-historical scale, to the environment. Both theoretical and practical population genetics usually pass by this higher systemic level of population organisation. And this is why the evolutionary theme predominates in discussions of the results of investigating elementary population groups and, all the more so, when discussing random samplings from historically undefined total populations. It is precisely the study of the system of partially isolated populations that makes it possible to distinguish the different evolutionary qualities of the consequences of gene dynamics for populations at different structural and chronological levels of the system's organisation.

We would hate to leave the impression of a hastily thrown together theoretical generalisation of the results described here. Empirical distributions of gene frequencies are, indeed, approximated by theoretical distributions of the stationary type; yet are there any other indications of lasting genetical stability of the Siberian population system? We shall answer this question in the next section.

On the duration of stable genetic equilibrium maintained by a population under the conditions of North Asia

A discussion of the results of analysing the stationary state of gene frequency distribution in North Asia follows from the well-known formulations of the spatial distribution in a series of populations being equivalent to the distribution in a series of generations of a single population and the distribution of many independent genes. It is on this principle of ergodicity that the next part of the discussion is based. The conclusion about the genetic stability of the system, made earlier on the basis of the established stationary quality of gene frequency distributions, can be verified, first of all, in regard to gene frequency $\bar{q}$ in the system. Let us make the following assumption: given common descent, the gene concentration in the contemporary system of North Asian populations is a concentration of the genes of the initial ancestral population. Such an assumption is based on the mathematical theory of random gene drift and is actively used in reconstructing the evolution of human populations and in assessing the degree of their genetic divergence. The divergence between two populations is assessed relative to the mean gene concentration $(\bar{q})$ which is assumed to be the gene frequency q_0 at the moment of the separation of populations. Though this assumption agrees with the mathematical model of divergence by random gene drift, it cannot, as a rule, be verified by a specific investigation of a natural population. In the absence of such confirmation, the use of $\bar{q}$ instead of q_0 in an analysis of actual natural populations is very arbitrary and involves the superficial extrapolation of an abstract model to real material. The experience of population studies in North Asia supports the view that only in a study planned and carried out in the framework of a long-standing vast population system can the $\bar{q}$ observed today in the system really be the same as the frequency in the population that founded the system. In the same way, only under conditions of a system of populations, is the ergodicity theorem fulfilled and, the gene distribution over the historical ensemble of populations be really equivalent to distribution in time.

The above is confirmed by data on the distribution in Siberian populations which have discretely varying characters of cranial anatomy, or 'cranial anomalies' as they are traditionally designated in the Russian literature (Rychkov & Movsesyan, 1972; Rychkov, 1973; Movsesyan, 1973, 1975). The hereditary component of these characters is quite large (Berry, 1963). Thanks to the archaeological expeditions of A. P. Okladnikov (1950, 1956, 1974), the osteological remains of the population which lived along the Angara and Lena rivers and in the Trans-Baikal area are known to us. They belong to different stages of the Neolithic epoch from the second to the fifth millenium B.C. As Okladnikov (1950, 1955) points out, the Neolithic cultures of Siberia reveal signs of the separation of large ethnic and tribal groups to which the roots of the modern Siberian peoples go back.

95

Yu. G. Rychkov and V. A. Sheremetyeva

Thus, we are dealing with a population in a space–time continuum, characterised by an area equal to 785000 km² and going back in time over 4000 years, i.e. over 160 generations. This Neolithic Siberian population, in turn, is separated from contemporary populations by no less than another 160 generations.

As Fig. 3.17 shows not one of the contemporary Siberian populations has a frequency distribution of the 12 independent cranial anomalies that is close to the distribution in any of the known Neolithic Siberian populations. It is all the more striking, therefore, that the total Neolithic population and the total contemporary population of Siberia are characterised by nearly identical distribution of characters. The same result is obtained by comparing also the quantitative craniometric characters (Rychkov, 1969).

Thus, we may be convinced of the genetic stability of the total Siberian population, at least in the interval of the last seven millenia. Therefore, the system of ancient isolated populations of North Asia, at least throughout 280 generations, preserves the entire initial genetic information, because this information, being constantly reorganised according to populational components, ensures the stability of the entire system in the natural and historical conditions of the North Asian subcontinent.

We could go further and show that the stationary state revealed when analysing gene frequencies in the modern population of Siberia is reflected in other parameters besides the constancy of $\bar{q}$ in time. Already in the Neolithic period the frequency variances of cranial anomalies reached a level close to the modern one. The mean variance values for 34 independent characters are compared below (Movsesyan, 1973). Taking into account the low frequency of a character, which is typical of anomalies, the angular transformation of frequency was used:

Population of Siberia

Neolithic	$\sigma_\theta^2 = 0.0281$	d.f. = 33
Modern	$\sigma_\theta^2 = 0.0321$	d.f. = 33
Variances ratio	$F = 1.14$	$P = 25\%$

Thus, the stationary state of the gene frequency distributions of modern Siberian populations scattered in space actually reflects the stationary character of the dynamics of gene frequency distributions in the generations. The property of the space–time continuum of the stationary process is used and confirmed by an example that would seem at first glance to be least suitable for this purpose – the example of populations isolated both in space and in time. Here it proved possible not only to assume theoretically that such a parameter of a modern system as mean gene frequency $\bar{q}$ is pre-set by the corresponding parameters of the ancestral population, but also to make certain that the assumption was true at least for the time span from the Neo-

96

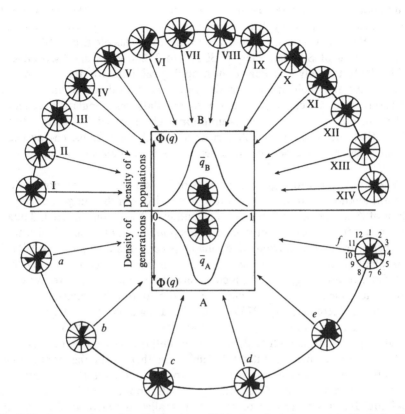

Fig. 3.17. A comparison of the distributions of 12 independent characters in Neolithic (A) and contemporary (B) populations of Siberia, demonstrating that the unweighted frequencies $\bar{q}_t$ in the contemporary population system are good assessments of $\bar{q}_0$ in the Neolithic ancestral population, $t \approx 200$ generations. In the centre of the Figure there is a schematic diagram of the theoretical justification of this comparison. *Contemporary populations*: I, Asian Eskimos; II, the Chukchi of the Chukotka Peninsula; III, the Aleuts; IV, the Negidals of the Amur River basin, V, the Ulchis of the Lower Amur; VI, the Tungus of Eastern Siberia; VII, the Buryats of the Baikal Area; VIII, the Mongols; IX, the Tuvinians of the Sayan Plateau; X, the Telengets of the Altai Mountains; XI, the Khakassians of Minusinsk Hollow, the Enissey river; XII, the Selkups of Western Siberia; XIII, the Mansi of the Eastern slopes of the Urals; XIV, the Khants of the Ob river basin. *Neolithic populations*: *a*, 'Serov' culture, Angara river; *b*, 'Kitoi' culture, Angara river; *c*, 'Glazkov' culture, Angara river; *d*, 'Serov' culture, Upper Lena; *e*, 'Glazkov' culture, upper Lena; *f*, 'Glazkov' culture, Eastern Baikal area. Frequency $q = 0$ in centre of circle; $q = 0.3$ on perimeter. 1–12 characters, a list of which can be found in Rychkov & Movsesyan, 1972; Movsesyan, 1973; Rychkov, 1973.

lithic to the present. But if, as follows from Fig. 3.17, the polygon of mean character frequencies in the modern Siberian population represents a good reconstruction of a similar polygon of the Neolithic population, the question arises as to what still more ancient stage in the history of Siberia's indigenous

97

population is reconstructed by the mean frequencies of characters established for the Neolithic epoch by direct observation. The naturally assumed answer is the Palaeolithic epoch, which, according to archaeological and radiocarbon dating of Siberian sites, is 30000 to 35000 years ago (Mochanov, 1969). Unfortunately the Palaeo-anthropological data for this epoch are fragmentary and their analysis in the above-mentioned way is impossible. However, here we can again return to the data on population genetics and recall the well-established assumption, substantiated by many disciplines, that the Siberian and American ancestral populations, having been at one time a single whole, were separated from Palaeolithic times onwards.

When, only a decade ago, the genetics of Siberia's indigenous population was still unknown, nothing but indirect comparisons of the indigenous population of America with the peoples of Southeast Asia could be used. Today we can resort to a direct comparison of the population of America with that of Siberia, through and from which man's penetration into the New World took place (Fig. 3.18). Such a comparison has been carried out for 11 loci including, in all, 28 genes, of which 17 are independent. The two systems' unweighted mean gene frequencies are compared (Sheremetyeva, 1973; Rychkov & Sheremetyeva, 1976b). For evaluating $\bar{q}$ on the American continent we used the tables compiled by Post, Neel & Schull (1968) for Amerindian populations, as well as data by other authors concerning populations of the American and Greenland Eskimos. In their turn, the populations of North Asia are taken into account throughout the entire subcontinent from the Urals to the Chukotka peninsula, populated by the Asian Eskimos. Just as in the case of cranial anomalies, the methods of building a gene frequencies polygon was used, the only difference being that the frequency $q = 0$ is placed on the perimeter, and $q = 1$ in the centre of the circle.

If one takes into account that each of the polygons compared may change in 17 independent directions on account of 11 independent combinations of natural selection, mutation, gene migration and random gene drift pressures, the similarity of the genetic structures of the peoples of Siberia and America becomes especially convincing.

It is all the more striking when compared with similar polygons (Fig. 3.18) for two Aleut populations of the Komandorski Islands, which have separated recently, in the accurately known period between 1825 and 1828. Before that the Komandorski Islands were uninhabited, and the Aleuts settled on them with the assistance of the Russian-American Company, a move which is well documented. Owing to this, the Aleut populations of the Komandorski Islands afford a good yardstick for considering the problems of the comparative genetics of the peoples of Siberia and America. The visual similarity of the polygons of each of the pairs is confirmed by a quantitative measure in

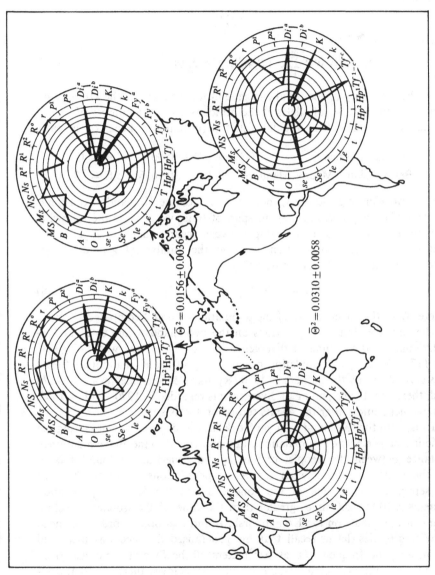

$\bar{\Theta}^2 = 0.0156 \pm 0.0036$

$\bar{\Theta}^2 = 0.0310 \pm 0.0058$

Fig. 3.18. Polygon of $\bar{q}$ genes frequencies, characterising the genetic structure of the North Asian and American population systems in comparison to the two Aleut populations of the Komandorski Islands. $\bar{\Theta}^2$ = mean square of genetic distance in radians.

Yu. G. Rychkov and V. A. Sheremetyeva

assessing the mean square of the angular distance (Θ) by the set of genes such that

$$\overline{\Theta}^2 = n^{-1} \sum_{j=1}^{m} \theta_j^2,$$

where

$$\theta_j = \text{arc} \left(\cos \sum_{i}^{n_j} \sqrt{(q_{ij}^k q_{ij}^{(l)})}, \ n = \sum_{j=1}^{m} (n_j - 1), \right) \tag{3.7}$$

q_{ij}^k = the frequency of the ith allelic gene in the jth locus of the kth population; n_j = the number of alleles at the jth locus; m = the number of loci.

The mean squares of the angular distances (in radians) of each of the pairs are:

Siberia–America $\qquad\qquad \overline{\Theta}^2 = 0.0310 \pm 0.0058$, d.f. = 17

Bering Island–Medny Island $\overline{\Theta}^2 = 0.0156 \pm 0.0036$, d.f. = 18

By approximating the divergence of Aleuts of the Komandorski Islands by the diffusion process in a small span of time, we can use the following formula expressing the relationship between mean square of angular distances with effective population size and the number of generations (t) (Malyutov, Passekov & Rychkov, 1972)

$$t = \overline{\Theta}^2 [(8N_{e_1})^{-1} + (8N_{e_2})^{-1}]^{-1} \tag{3.8}$$

where N_e = the effective size of the populations 1 and 2 compared.

The effective sizes of populations on the two islands came to 81 and 70 individuals, and the time of divergence in the generations was 4.67 ± 1.08 (or 117 ± 27 years). A population genetic study of the Komandorski Aleuts was carried out by the authors in 1970 (Rychkov & Sheremetyeva, 1972a, b) and, therefore, the beginning of divergence is regarded as $1970 - 117 = 1853$, i.e. one generation after the Aleuts settled on the Islands.

Thus, with the 'Komandorski yardstick' at our disposal for measuring the genetic distance between populations, and applying it to the above-mentioned distance between the peoples of Siberia and America, we found this to be the same as that between the Aleut populations; whereas the time of separate development of the American and the Siberian population systems is 100 to 200 times greater than the time of the isolated development of the Aleuts on the Komandorski Islands. In other words, the mean gene frequencies (let us recall that the generalised distance was measured by mean gene frequencies), as parameters of the Siberian and American population systems, remain stable in time. It follows, therefore, that $\bar{q}$ represents a good reconstruction of the genetic structure of the two ancestral populations at the beginning of the peopling of Siberia and America. Moreover, thanks to the property of lasting genetic stability of the systems, it proved possible, from comparing the modern parameters of the two systems, to suggest that at the dawn of man's pre-history, in the epoch of the upper

100

Palaeolithic, the Siberian and American ancestral populations were genetically close to one another, like neighbouring populations of a single people, as for example, the Aleut populations of the Komandorski Islands. Being unable to estimate the effective size of the total American population system we could not use equation 3.8 for confirming the previous thesis by assessing the time of divergence. This, however, may be done by another, equally direct method, if one takes into account one of the preceding conclusions – that the genetic stability of a population system is achieved through genetic dynamics, i.e. evolution of populations that are the system's structural components. With such a relation between the process of genetic evolution at different structural levels of the system, the stability in time of the mean gene frequencies within the system, and the system's internal genetical differentiation, must be inter-related. Using the results of gene frequency variance analysis described above pp. 74–79, the process of internal differentiation of the Siberian population system, as distinct from a simple process of random gene drift, may be expressed by the formula

$$f_0 = f_e(1 - e^{-t/2N}), \tag{3.9}$$

where f_0 stands for the observed variance value in the modern system, while f_e is the theoretically expected value, at equilibrium, of random gene drift and gene migrations.

If we assume that the migration structure of the Siberian population system is developing towards the stepping-stone model type, at its limit, then $f_e = 0.0974$ according to the earlier predicted theoretical value. If, on the other hand, we assume that the development limit of the structure is a type intermediate between the island and stepping-stone ones, then $f_e = 0.0864$ is the mean value of standardised variances predicted on the basis of both models.

Having the empirical value $f_0 = \bar{f}_w = 0.810 \pm 0.0071$ and using equation 3.9, we obtain the value of $t = 777 \pm 68$ (or 19419 ± 1702 years) for the stepping-stone model and $t = 1209 \pm 106$ (or 30221 ± 2649 years) for the population structure of the intermediate type. Both time assessments take us back to the Upper Palaeolithic, the first nearer the end of that epoch and the second to its middle. Both time assessments agree with the datings made on the basis of other data and methods: (1) dating of the concluding stages in the separation of the Siberian and American ancestral populations in the area of the ancient Bering landbridge, and (2) dating of the beginning of this process. From that time until now the genetic distance between the evolved population systems of Siberia and America did not exceed the distance between neighbouring groups belonging to the same people. The genetic dynamics of an elementary population, taking place under the joint effect of evolutionary forces, natural selection included, but within the boundaries of a population system, has indeed evolved into a stationary process.

Yu. G. Rychkov and V. A. Sheremetyeva

This genetic excursion into the past has led us to the truism: the past lives on in the present. But this past is untraceable in every individual and, even more so, in an isolated population, and the number of such elementary populations is without end. This past really exists only in a system. And it survives all the longer, the more complicated the population system's structure. Moreover, as we have seen, this past pre-sets and determines the system's properties. For thousands of years a population system is subjected to the pressure of random and systematic evolutionary forces, but distributes these pressures throughout its structural components which retain some degrees of freedom of genetic reorganisation. Even in such extreme environmental conditions as those of North Asia, the system as a whole has not made a single evolutionary step in the sense of the contemporary genetic interpretation of evolution described at the beginning of this review.

Thus, having examined simultaneously the process of genetic variability and its consequences at the lower and higher structural levels of a population's system, we may conclude that a system's highest level depends to a much lesser degree on the state of the environment, whereas a lower level population, in agreement with population genetics theory, depends greatly on the state of the environment and constantly forms and changes its genetic structure under environmental influence. A population of the lower level is actually included in the Markovian process, in which its future changes are independent of the memory of past events. It is precisely a population system that acquires such a memory, as a result of which the initial stages of evolution can be easily discerned in the present-day system's genetic parameters.

The formation of the internal structure of a population, the emergence of the most complex population system from an elementary population unit in combination with the continuously proceeding genetic reorganisation of its components, appears as a fundamental property of the population level in the organisation of life that ensures effective adaptation and, thereby, stability in the fluctuating conditions of the environment.

Conclusions

The problem 'population structure and human variability' contains a concealed contradiction in its very formulation, for the concept of structure brings associations of something pre-set and permanent, while the concept of variability implies something progressive. Actually, the structure of a population develops with passage of time and so the results of the variability process differ for the different chronological levels of a populations' structure. Hence, the outcome of studying this problem will depend on the level of the population structure at which the investigation is carried out. The historical nature of a population and the historical nature of a population structure is particularly obvious in man. The desire to carry out the investigation at the simplest level of the structure – at the level of 'villages' and 'isolates' – turns out, after all, to be rather shallow in regard to length of time, as reflected in man's biological and, in particular, genetical variability. The difficulty of investigating a sufficiently large number of such elementary populations prompts a purely statistical approach in which populations are replaced by random samplings from an unknown general aggregate, which may disorganise the study altogether. In actual fact, it is none other than time itself which, despite us and for us, and according to its own plan, makes samplings, such as populations, and organises them in aggregates (population systems). The task, therefore, is to reveal the plan according to which such samplings (populations) are organised and to carry out, on that basis, studies into the variability in a population as a natural-historical structure. As such, population system is a time function, whereas an elementary population, whether a 'village' or an 'isolate', etc., represents merely a point in the developmental process of the system.

It is according to the system of population points that this study of man's genetic variability in North Asia has been carried out. When investigating variability at the lowest levels of population structure we encountered all the factors of genetic dynamics expected beforehand and the consequences of existence under conditions of spatial isolation and an extremely severe natural environment. The effects of random gene drifts, of various types of natural selection and a weakly developed gene migration structure were found to be intricately intertwined and reflected in the heterogeneity of gene frequency variances of different loci, in different types of spatial gene distributions and, lastly, in different types of statistical distributions. It was, however, against this dynamic background of the simplest level of population structure that it proved possible to observe with particular clarity the consequence to which the seemingly boundless variability arising in the foundation of the system leads to the structure's higher levels. This consequence appeared as a strikingly stable gene concentration in the population system as a whole. The stability was so high that, at mean gene and character frequencies in Siberia's contemporary indigenous population, it was easy to recognise not only a close conformity to Siberia's Neolithic population but also that Palaeolithic ancestral population associated with which is man's discovery and peopling of America.

To Moscow University students and post-graduate students D. Biryukov, S. Borodina, T. Volkova, V. Gorshkov, V. Groshev, E. Gubenko, O. Zhukova, N. Zavodnik, S. Yefimova, I. Kozlova, N. Klevtsova, V. Kondik, A. Lukyanov, A. Moskalenko, I. Perevoschikov, V. Spitzyn, K. Sapata, T. Tausik, F. Funes, Yu. Chaikovski and A. Shuvayaeva the authors are indebted for assistance in the work of the expeditions they had directed over the years. Appreciation is extended to student O. Davydov for assistance in preparing the section on analysis of gene frequency variances and to post-graduate student V. Kovalenko for her assistance in preparing the section on anthropometric characteristics.

Yu. G. Rychkov and V. A. Sheremetyeva

Thanks are due to A. G. Bashlai of the Moscow City Blood Transfusion Station, Professor A. M. Umnova of the Central Institute of Blood Transfusion, Pofessor M. A. Bronnikova, Professor M. I. Potapov and L. K. Arzhelas of the Institute of Forensic Medicine, for reagents provided and methodological advice given. The authors gratefully acknowledge the contribution of foreign scientists: Dr A. E. Mourant (Britain), Dr M. Layrisse (Venezuela), Dr A. Steinberg (USA), who provided serum reagents for serological work during expeditions. The 1970 expeditions to Chukotka and the Komandorski Islands and of 1971 to the basin of the Podkamennaya Tunguska river were carried out with partial financial support of the World Health Organization, to which appreciation is extended.

The authors are particularly indebted to Professor Z. I. Barbashova, Chairman of the HA Section, USSR National Committee for the International Biological Programme, for her invaluable help in organisation. Last but not least, appreciation is extended to B. M. Meerovich, who translated this work into the English language.

References

Alexeyev, V. P., Benevolenskaya, Yu. D., Gokhman, I. I., Davydova, G. M. & Zhomova, V. K. (1968). Anthropological studies on the Lena. In Russian. *Sovetskaya etnografiya*, no. 5.

Alexeyeva, T. I., Volkov-Dubrovin, V. P., Golubchikova, Z. A., Pavlovski, O. M., Smirnova, N. S. & Spitzin, V. A. (1972). An anthropologica linvestigation of the forest Nentsy (morphology, physiology and population genetics). In Russian. *Voprosy antropologii*, **41**, 19–35.

Alexeyeva, T. I., Volkov-Dubrovin, V. P., Pavlovski, O. M., Smirnova, N. S. & Shekochikhina, L. K. (1970). Anthropological surveys in the Trans-Baikal region and the problem of human adaptation (morphology, physiology and population genetics). In Russian. *Voprosy antropologii*, **36**, 3–20.

Allison, A. E. (1955). Aspects of polymorphism in man. *Cold Spring Harbor Symposia on Quantitative Biology*, **20**, 239–55.

Berry, R. J. (1963). Epigenetic polymorphism in wild mouse populations. *Genetical Research*, **4**, 193–220.

Cavalli-Sforza, L. L. & Bodmer, W. F. (1971). *The Genetics of Human Populations*. San Francisco: Freeman & Co.

Debetz, G. F. (1951). Anthropological studies in the Kamchatka region. In Russian. In *Trudy instituta etnografii ANSSSR*, Novaya seriya, **17**, Moscow: USSR Academy of Sciences Press.

Debetz, G. F. & Trofimova, G. A. (1941). The West-Siberian Expedition of 1937. In Russian. In *Kratkiye soobshcheniya o nauchnykh rabotakh NII i Muzeya antropologii za 1938–1939*. Moscow.

Dolgikh, B. O. (1960). The gentile and tribal composition of Siberian peoples in the 17th century. In Russian. In *Trudy instituta etnografii ANSSSR*, Novaya seriya **55**. Moscow: USSR Academy of Sciences Press.

Galperin, E. A. & Zhitomirski, V. K. (1967). Blood groups in the Evenks. In Russian. *Voprosy antropologii*, **27**, 140–1.

Glass, B. (1956). On the evidence of random genetic drift in human populations. *American Journal of Physical Anthropology*, **14**, 541–55.

Glass, B. & Li, C. C. (1953). The dynamics of racial intermixture – an analysis based on the American Negro. *American Journal of Human Genetics*, **5**, 1–20.

Gorokhov, V. G. (1971). Distribution of the ABO blood groups system among Khakassian people. In Russian. *Voprosy antropologii*, **37**, 136–9.

Kimura, M. & Weiss, G. H. (1964). The stepping stone model of population structure and the decrease of genetic correlation with distance. *Genetics*, **49**, 561–76.

Korostelyov, A. P. & Malyutov, M. B. (1975). On the estimation of genotype fitness by using stationary gene frequency distribution values. In *Applications of Statistical Methods in Problems of Population Genetics. Collected Papers*, ed. M. B. Malyutov, pp. 45–51. In Russian. Moscow: Moscow State University Press.

Levin, M. G. (1958*a*). Blood groups among the Chukchi and the Eskimo. In Russian. *Sovetskaya etnografiya*, **5**, 113–16.

Levin, M. G. (1958*b*). Ethnic anthropology and problems of ethnogenesis of the peoples of the Far East. In Russian. In *Trudy instituta etnografii ANSSSR*, Novaya seriya, **36**, Moscow: USSR Academy of Sciences Press.

Levin, M. G. (1959). New materials on the blood groups among the Eskimo and the Lamuts. In Russian. *Sovetskaya etnografiya*, **3**, 98–9.

Malyutov, M. B., Passekov, V. P. & Rychkov, Yu. G. (1972). On the reconstruction of the evolutionary trees of human populations resulting from random genetic drift. In *The Assessment of Population Affinities in Man*, ed. J. S. Weiner & J. Huizinga, pp. 48–71. Oxford: Clarendon Press.

Mochanov, Yu. A. (1969). The earliest Stages in the peopling of North-East Asia and Alaska (to the question of the first human migrations to America). In Russian. *Sovetskaya etnografiya*, 1969, pp. 79–86.

Mourant, A. E. (1954). *The Distribution of the Human Blood Groups*. Oxford: Blackwell.

Mourant, A. E., Kopeć, A. C. & Domaniewska-Sobczak, K. (1958). *ABO Blood Groups*. Oxford: Blackwell.

Movsesyan, A. A. (1973). Some aspects of the population genetics of modern and ancient populations of Siberia. In Russian. *Voprosÿ antropologii*, **45**, 77–84.

Movsesyan, A. A. (1975). Anatomical peculiarities of cranial structure and their application in the genetical–anthropological analysis of modern and ancient populations of Siberia. Author's Dissertation Abstract. Moscow.

Neel, J. V. & Ward, R. H. (1972). The genetic structure of a tribal population; the Yanomama Indians. vi. Analysis by *F*-Statistics (including a comparison with the Makiritare and Xavante). *Genetics*, **72**, 639–66.

Nei, M. (1965). Variation and covariation of gene frequencies in subdivided populations. *Evolution*, **19**, 256–8.

Okladnikov, A. P. (1950). The Neolithic and the Bronze Age in the Baikal Area. In Russian. In *Materialy i issledovania po arkheologii SSSR*, issue **18**, part I.

Okladnikov, A. P. (1955). The Neolithic and the Bronze Age in the Baikal Area. In Russian. In *Materialy i issledovaniya po arkheologii SSSR*, issue **43**, part 3.

Okladnikov, A. P. (1974). *Neolithic Monuments on the Angara*. In Russian. Novosibirsk: 'Nauka' Publisheis.

Patkanov, S. K. (1911). Statistical data showing tribal composition of the Siberian population, language and clans of national minoiities (based on data from special processing of the 1897 Census materials). In Russian. *Papers of the Imperial Russian Geographic Society, Department of Statistics*, ed. V. V. Morachevsky, vol. **11**, issue 2. St Petersburg.

Patkanov, S. K. (1912). Statistical data showing tribal composition of the Siberian population, the language and clans of national minorities (based on data from special processing of the 1897 Census materials). In *Papers of the Imperial Russian Geographic Society, Department of Statistics*, ed. V. V. Morachevsky, vol. **11**, issue 3, St Petersburg.

Post, R. H., Neel, J. V. & Schull, W. J. (1968). Tabulations of phenotype and gene frequencies for 11 different genetic systems studied in the American Indian. In *Biomedical Challenges Presented by the American Indian*. Scientific Publication no. 165 (September 1968). Washington, D.C.: Pan American Health Organization.

Roginski, Ya. Ya. & Levin, M. G. (1963). *Antropologia*. In Russian. Moscow: Higher School Press.

Rychkov, Yu. G. (1965a). Some aspects of serological examinations in anthropology. In Russian. *Voprosy antropologii*, **19**, 95–105.

Rychkov, Yu. G. (1965b). Peculiarities of serological differences in Siberian peoples. In Russian. *Voprosy antropologii*, **21**, 18–34.

Rychkov, Yu. G. (1968). Reaction of populations to isolation. In Russian. In *Problemy evolyutsii*, vol. 1, pp. 212–36. Novosibirsk: 'Nauka' Publishers, Siberian Department.

Rychkov, Yu. G. (1969). Some approaches to Siberian anthropology based on population genetics. In Russian. *Voprosy antropologii*, **33**, 16–33.

Rychkov, Yu. G. (1973). The system of ancient human isolates in Northern Asia from the genetic aspects of evolution and stability of populations. In Russian. *Voprosy antropologii*, **44**, 3–22.

Rychkov, Yu. G. (1975). Detecting and evaluating the pressure of natural selection on ABO blood groups' genotypes from empirical data on the genetic structure of human populations in Northern Asia. In *Applications of Statistical Methods in Problems of Population Genetics. Collected Papers*, ed. M. B. Malyutov, pp. 19–44. In Russian. Moscow: Moscow State University.

Rychkov, Yu. G. & Borodina, S. R. (1969). Hypersensitivity to phenylthiourea (PTC) in one of the isolated populations of Eastern Siberia. A hypothesis of inheritance. In Russian. *Genetics*, **5**, 116–23.

Rychkov, Yu. G. & Borodina, S. R. (1973). Further genetical research into man's hypersensitivity to phenylthiocarbamide (experimental, population, family evidence). In Russian. *Genetics*, **9**, 141–52.

Rychkov, Yu. G. & Movsesyan, A. A. (1972). A genetical anthropological analysis of the distribution of cranial anomalies in the mongoloids of Siberia in connection with the problem of their origin. In Russian. In *Transactions of the Moscow Society of naturalists*, **58**, part 3, 114–32.

Rychkov, Yu. G., Perevozchikov, I. V., Sheremetyeva, V. A., Volkova, T. V. & Bashlai, A. G. (1969). To the population genetics of Siberia's indigenous population. The Eastern Sayan mountains. In Russian. *Voprosy antropologii*, **31**, 3–32.

Rychkov, Yu. G., Rusakova, O. L., Rappoport, M. P. & Sheremetyeva, V. A. (1973). Factors of genetic differentiation in the population system of the North Asian aboriginal stock. 1. Estimation of the integrative adaptive values of ABO blood groups' genotypes in the Siberian native population. In Russian. *Genetics*, **9**, 136–45.

Rychkov, Yu. G. & Sheremetyeva, V. A. (1972a). Population genetics of the Aleuts of the Komandorski Islands: problems of history of the peoples and the adaptation of the population in ancient Beringia. I. In Russian. *Voprosy antropologii*, **40**, 45–70.

Rychkov, Yu. G. & Sheremetyeva, V. A. (1972b). Population genetics of the Aleuts of the Komandorski Islands: problems of the history of the peoples and adaptation of the population in ancient Beringia. II. The Komandorski yardstick for studying genetic differentiation in the North Pacific Basin. *Voprosy antropologii*, **41**, 3–18.

Rychkov, Yu. G. & Sheremetyeva, V. A. (1972c). Genetics of North Pacific populations related to problems of the history and adaptation of peoples. III. The Eskimo and Chukchi populations of the Bering Sea Coast. In Russian. *Voprosȳ antropologii*, **42**, 3–30.

Rychkov, Yu. G. & Sheremetyeva, V. A. (1974). Factors of genetic differentiation of the population system of the North Asian aboriginal stock. II. Structure of gene migration in Siberia and the adaptive landscape of Northern Asia in the space of ABO blood groups' gene frequencies. In Russian. *Genetika*, **10**, 147–59.

Rychkov, Yu. G. & Sheremetyeva, V. A. (1976a). Population genetics of the Komandorski Islands Aleuts, and its relevance to problems of adaptation of ancient Beringian Aborigines. *Proceedings of the 9th International Congress of Anthropological and Ethnological Sciences*, August-September 1973, Chicago, USA. Paris: Mouton (in press).

Rychkov, Yu. G. & Sheremetyeva, V. A. (1976b). Genetical and anthropological aspects of the problem 'man and environment' in the region of Beringia. In *Proceedings of the 2nd Symposium on the Role of the Bering Land Bridge in the History of Holarctic Floras and Faunas in the Late Cenozoic*, May 1973, Khabarovsk, USSR. Vladivostock.

Rychkov, Yu. G. & Sheremetyeva, V. A. (1976c). Genetic investigation of the Circumpolar populations in connection with the problem of human adaptation. In *Resources of the Biosphere* (Synthesis of the Soviet studies for the International Biological Programme), vol. 3, *Human Adaptation*, ed-in-chief, Z.I. Barbashova, pp. 10–41. In Russian. Leningrad: 'Nauka'.

Rychkov, Yu. G., Tauzik, N. E., Tauzik, T., Zhukova, O. V., Borodina, S. R. & Sheremetyeva, V. A. (1974a). Genetics and anthropology of the Siberian taiga-hunting and reindeer-breeding populations (the Evenks of Central Siberia). I. Tribe structure, subisolates and inbreeding in the Evenk population. In Russian. *Voprosȳ antropologii*, **47**, 3–26.

Rychkov, Yu. G., Tauzik, T., Tauzik, N. E., Shukova, O. V., Borodina, S. R. & Shermetyeva, V. A. (1974b). Genetics and anthropology of the Siberian taiga-hunting and reindeer-breeding populations (the Evenks of Central Siberia). II. Effective size, time and spatial structure of the population and gene migration intensity. In Russian. *Voprosȳ antropologii*, **48**, 3–18.

Seleznyova, V. I. (1974). On the work of the anthropological team of the 1972 Siberian ethnographic expedition in Tuva, Chair of Ethnography and Anthropology, Leningrad State University. In Russian. In *Novoye v etnographicheskikh i antropologicheskikh issledovaniyakh v 1972 godu*, ed. L. A. Tultseva, T. V. Lukyanenko & A. E. Ter-Sarkisyants, pp. 138–47. Moscow: Press of the Institute of USSR History, USSR Academy of Sciences.

Sheremetyeva, V. A. (1971). To the population genetics of the indigenous peoples of Siberia (Western Siberia). In Russian. In *Vestnik Moskovskogo Universiteta*, **2**, 97–8.

Sheremetyeva, V. A. (1973). Population genetics of the North-East Asian Peoples, related to problems of ethnic anthropology. In Russian. Dissertation, Moscow State University.

Sheremetyeva, V. A. (1975). A genetic reconstruction of the evolution of human populations in the North Pacific Basin and the 'Proximity aberration' phenomenon. In *Applications of Statistical Methods in Problems of Population Genetics. Collected Papers*, ed. M. B. Malyutov, pp. 52–62. In Russian. Moscow: Moscow State University.

Shluger, S. A. (1940). The Nenets anthropological type in relation to the ethnogenesis of northern peoples. In Russian. Dissertation, Moscow State University.

Yu. G. Rychkov and V. A. Sheremetyeva

Spitzin, V. A. (1967). The distribution of haptoglobin and some other inheritable factors in North-Eastern Siberia. In Russian. *Voprosȳ antropologii*, **25**, 62–9.

Spitzin, V. A. (1976). The polymorphism of blood serum proteins and some blood enzymes in the aboriginal population of Siberia and some other adjacent territories. In *Proceedings of the 9th International Congress of Anthropological and Ethnological Sciences*, August-September 1973, Chicago, USA. Paris: Mouton (in press).

Spitzin, V. A. & Irissova, O. V. (1973). The ethnogeographical aspect in the study of the group-specific component (Gc). In Russian. *Voprosȳ antropologii*, **45**, 85–93. USSR, 1959. Census Data, 1963. Moscow State Statistical Board Publishing House.

Voronina, V. G. (1975). A genetical–anthropological characteristic of the indigenous population of the Maritime Territory. I. Population-genetical data on the peoples of the Maritime Territory. In Russian. *Voprosȳ antropologii*, **50**, 46–66.

Wright, S. (1931). Evolution in Mendelian populations. *Genetics*, **16**, 97–159.

Wright, S. (1943). Isolation by distance. *Genetics*, **28**, 114–38.

Wright, S. (1970). Random drift and the shifting balance theory of evolution. In *Mathematical Topics in Population Genetics*, ed. K. Kojima, pp. 8–31. Berlin: Springer-Verlag.

Yarkho, A. I. (1947). *The Altai-Sayan Turks*. Abakan: Khakass Regional Publishing House.

Zhinkin, V. I. (1927). On the blood groups and the racial-biochemical index in the Avinsk Buryats. In Russian. In *Zhizn Buryatii*, pp. 7–9.

Zhukova, O. V., Bronnikova, M. A., Rychkov, Yu. G. & Sheremetyeva, V. A. (1975). The genogeography of hereditary polymorphism of ABH-antigen secretion in North Asian peoples. In Russian. *Voprosȳ antropologii*, **49**, 15–23.

Zilberman, S. A. (1975). Collection of computer programmes on population and molecular genetics. In *Applications of Statistical Methods in Problems of Population Genetics. Collected Papers*, ed. M. B. Malyutov, pp. 74–83. In Russian. Moscow: Moscow State University.

Zolotaryova, I. M. (1968). Blood group distribution of the peoples of Northern Siberia. In *Proceedings of the 7th International Congress of Anthropological and Ethnological Sciences*, August 1964, Moscow, pp. 502–8. Novosibirsk: 'Nauka' Publishers.

4. Man in the tropics: the Yanomama Indians

JAMES V. NEEL, MIGUEL LAYRISSE & FRANCISCO
M. SALZANO

As the objectives of the studies on human adaptability to be encouraged by the International Biological Programme began to emerge, one of the populations of unusual relevance to the kinds of studies envisaged was the Amerindian of South America. He represents one of the few remaining opportunities to study relatively unacculturated, primitive man; the term 'primitive' being used here designates a society which is preliterate, employs very simple agricultural and 'manufacturing' techniques, and is primarily organized around concepts of kinship. Since the term 'primitive' seems pejorative to some, we reiterate once again our recognition of the highly sophisticated interplay of the human mind under these conditions, and the repeatedly demonstrated tendency of the Western World to confuse technological knowledge with intellectual superiority. Although disturbed in both obvious and subtle ways by post-Columbian developments, some of the tribes of South America nevertheless still appear to retain the essential features of primitive tribal organization. The strenuous efforts being made by the nations of South America to develop their interior regions impart an air of urgency to studies of these people: it is inevitable that within another generation, although some of the indigenous populations of South America may remain genetically intact, they will all have been subjected to so many extraneous influences that aboriginal population structure and environmental adaptations will have undergone major disturbances.

The scientific challenges presented by these Amerindian populations were apparent to the authors of this chapter well before the advent of the IBP, Layrisse's publications on Amerindian populations dating back to 1955, Salzano's to 1960, and Neel's to 1964. A collaborative program involving these three investigators was in fact already underway in the mid-1960s, with many of the principal objectives enumerated in 1964 (Neel & Salzano, 1964). The IBP provided an opportunity to strengthen and expand that program.

Our major objectives may be described as follows:

Tribal structure

Although during all but a small fraction of man's evolutionary history the basic population unit has been the band or tribe (as it was for his primate

James V. Neel and others

ancestors before that), the structural and demographic parameters for such groups remain poorly defined. At a simple descriptive level, we wished to improve upon existing knowledge.

Genetic constraints

Given adequate descriptive data, our next objective has been to attempt to understand the constraints this structure and demography place on the evolutionary process. Our present gene frequencies arose subject to these constraints and can only be understood in the light of such knowledge.

Environmental pressures

Although we all pay lip service to the principle that the nature of natural selection has changed, there are precious few hard data on how it has changed. We hoped by a series of judicious probes to develop some insights into the biomedical pressures on primitive man, and, by inference, how these changed with civilization.

Human evolution

Finally, given the uniqueness of Amerindian origins and spread (small numbers reaching a large area by a defined route), we saw in the Amerindian a most unusual testing ground for some of the newer techniques of evaluating human differentiation and evolution. A particular effort has been directed towards a classical question of anthropology – to what extent is the picture of intergroup differentiation yielded by one metric confirmed by another.

During the years of the IBP (1970–74), our teams, separately and together, have made observations on some eight South American tribes (Yanomama, Cayapo, Makiritare, Warao, Macushi, Wapishana, Krahó, and Moro), but the bulk of the effort has been expended on the first-named three. In the following, we propose because of limitations on the length of this article to restrict the discussion to the results of studies of the Yanomama, the one tribe on which all the authors have collaborated, but to refer to our work on other tribes where relevant. Because of the overlapping objectives of successive periods of field work, and the continuing elucidation of some of our early findings, it is simply impossible to restrict this narrative entirely to work accomplished within the time span of IBP. Layrisse first reported on the blood types of the Yanomama in 1962, and the intensive studies of the Yanomama herein described began in 1966; the studies of the Cayapo began in 1965 and of the Makiritare in 1966. This has been the work of many people; a list of colleagues can be found in the Appendix, pp. 141–2.

The following few pages may to some extent capture, of necessity in a some-

110

what telegraphic style, the essence of our studies of population structure and physiological adaptation among the Indian. However, because of the nature of this assignment, we exclude from this chapter consideration of a matter very much on our minds: how to minimize the toll usually exacted by the impact of what we are pleased to call civilization. We submit, however, our hope that both through our efforts in the field and through reports to Governmental Health Services, Commissioners of Indigenous Affairs, and Mission Headquarters, we have discharged the ethical obligations of the biomedical scientist working among exceptionally vulnerable peoples.

The Yanomama

There are estimated to be some 12000 Yanomama, whose distribution is shown in Fig. 4.1. The center of their distribution is the Parima Mountains which at longitude 64° W form the boundary between Brazil and Venezuela between latitude 2° 30′ N and 4° 30′ N. Within the past century the Yanomama appear to have undergone a significant centrifugal expansion which has increased the area they occupy by a factor of two to three. Although they had fleeting contacts with some of the early explorers of this region (cf. Koch-Grünberg, 1917), semi-permanent contacts with non-Indians date back only to the early 1950s. There are still Yanomama villages not yet reached by non-Indians, but this isolation is vanishing rapidly: the Brazilian 'perimeter highway' west from Boa Vista, Roraima Territory, cuts through a corner of the Yanomama distribution until now only accessible by arduous river travel or small planes able to use the few airstrips maintained by isolated missions.

Some unusual features of the Yanomama and their explanation: long relative isolation

The Yanomama were chosen as the subjects for this multi-disciplinary study essentially because they were known to be *relatively* undisturbed and to occur in the numbers necessary to yield significant bodies of data – assuming successful field work! Early in the course of the investigations, it became apparent that this was a most unusual group of Indians who, as their story unfolded, presented exceptional opportunities for genetic research. The unique constellation of characteristics found in the Yanomama includes the following:

Culture

Their material culture is much simpler than that of the surrounding Carib and Arawak groups. Characteristically the entire village lives in a single circular house of 'lean-to' type construction, open at the center. Clay pots (already largely replaced by trade pots) and basketry are of simple construc-

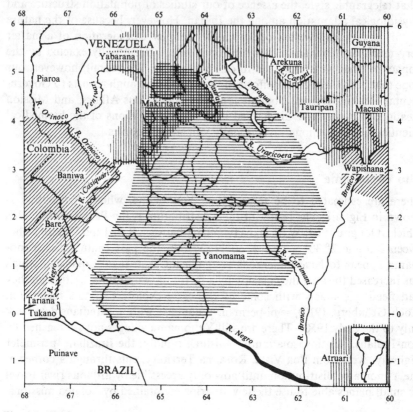

Fig. 4.1. The location of the Yanomama.

tion and design. Hammocks are usually made of longitudinal bark strips, more rarely of native cotton. The cooking banana (*Musa paradisica*) is the staple crop; corn and mandioca are raised only as very recent introductions. If as some authorities surmise the cooking banana was introduced from the Old World in post-Columbian times, then until several hundred years ago the Yanomana would have been largely hunter–gatherers. A sylvan rather than a riverine people, they have only recently learned to make dug-out canoes. They do not make fermented beverages. Characteristic practices and beliefs include cremation of the dead, drinking of the ashes of their bones, nasal insufflation of hallucinogenic drugs, an ever-present cud of tobacco between lower lip and teeth, and organized, duel-type fighting (Chagnon, 1968, 1974; Migliazza, 1972).

112

Language

This is not readily identified with any of the principal language families of
South America. Spielman, Migliazza & Neel (1974) have attempted to apply
glottochronology to estimating elapsed time since the ancestors of the Yano-
mama separated from other tribes. Unfortunately, data exist for only a very
modest sample of the indigenous languages of Central and South America.
The percentage cognates with Yanomama in the Swadesh 100-word list are
Shipibo (Peru), 27%; Warao (Orinoco Delta, Venezuela), 27%; Guaymí
(Panama), 25%; Macushi (Northern Brazil), 25%; Makiritare (or Ye'kuana)
(Southern Venezuela), 23%; Trio (Northern Brazil), 21%. Given the im-
precision of these figures as measures of language similarity, no one language
so far studied stands out on the basis of cognates as 'significantly' more
similar to Yanomama than the others. The two 'closest' languages (Shipibo,
Warao) imply a time of separation from Yanomama of 1500 to 3000 years,
when standard word-retention rates are used as the basis for calculation.
These times are very approximate; the indisputable fact is that thus far the
Yanomama language has no close relatives.

Physical traits

They are a small, often lightly built, almost Pygmoid people. In our series of
391 adult males, mean stature was 153.1 cm, with a standard deviation of
4.8 cm (Spielman *et al.*, 1972). By contrast, the Cayapo males have an average
height of 165.4 ± 5.1 cm (da Rocha & Salzano, 1972). Of the 28 different
South American tribes represented in Comas' (1971) summary, only two
(Yupa and Kuaiker) are shorter in average stature. Yanomama dental
morphology is distinguished by the fact that the frequency of occurrence of
six cusps on the first mandibular molars (50.5%) greatly exceeds that observed
in any other Amerindian tribe to date (Brewer-Carias, LeBlanc & Neel, 1976).
In comparison with such tribes as the Xavante (Neel *et al.*, 1964) or the
Cayapo (Peña, Salzano & da Rocha, 1972), their dermatoglyphics are unusual
in a low frequency of whorls and a high frequency of arches, and a low frequency of thenar/first interdigital (I and Ir) palmar patterns (Rothhammer,
Neel, da Rocha & Sundling, 1973).

Genetic markers

Typings have thus far been carried out for 28 serological or biochemical
genetic systems (Gershowitz *et al.*, 1972; Weitkamp *et al.*, 1972; Weitkamp
& Neel, 1972; Tanis, Neel, Dovey & Morrow, 1973; Ward, Gershowitz,
Layrisse & Neel, 1975; Gershowitz & Neel, unpublished). All 3416 Yanomama typed to date have been blood group O, Kell(-) and Gm(ag) or (axg),

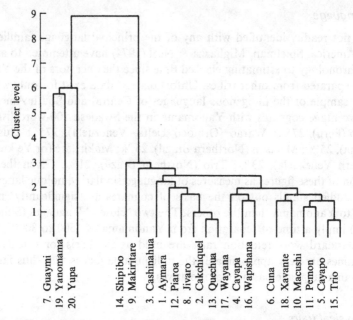

Fig. 4.2. A 20-tribe dendrogram derived from a matrix of genetic identities by an agglomerative clustering process using the 'median' method of Lance & Williams (1967). Hierarchic non-overlapping clusters have been produced in sequential fashion by joining clusters when the distance between their centroids is a minimum. The numbers correspond to those given in Fig. 4.3. Further explanation in Ward *et al.* (1975).

strongly suggestive of a complete lack of admixture with Caucasoids or Negroids. In the heartland of the tribal distribution, neither the *a* allelle of the Diego system nor the *A* allele of the acid phosphatase system are present, although these have been observed in most of the surrounding tribes. It is presumed that until the recent tribal expansion, these alleles were absent from the tribe. In several villages, the precise manner of introduction of these genes has been documented (Chagnon *et al.*, 1970). On the other hand, they possess a 'private' polymorphism of serum albumin (Yanomama-2) thus far not encountered in any of the surrounding tribes (Tanis, Ferrell, Neel & Morrow, 1974). In a recent analysis of the genetic relationship of the Yanomama with 19 other tribes of Central and South America, intertribal genetic distances were derived by the method of Edwards & Cavalli-Sforza (1963; Cavalli-Sforza & Edwards, 1965, 1967; Edwards, 1971) on the basis of six systems: Rh, MNSs, Kidd, Duffy, Diego and haptoglobin (Ward *et al.*, 1975). The average pairwise distance between all tribes was 0.41 but the average distance of the Yanomama from each of the other tribes was 0.53 genetic units; only the Guaymí slightly exceeded this value (0.57). In a genetic network, the Yanomama, Guaymí, and Yupa stand sharply apart from the other 17 (Fig. 4.2) (Ward *et al.*, 1975).

Fig. 4.3. The geographic location of 20 relatively pure South and Central American Indian tribes together with contours indicating the deviation from the multi-tribal average or centroid. The units of deviation are measured in the stereographic approximation for a genetic distance based on six loci (Rh, MNSs, Kidd, Duffy, Diego, and haptoglobin); each contour corresponds to 0.030 units. Further explanation in Ward *et al.* (1975). The tribes are 1, Aymara; 2, Cakchiquel; 3, Cashinahua; 4, Cayapa; 5, Cayapo; 6, Cuna; 7, Guaymí; 8, Jivaro; 9, Makiritare; 10, Macushi; 11, Pemon; 12, Piaroa; 13, Quechua; 14, Shipibo; 15, Trio; 16, Wapishana; 17, Wayana; 18, Xavante; 19, Yanomama; and 20, Yupa. Note the anomalous position of the Yanomama.

An alternative approach to defining the unusual position of the Yano-mama has recently been developed by Ward *et al.* (1975), an approach which at the same time may bring a greater order into the genetic relationships of the tribes of South America than is apparent from a single dendrogram. A multi-variable mean genetic distance (based on the above-mentioned six systems) of each of the 20 tribes from the centroid of their genetic hyperspace (defined as the vector of unweighted mean gene frequencies) has been obtained. This value was then assigned to the geographic location of that tribe and the existence of a geographic pattern tested by drawing contour lines at intervals of 0.03 units from a centroid. A regular pattern emerges; only two tribes violate it: the Yanomama and the Guaymí (Fig. 4.3). Our working hypo-thesis is that this regularity reflects the pattern of dispersion of the Amerindian through Central and South America; this general explanation is consistent with a number of different specific schema for the spread of the early Amerindian. Twenty points are scarcely adequate for such an elaborate plot; more data are needed badly. However, it is unlikely that these additional data can erase the distinct positions of the Yanomama and Guaymí.

All these facts are best explained by the hypothesis that the ancestors of the Yanomama – probably a relatively small group – found their way into the Parima Mountain Range several thousand years ago and have remained in virtual isolation ever since. We do not mean to imply they are undisturbed (see Neel, 1972). If this hypothesis – which of course could not be anticipated at the outset of field work – is correct, then the Yanomama are most unusual among human populations, and provide an exceptional opportunity not only to observe the genetic structure of primitive man but also to explore genetic inference procedures best carried out on relatively undisturbed peoples.

Some focal points of genetic interest

We turn now to a consideration of some of the chief points to emerge from the Yanomama studies.

Demography

Most of human evolution occurred when man was living in tribes sub-divided into smaller aggregates varying in size from single, extended families to villages of several hundreds. Any attempt at a precise mathematical treatment of human evolution must be modeled after such populations. Demographic data on such groups are sparse (cf. Salzano, 1972, for a review of Amerindians and Eskimos). This led us to direct a major effort towards (1) the development of fertility rates through physical examinations and tests of urine, (2) the estimation of the age pyramid, (3) the derivation of a mortality schedule (which would include infanticide), (4) the estimation of population

Table 4.1. *An estimation of the annual frequency of pregnancy by age group on the basis of physical examinations and urine tests. As an illustration, of 100 women aged 15–19 years, one expects 29.4 to become pregnant during any single year*

	Age class (years)					
	15–19	20–24	25–29	30–34	35–39	40–49
Palpation rates	0.172	0.205	0.426	0.385	0.271	0.039
Urine test rates	0.416	0.328	0.295	0.342	0.141	0.088
Average of tests	0.294	0.267	0.361	0.363	0.206	0.064

growth through an analysis of areal expansion, and, from all this, the generation of Model Life Tables (Neel & Weiss, 1975). Table 4.1 presents the estimated fertility rates, and Tables 4.2 and 4.3 Model Life Tables for males and females, respectively. Pertinent facts concerning Yanomama demography are as follows: the growth rate is currently between 0.5 and 1% annually. Infant and childhood mortality rates from natural causes are about 30%, and female infants are subjected to an initial 25% infanticide rate above what males experience (perhaps 5%). Survivorship to age 15 is about 50%, although somewhat higher in males than in females because of infanticide. In general, female mortality rates, after infanticide, are slightly lower than are those for males (which is commonly observed in primitives). The life expectancy at birth is about the same for males and females, being about 20 years. For those surviving to age 15, there is an expectation of 20–30 more years of life. Between 45 and 50% of the total population are under age 15, with 7–10% over age 50.

Fertility measures show that the annual number of births per fertile female is about 0.25. The mean number of full-term births for females who reach at least age 15 is 5.7, and the mean family size for those surviving beyond menopause is approximately 8.2 births; only 35% of all females who reach age 15 will live to age 50. The mean age for reproduction is 26.9 years and the generation length is 26.6 years. These figures differ because the population is growing; the generation length is the time required for a cohort of women to reproduce themselves R_0 times, where R_0 is the net reproduction rate. R_0 is 1.25, that is, each female born will on average replace herself 1.25 times. Each adult must care for 1.22 dependents. These values were all computed in a straightforward manner, using the appropriate model rates of birth and death.

The crude birth rate of the females is 0.059, and 0.056 for males, and the crude death rates are 0.050 and 0.048, respectively. By comparison, crude birth and death rates for some modern nations are, after Keyfitz & Flieger

Table 4.2. *The most satisfactory male Life Table for the Yanomama*

The first column is the age in years at the beginning of each age class; the second column, $Q(x)$, is the mortality schedule or the chance that those who reach the age class will die before reaching the next age class. $P(x)$ is the chance of surviving the age class, and is $1 - Q(x)$. $l(x)$ is the number of survivors left at the beginning of the age class out of every 100 born. $L(x)$ is the number of person-years lived in the age class per 100 born; $T(x)$ is the total number of person-years left to be lived from the beginning of the age class until all who are alive then have died. $E(x)$ is the life expectancy of those surviving to the beginning of the given age class, and is the mean number of years left to them. $C(x)$ is the proportion of the total population which is in the given age class, and this is the age distribution which was fitted in choosing these Life Tables.

Age	$Q(x)$	$P(x)$	$l(x)$	$L(x)$	$T(x)$	$E(x)$	$C(x)$
0	0.267	0.733	100	83	2147	21.5	4.6
1	0.160	0.840	73	262	2065	28.2	14.3
5	0.110	0.890	62	291	1802	29.3	15.3
10	0.088	0.912	55	262	1512	27.6	13.2
15	0.148	0.852	50	231	1250	25.0	11.2
20	0.152	0.848	43	197	1018	23.9	9.1
25	0.156	0.844	36	166	822	22.7	7.4
30	0.160	0.840	30	140	655	21.5	6.0
35	0.164	0.836	26	117	515	20.1	4.8
40	0.168	0.832	21	98	398	18.5	3.8
45	0.173	0.827	18	81	300	16.8	3.0
50	0.177	0.823	15	67	218	14.8	2.4
55	0.213	0.787	12	54	151	12.5	1.9
60	0.282	0.718	10	41	97	10.2	1.3
65	0.362	0.638	7	28	56	8.2	0.9
70	0.470	0.530	4	17	28	6.4	0.5
75	0.612	0.388	2	8	11	4.9	0.2
80+	1.000	0.000	1	3	3	3.8	0.1

(1968), United States, 0.021 and 0.010; Japan, 0.017 and 0.007; Fiji, 0.038 and 0.006; India, 0.041 and 0.019 (for years 1960–65). Thus the Yanomama rates are quite high by national standards, although they are typical for primitive populations (Weiss, 1973). These figures give some gross measures of the Yanomama's general demographic experience, if they were to live under a stable condition as reflected by the available data. In order to facilitate comparison of these data on the Yanomama with similar data for other peoples, such statistics as Fisher's Reproductive Value (1929), Crow's Index of Total Selection (1958), and Weiss' Index of Growth Regulation (1972) have been computed, and will be found in Neel & Weiss (1975).

Two points about these findings seem to us particularly noteworthy: (1) There is substantial control of the rate of introduction of new life to the tribe, achieved through a combination of intercourse taboos for prescribed periods following childbirth, induced abortion, and infanticide. There is clearly room for significant reproductive compensation for the death of a

Table 4.3. *The most satisfactory female Life Table for the Yanomama*

The first column is the age in years at the beginning of each age class; the second column, $Q(x)$, is the mortality schedule or the chance that those who reach the age class will die before reaching the next age class. $P(x)$ is the chance of surviving the age class, and is $1 - Q(x)$. $l(x)$ is the number of survivors left at the beginning of the age class out of every 100 born. $L(x)$ is the number of person-years lived in the age class per 100 born; $T(x)$ is the total number of person-years left to be lived from the beginning of the age class until all who are alive then have died. $E(x)$ is the life expectancy of those surviving to the beginning of the given age class, and is the mean number of years left to them. $C(x)$ is the proportion of the total population which is in the given age class, and this is the age distribution which was fitted in choosing these Life Tables.

Age	$Q(x)$	$P(x)$	$l(x)$	$L(x)$	$T(x)$	$E(x)$	$C(x)$
0	0.430	0.570	100	52	1982	19.8	3.2
1	0.118	0.882	57	210	1920	33.7	12.7
5	0.066	0.934	50	243	1710	34.0	14.2
10	0.042	0.958	47	230	1467	31.2	12.8
15	0.130	0.870	45	210	1237	27.5	11.2
20	0.134	0.866	39	182	1027	26.3	9.3
25	0.137	0.863	34	158	845	24.9	7.8
30	0.140	0.860	29	136	687	23.5	6.4
35	0.143	0.857	25	117	551	21.9	5.3
40	0.147	0.853	22	100	435	20.2	4.3
45	0.150	0.850	18	85	335	18.2	3.5
50	0.154	0.846	16	72	250	16.0	2.9
55	0.188	0.812	13	60	178	13.4	2.3
60	0.251	0.749	11	47	118	11.0	1.7
65	0.327	0.673	8	34	71	8.8	1.2
70	0.432	0.568	5	21	37	6.9	0.7
75	0.570	0.430	3	11	16	5.2	0.4
80+	1.000	0.000	1	5	5	3.8	0.2

child. Mathematical treatments of selective equilibria which do not allow for this phenomenon may be substantially in error. (2) Clearly this rate of population increase cannot have characterized the Amerindian since his arrival in the Americas some 20000–30000 years ago. Of the alternative growth patterns – a lower sustained rate or periodic reductions in population through natural disaster, disease, or war, we favor the latter. This has important implications concerning the probability of identity by descent for randomly chosen pairs of alleles (see Inbreeding, below).

Microdifferentiation and population structure

Early in the course of the work we were impressed by the large differences in gene frequencies between villages (Arends *et al.*, 1967). For instance, for 47 villages in which more than 30 persons (60 alleles) were sampled, gene frequency ranges such as the following were observed: *MS*, 0.00 to 0.49; *Ms*,

Table 4.4. *Intratribal six-loci genetic distances for four relatively undisturbed Indian tribes compared with similar intertribal relationships*

Tribe	Reference	Number of villages	Smallest distance	Largest distance	Mean
Yanomama	Neel & Ward, 1970	7	0.144	0.537	0.330
Makiritare	Neel & Ward, 1970	7	0.158	0.588	0.356
Cayapo	Salzano, Neel, Weitkamp & Woodall, 1972	4	0.190	0.275	0.248
Xavante	Salzano, Neel, Weitkamp & Woodall, 1972	3	0.105	0.231	0.178
20 tribes	Ward *et al.*, 1975	20	0.176	0.712	0.409

0.20 to 0.81; R^1, 0.60 to 1.0; Fy^a, 0.22 to 0.84; and Jk^a, 0.20 to 0.84. This variation is conveniently quantified in two ways. The first is by the F_{ST} statistic of Wright (1943, 1965), defined as the correlation between random gametes within subdivisions, relative to gametes of the total population, and measured as $\sigma_p^2/\bar{p}\bar{q}$ where $\bar{p}$ and $\bar{q}$ represents the weighted means, and σ_p^2 the weighted sum of the squared deviations of the individual subpopulation gene frequencies from the mean gene frequency, divided by the number of sub-populations. For 15 genetic systems studied in 47 Yanomama villages, the values were: MS, 0.095; Ms, 0.122; NS, 0.041; Ns, 0.111; R^z, 0.071; R^1, 0.067; R^2, 0.081; R^o, 0.063; P^1, 0.071; Fy^a, 0.070; Jk^a, 0.072; Le^a, 0.041; Hp^1, 0.069; Gc^1, 0.066; PGM^1, 0.038; $Mean$, 0.073 (Neel & Ward, 1972; Ward *et al.*, 1975). (The Di^a and Ph^A alleles have been excluded from this calculation because of the evidence that they are recent additions to the tribe as a result of admixture.) Salzano (1975) has recently computed intertribal F_{ST} values based on data on 11 alleles (MS, Ms, NS, Ns, R^1, R^2, R^z, Fy^a, Jk^a, Hp^1, and Di^a) for 29 tribes of South America. The Yanomama *intratribal* values were higher than these *intertribal* values for six alleles (MS, Ms, Ns, R^2, Jk^a, and Hp^1). Useful summaries of the variability of blood group gene frequencies in South American tribes will be found in Layrisse & Wilbert (1966); Salzano (1968); Post, Neel & Schull (1968); and Layrisse (1971).

A second way of expressing the intervillage variability is by the use of a genetic distance function, as mentioned earlier, for the quantification of intertribe differences, and again based on the Rh, MNSs, Duffy, Kidd, Diego, and haptoglobin loci. Table 4.4 presents such distances for four of the tribes we have studied. We draw two tentative conclusions: (1) the mean genetic distances between the smaller and probably less permanent villages of the Yanomama (and Makiritare) are greater than those between the larger and probably more permanent villages of the Gê-speaking Xavante and Cayapo, and (2) more surprisingly, the genetic distances between pairs of Yanomama villages are almost as great as those between randomly chosen pairs of South American tribes.

The origin of these village differences is thought to result primarily from the manner in which new villages come into being, aided subsequently by genetic drift. New villages (populations) do not arise by some random sampling of a large 'parental' gene pool, but by a fissioning of a pre-existing village, usually to some extent along lineal lines. Our published documentation of that statement involves an example from the Xavante Indians (Salzano, Neel & Maybury-Lewis, 1967), but the same is clearly true for the Yanomama. Thus as an Indian population expands, its relatively small demes come into being through a succession of non-random samplings of the 'gene pool'. Migration between demes would be expected to reduce these original differences, but it has thus far not been feasible to develop matrices for the Yanomama migration, which would permit a treatment of this problem. Furthermore, even if such matrices did exist, it would be a meaningless exercise to power them in such a way as to estimate the number of generations necessary to erase these differences. This is because village alliances (and so migration matrices) are subject to sudden change and because the dictates of war and disease also result in village fusions, which probably occur with sufficient frequency to equal in importance small-scale intervillage migrations. Many of the mathematical treatments of classical population genetics involve the concept of genetic equilibrium. The net result of the Yanomama structure is to ensure that the approach to equilibrium at the deme level is so constantly disturbed that 'equilibrium' can be only fleetingly, if ever, attained for any one system in a village, and never for more than a relatively few systems at any one time. Those who like to calculate the 'genetic load' created by departure from equilibrium will find rich material in our data!

We feel there is an additional genetic significance to these findings. It seems quite likely that in the past, new tribes (whether of Indians, Caucasoids, or Negroids) originated from old by a budding-off from the periphery of the tribe, followed by the migration of the derivative group to an area sufficiently far from its ancestors that it could begin to develop its own cultural and breeding integrity. If the Yanomama are a guide, the sampling of the gene pool which occurred at the origin of a new tribe was highly biassed and far from random. Social structure thus sets the stage for isolated demes with unusual combinations of gene frequencies and this may play a role in the rapid evolution which seems to have characterized our species (Neel & Ward, 1970).

Despite these dispersive tendencies, there is clearly defined genetic 'structure' when we consider the Yanomama as a whole. This emerges from two different types of analysis. Firstly, Ward (1972), from genetic determinations on 11 loci for 2416 Yanomama distributed among 37 villages, derived a pairwise genetic distance matrix for these villages, using for this purpose the statistic of Edwards & Cavalli-Sforza (1964; Cavalli-Sforza & Edwards, 1967). A genetic network (dendrogram) was then developed, by the procedure of deriving a minimum net length by the algorithm of Cavalli-Sforza & Edwards (1967) for a constellation of relationships defined by the cluster-analysis technique of Edwards & Cavalli-Sforza (1965). The resulting dendrogram exhibited principal subdivisions which corresponded rather well to geographical blocks of Yanomama villages (cf. Fig. 5 of Ward, 1972; also Fig. 7 of Neel *et al.*, 1972). Furthermore, this dendrogram, as well as a similar dendrogram developed for a smaller number of Makiritare villages (Ward & Neel, 1970) corresponded satisfactorily to the known ethnohistory of the villages. Secondly, Ward & Neel (1974) have investigated the accumulated data for the occurrence of clines in gene frequency. With the statistical techniques of Smouse & Kojima (1972) and Smouse (1974), significant clines can be demonstrated for the following alleles among the 15 for which F_{ST} values were given above: *MS*, *Ms*, *NS*, *Ns*, R^z, R^1, R^2, R^o, and Jk^a. The significance of the clines has been determined by deriving the logistic regression of best fit, obtaining the component of the variation in village frequencies attributable to the cline and the residual component, dividing each by the appropriate degrees of freedom, and treating the ratio of the two as an asymptotic F distribution. This is a conservative approach. The unusual feature (mentioned above), that the Yanomama appear until recently to have lacked two genetic polymorphisms present in most of the surrounding tribes (involving the Diego and acid phosphatase systems) but do possess a 'private' genetic polymorphism (of serum albumin), provides especially favorable material for the analysis of the interplay of admixture, selection, and population dynamics in the etiology of these clines. It is concluded that it is unlikely that selection makes a major contribution to the clines; that the role of admixture varies considerably with the locus but on balance is minor; and that the clines, given the fission process described above, are probably primarily related to population fissioning and the recent population expansion noted under demography. Admittedly the last of these is a difficult hypothesis to test, but yet it is felt that the possible role of population expansion in the etiology of genetic clines has received insufficient attention in the past.

Inbreeding

Because of the short average life span and the lack of written records, the Yanomama are not suitable for the estimation of inbreeding. They have a prescriptive bilateral cross-cousin marriage rule. In a complex of four villages known as the Namoweitedi, studied with particular care by Chagnon (1968), there were only 37 marriages in which all four grandparents of both spouses were known; 13 of these marriages (35.1%) involved first cousins. This implies a high rate of inbreeding, but it was impossible to begin to reconstruct, satisfactorily, more remote loops of relationship. To meet this and other problems, a Monte Carlo type computer simulation model of the Yanomama has been developed for which the input is the above-mentioned complex of four villages (MacCluer, Neel & Chagnon, 1971). The model allows for birth, death, marriage, and intervillage migration on the basis of Yanomama rules and schedules. Genes may be introduced and their fate followed. There were recognized imperfections in the model, such as no allowance for exchange with the outside world (closed population) or for divorce, which are being rectified, but the preliminary results are already of interest. The computer program was restricted to reconstructing inbreeding loops for the proximal seven generations; at the end of a 400 year run, the coefficient of inbreeding was 0.034, and it appeared that if all loops could be reconstructed, the demonstrable F might approximate to 0.05. Results with the improved model suggest that even this high value may be an underestimate (Li, unpublished).

Earlier we suggested that it was probably the fate of most Indian populations to be decimated periodically by war, disease, or natural disaster. We presume that this decimation was non-random, i.e. that because of genetic factors, there was a familial correlation in survival, so that the population was periodically cut back to a smaller group of related persons. The implication of this cyclical fluctuation in population numbers for the accumulation of inbreeding (more accurately, identity of genes by descent) is currently under investigation through simulation.

A comparison of distance matrices based on different sets of biological data

A question of long standing in biology is the extent to which a triangular, pairwise distance matrix based on one set of characteristics – say, anthropometrics – will in its essential details correspond to a comparable matrix based on different characteristics – say, gene frequencies. A major objective of the Yanomama studies has been to examine this question, using matrices based on geographic distance, genetic markers (11 systems), anthropometrics (12 measurements), dermatoglyphics (9 traits), and dental morphology (6 traits). The data necessary to the treatment will be found in the papers cited earlier as documenting the unusual features of the Yanomama. As noted above, genetic matrices were derived by the technique of Cavalli-Sforza & Edwards (1965; Edwards & Cavalli-Sforza, 1967), while the anthropometric and dermatoglyphic distances are based on Mahalanobis' D^2 (Mahalanobis, Majumdar & Rao, 1949) and the dental on the use of Pearson's Coefficient of Racial Likeness (1926). The recent expansion of the Yanomama and their marked micro-differentiation render them unusually favorable material for such an enquiry. On the other hand, the smallness of the sample size for most villages and the impossibility of obtaining completely overlapping data sets detract from the rigor of the approach possible.

123

We have already commented on the correspondence which Ward (1972) observed between genetic clusters and geographic clusters, based on 11 genetic loci studied in 37 Yanomama villages. Spielman (1973) then extended this type of comparison to involve anthropometric, genetic, and geographic matrices, and certain derivative representations, such as dendrograms. Next, Neel, Rothhammer & Lingoes (1974) added dermatoglyphic traits to the comparison, and now a comparison of matrices based on dental and on genetic traits is also available (Brewer-Carias *et al.*, 1976). All told, four different types of comparison have been utilized, but not all types have been applied to each contrast. The comparisons are

(1) an examination of the rank–order correlation between corresponding elements of two matrices;

(2) an examination of the fit of one data matrix to another under an optimal choice of a central dilation and a rigid motion, as advocated by Schönemann & Carroll (1970);

(3) a comparison of the correspondence of the dendrograms best representing one set of data with those representing another;

(4) a comparison of data matrices derived by the non-metric approach of monotone distance analysis (MDA) developed by Lingoes (1965, 1971), Guttman (1968), Lingoes & Roskam (1971) and Roskam & Lingoes (1970).

A standard problem in the past has been to introduce the requisite degree of objectivity into the comparisons. This has been facilitated by a number of empirical developments:

(1) Lingoes (1973) has shown empirically that in a triangular matrix of pairwise distances between n populations, containing $(n)(n-1)/2$ matrices, the distribution of τ, the rank order correlation of Kendall (1962), corresponds satisfactorily with that expected with $[n(n-1)/2]-1$ degrees of freedom.

(2) Lingoes & Schönemann (1976) have recently provided a measure of fit (S) between pairs of matrices that is invariant under separate stretchings and order of fittings and lies in the range (0, 1). S has no known theoretical distribution, but Spielman (1973) has developed a method for empirically generating the distribution of S when the similarities between the two matrices to be fitted to one another arise by chance alone (i.e. when the positions of the populations are the product of a random number generator). The vector means and variances of the randomly generated matrices must of course match those of the real set. A random distribution for S can be generated for each of the comparisons of interest by fitting pairwise, for some predetermined number of times (minimum of 100), the simulated matrix for one kind of real data with the simulated matrix for a second kind of real data.

(3) To assess how well representations as dendrograms of two different matrices agree, Spielman (1973) restricts attention to that situation where it is practical to derive all possible dendrograms (for us, with present computer

124

facilities, 7 populations, yielding 945 possible dendrograms). One then determines how often among some arbitrary number of 'best' representations – say, 50 – the same topologies are found among the best 50 when a different set of measurements is employed. Empiric probabilities for the number in common to be expected have been developed by using a random number generator to repeatedly locate seven populations in a unit hypercube and then generating pairs of sets of dendrograms to determine how often among the best 50 dendrograms for one of these sets the same dendrogram was found in the best 50 of the other set. This provides an empirical basis for establishing approximate probability levels of any desired degree.

(4) Finally, MDA, as mentioned above, generates a data matrix, and the fit of two such matrices can be tested as under (2) above.

A detailed discussion of the results of these various comparisons is beyond the scope of this review. In general, the correspondences observed with all four of the methods of comparison are significant, i.e. would be expected by chance in less than 5% of comparisons involving matrices where there was no basic similarity, although there are some notable exceptions. (As the extreme example, Spielman (1973) obtained a *negative* correlation of -0.25 between genetic markers and corresponding entries in a Yanomama seven-village matrix based on anthropometric traits. Limited comparisons among the Cayapo also did not always yield a consistent picture (Salzano, 1976).) The level of significance was almost as great where dermatological and dental traits were concerned as where genetic markers and physical measurements were employed. Since all the traits are, of course, subject to mensurational or classification errors, and since samples did not overlap completely and we have demonstrated sizeable differences between distances based on successive samplings (Ward *et al.*, 1975), we infer that the 'true' correspondence between these matrices should be *highly* significant. The validation of this inference is best pursued under circumstances where sampling difficulties are not as great as in the Yanomama. If this point of view is upheld, and if the similarities in matrices do not result from unequally distributed exogenous influences (an unlikely prospect), then on an empirical level it is clear that the different ways of comparing such data sets are more or less equivalent; but on the theoretical level, it may be surmised that these matrices are accurate indices to biological differentiation and that the genetic systems underlying the various traits are all sufficiently complex (multigenic) that in a statistical sense they are similarly affected by the meiotic partitionings.

Attempts to study selection and mutation

Despite the current flurry of interest in the possibility that a large fraction of mutations may be essentially neutral in their phenotypic effects, there is little doubt that most of the distinguishing features of man are the product of natural selection occurring under conditions much more similar to Yano-mama-type conditions than London, Tokyo, or New York City-type con-ditions. It follows that to understand our biological selves, we must under-stand the interaction of mutation and selection under the former conditions. The various methods of searching for evidence of natural selection may be classified as direct or indirect approaches (Bodmer, 1973; Neel, 1975). In the direct approach, one attempts to establish a direct relationship between genotype and indicators of selection such as survival and/or fertility, distor-tion of genetic ratios, etc. Large numbers are required for a convincing demonstration of selective differentials of 1 or 2%, i.e. the kind of differen-tials which seem 'reasonable' (Neel & Schull, 1968). The Yanomama are not suitable for the accumulation of large series of this sort. The indirect approach, in contrast, seeks to document phenomenon which could be taken as evi-dence for selection, but often have other possible explanations as well. The same distinction can be drawn for studies of mutation: the direct approach involves searching for children with a genetic trait not present in either parent; the indirect approach involves generating an estimate of the mutation rate that should exist to maintain certain phenotypes, given assumptions concern-ing selection, equilibrium conditions, and stable population size (Neel, 1962). Again, the Yanomama are not very suitable for the direct approach. In this section, we summarize efforts to apply indirect approaches to the Yanomama.

Indirect approaches to selection

Is there heterogeneity in F_{ST} values?

In the absence of selection, the allele frequencies at each locus within demes are all subject to the same dispersive forces – lineal effect, drift, mutation – and the dispersion of allele frequencies in villages should be uniform across loci. If selection operates on the phenotypes associated with one or more loci, however, $\hat{F}_{ST}$ (estimated) values should be heterogeneous, village allele frequencies for the loci subject to selection either being relatively over- or under-dispersed. On the basis of the statistic developed by Lewontin & Krakauer (1973) there is no evidence for heterogeneity of $\hat{F}_{ST}$ for eight co-dominant genes in the Yanomama (M/N, S/s, C/c, E/e, Hp^1/Hp^2, Gc^1/Gc^2, Fy^a/Fy^b, and PGM^1/PGM^2) (Neel & Ward, 1972; Ward et al., 1975), and so no evidence for the operation of selection on these systems.

Is there heterogeneity in F_{IS} values?

Wright (1943, 1965) defined a statistic, F_{IS}, as the average over all sub-divisions of a population of the correlation between uniting gametes relative to those of their own subdivision. Mathematically it is estimated as

$$\hat{F}_{IS} = \sum_i w_i \hat{F}_i,$$

where w_i represents the appropriate weight of the ith deme and $\sum_i w_i$ equals 1, while

$$\hat{F}_i = 1 - \frac{H_i}{2N_i\hat{p}_i\hat{q}_i},$$

where H_i = number of heterozygotes in deme sample, N_i = total number of individuals in deme sample, $\hat{p}_i$ = frequency of A allele in deme sample, $\hat{q}_i$ = frequency of a allele in deme sample ($\hat{p}_i + \hat{q}_i = 1$). F_{IS} is thus a measure of heterozygote excess or deficiency. A deviation from the expected number of heterozygotes could be a product of selection. However, Neel & Ward (1972) (see also Jain & Workman, 1967) have discussed the many other factors in real, subdivided populations which could also produce heterozygote excesses or deficiencies, and have concluded that it is impossible with respect to a single locus to draw inferences concerning the action of selection. In the case of the Yanomama, the mean $\hat{F}_{IS}$ across loci is -0.012, not a clearly significant departure from zero. However, the same argument advanced concerning heterogeneity for F_{ST} values as an indication of the action of selection holds for F_{IS} values. There is no evidence for heterogeneity of $\hat{F}_{IS}$ values in the Yanomama for the same eight co-dominant genes enumerated above (Neel & Ward, 1972).

The HL-A alleles

Kimura & Crow (1964) have developed the relationship

$$n_e = 4N_e\mu + 1,$$

where n_e = effective number of alleles maintained, N_e = effective population size, μ = mutation plus introduction of new alleles through migration. Layrisse *et al.*, (1973) have determined the number of alleles present at the two 'loci' of the HL-A system (the HL and four loci) in 221 Yanomama. The minimum number of alleles in the first series is five and six in the second series. There is highly significant linkage disequilibrium between the alleles. It is argued that, given the population characteristics of the Yanomama, the maintenance of this number of alleles plus the linkage disequilibrium is indirect evidence for the action of selection upon these two loci. A weakness in this argument is the key role played in the formulation by N_e, whose estimation, as we shall discuss briefly below, presents many difficulties.

Indirect approaches to mutation

A special effort has been directed in these studies towards the accumulation of data concerning rare electrophoretically detectable variants at 15 loci (6-phosphogluconate dehydrogenase, phosphoglucomutase 1 and 2, lactate dehydrogenase, adenylate kinase, malate dehydrogenase, adenosine deaminase, peptidase A and B, phosphohexose isomerase, isocitrate dehydrogenase,

James V. Neel and others

hemoglobin A, transferrin, ceruloplasmin, and albumin). Although the bulk of the data involve the Yanomama, additional information is available for five other tribes (Tanis *et al.*, 1973). Thus far, 56237 system-determinations have been performed. An indirect estimate of mutation rate has been obtained from a formulation by Kimura & Ohta (1969), namely

$$\mu = \frac{I}{2N} \cdot \frac{1}{t_0},$$

where I = the average number of mutant alleles per locus among the loci sampled, N = number of individuals in one generation in the population in which I has been determined, and t_0 = average mutant survival time in generations for those mutants not going to fixation. By definition, I should not include those variants which will ultimately go to fixation, but this presumably very small fraction of the total cannot be excluded from any data set.

The final formula has an appealing simplicity: on the assumption of equilibrium, the mutation rate is equal to the mean proportion of different variants per locus ($I/2N$) times the fraction of these variants lost each generation ($1/t_0$). While the assumption of neutrality of mutant phenotypes is essential to this formulation, it is clear that the effect of the most probable departure from this assumption (deleterious mutants) will be to result in an underestimate. However, the estimation of these parameters presents many difficulties in practice. This is particularly true of t_0, which has been determined by computer simulation, and, for mutants whose recipient survives to adulthood in the first generation, has an estimated value of 4.7 (Li & Neel, 1974). Furthermore, the assumption of constant population size is clearly incorrect for most Amerindian tribes. The values which can be plugged into the equation are given in Table 4.5. The final average of the various possible estimates is approximately 3×10^{-5}/(locus/generation) for electrophoretically detectable mutants, and approximately 8×10^{-5}/(locus/generation) when allowance is made for electrophoretically 'silent' mutations. Albumin makes a disproportionate contribution to this estimate and the relatively high value is certain to receive very critical scrutiny – as indeed it should. In this connection, we note the discovery (see below) of an unexpectedly large amount of chromosomal damage in the Yanomama (Bloom, Neel, Tsuchimoto & Meilinger, 1973).

The rate of mutation is a basic biological parameter, the estimation of which in higher organisms has been quite unsatisfactory. At the moment there is room for disagreement as to order of magnitude (let alone factors of 2 or 3). The average for a series of a dozen dominantly inherited phenotypes in man is of the order of 10^{-5}/('locus'/generation) (Neel, 1962; Cavalli-Sforza & Bodmer, 1971). Some have argued that these estimates are biassed by the fact that the investigator tends to select for study, traits determined by unusually mutable genes (otherwise he would have difficulty assembling

128

Table 4.5. *A summary of the existing data on rare protein variants among the Indians studied in this project. Fifteen different proteins have been screened*

Tribe	Estimated total population	Approximate N	Total no. of system determinations	Average no. sampled per locus	Estimated adults sampled	No. of variants in adults[a]	$\%N$ sampled
Yanomama	12 000	5760	37 678	2512	1206	4	21
Makiritare	1500	720	8375	558	268	3	37
Cayapo	1500	720	4567	304	146	3	20
Piaroa	1000–2000	~720	1897	126	60	4	8
Macushi	4000	1920	2792	186	89	2	5
Wapishana	1000	480	928	62	30	3	6
Total			56 237	3748	1799		

[a] Includes the transferrin and phosphoglucomutase polymorphism when present.

material for a study) and that a better order of magnitude is 10^{-6} (Stevenson & Kerr, 1967; Cavalli-Sforza & Bodmer, 1971). There is a counter-argument to this, that the loci selected for study are those at which an unusually high proportion of code changes result in gross phenotypic effects, in which case the bias is much less (Neel, 1957; see also Morton, 1974). When electrophoretic variants are the basis for the study, bias for mutability in the selection of loci is much less likely, since the prime criterion for choice is technical, i.e. availability of an electrophoretic system clearly definining the protein. Attempts to improve this estimate will be the chief legacy of the IBP to the Ann Arbor laboratory.

The evolution of language

When it became apparent, in the course of the field work, that there were considerable linguistic differences between Yanomama groups, and as the case built up for the long-time isolation of the Yanomama, it seemed there was an unusual opportunity to determine whether there was a significant parallelism between linguistic and genetic micro-evolution. To this end, in each of seven villages or village-constellations dispersed throughout the Yanomama distribution, a lexicon of 750 entries was compiled, plus data on two sorts of grammatical rules: 15 syntactic rules and 23 phonological rules (Spielman *et al.*, 1974). Distance functions and matrices were developed, comparable to those based on gene frequencies. The congruence of the genetic and linguistic data was tested in two ways: (1) by the matrix superimposition technique of Schönemann & Carroll (1970), and (2) by a comparison of the number of dendrograms in common among the best 50 dendrograms yielded by each of the two data sets, using in either case the tests of significance developed by Spielman (1973) (see above). The correspondences were highly significant. While a parallelism of this sort does not of course prove similarity in mechanism, we suggest that the introduction and spread of linguistic traits may follow a set of rules very similar to those regulating the introduction and dissemination of genetic characteristics.

The concept of N_e

One of the goals of this program has been to determine whether details of genetic structure really do have an important bearing on genetic outcomes. If not, then much of the justification for studies such as these is removed. This issue is perhaps best approached through a consideration of N_e, *effective population size*, an entity which appears in a variety of genetic equations. It has already been established that on theoretical grounds, given defined population conditions, there is a difference between 'inbreeding effective number', 'variance effective number', and 'extinction effective number', but these differences have not seemed large (discussion in Crow & Kimura, 1970; Cavalli-Sforza & Bodmer, 1971; Kimura & Ohta, 1971). The Monte Carlo type simulation program of the Yanomama described briefly under 'inbreeding' (above) sheds new insights on how complex the 'black box' termed N_e really is. What corresponds most closely to 'variance effective N_e' has been determined for the simulated population in two ways, namely, from a formulation of Crow & Kimura (1972) and from the variance of the gene frequency obtained in a series of simulations, namely $N_e = pq/2V_p$, where V_p was the variance of gene frequency obtained by dropping a pair of alleles with initial frequencies $p = q(= 0.5)$ 100 times through a pedigree constructed from a 400-year simulation of this population (MacCluer & Neel, unpublished). The two estimates were 209 and 195, respectively. Now, $\bar{t}_0$ for a mutation introduced into a newborn child was 2.3 generations (the figure given earlier was for a mutant introduced in an adult, 'corrected' for some imperfections in the model). For reasons given in the original paper, we feel $\bar{t}_0$ has been underestimated by approximately one generation, and prefer for the present to work with a $\bar{t}_0$ of 3.3 for mutations introduced at birth and 4.7 for mutations introduced into mature adults (aged 15–19 years). A $\bar{t}_0$ of 3.3 would be expected in a population whose N_e was 55 (Kimura & Ohta, 1971). Without this empiric demonstration, the tendency would have been to insert the 'variance effective N_e' in formulations involving mutant loss. It will be of great interest to determine the 'mutant loss' and 'number of transmission' N_es for populations with other mortality and mating patterns. At the least, while one can challenge details of the simulation, it seems likely that N_e can assume widely different values according to the phenomena in question. Attempts to improve these estimates of N_e and $\bar{f}_0$ are another ongoing activity in this program.

Some biomedical observations

The foregoing discussion has been primarily directed at defining the genetic structure within and upon which the forces of evolution operated in tribal societies. Our Integrated Research project has included a number of 'probes' aimed at evaluating the nature of these evolutionary pressures, i.e. the pressures to which tribal man must have adapted under tropical conditions. A word must be said about the strategy dictated by the circumstances. There are no laboratory facilities available in the field; to our knowledge there has never been an autopsy on a Yanomama. Thus, a variety of investigations of great interest simply cannot be performed. The strategy has perforce been to base the 'probes' on blood, urine, and stool samples which can be collected in the field and analyzed under favorable conditions in base laboratories.

Many of these studies are still in progress. This fact, plus the space limitations of this review, lead us to restrict this section to scarcely more than an enumeration of activities in this area.

High incidence of cold agglutinins of anti-I^T specificity

A cold auto-agglutinin that reacts at relatively high titers with human blood group O red cells from both adults and neonates (umbilical cord blood samples) was found in 88 of 90 Yanomama tested (Layrisse & Layrisse, 1968). From a study of the relative agglutinability of the two types of red cells, it was concluded that the antibody should be classed as anti-i in 12 instances but anti-I^T in 76 instances. Sera from 30 members of another group of Venezuelan Indians (the Warao) and from 50 healthy residents of Caracas did not react similarly. The physical characteristics of the antibody identified it as an IgM globulin. The cold agglutinin titers did not correlate with the titer of the serum against *Trypanosoma cruzi* or a battery of antigens of the arbovirus group. The most comparable finding is that of Booth, Jenkins & Marsh (1966) in sera from New Guinea Melanesians. Salzano, Steinberg & Tepfenhart (1973) have reported that the sera of 116 of 440 Cayapo Indians showed antibody activities against globulin-sensitized red blood cells, a surprisingly high frequency. This is apparently a different phenomenon from that reported by the Layrisses. The significance of this unusual antibody, as well as the antiglobulin activity, remains unknown; the subject is a fertile field for investigation.

131

James V. Neel and others

Chromosomal damage

In approximately 2% of mitotic preparations from cultured leucocytes of normal individuals, one or more chromosomes will exhibit evidence of gross damage – usually a chromatid or chromosome break, more rarely a dicentric chromosome or an apparent translocation. Such damage presumably results in part from noxious agents in the environment. The Yanomama presented an excellent opportunity to develop a baseline applicable to a situation free from the contaminants of civilization. Cytogenetic studies were performed in 1969, 1970, and 1971 on the Yanomama and on a neighboring tribe, the Piaroa (Bloom *et al.*, 1973). The studies in 1969 involved 49 Yanomama from two villages, of whom 13 had 23 leucocytes (out of 5165) with extensive chromosome breakage and rearrangement. Cells with such complex damage were not encountered in subsequent years, even (in 1971) in the original two villages. This was accompanied by a decline in the percentage of cells with *any* aberration, from 4.2 in 1969 to 2.6 in 1970 to 1.4 in 1971. Members of the expedition served as controls: 1.5% of 1450 cells showed some damage, but none the extensive picture encountered in the Yanomama. It was concluded that the original findings resulted from the exposure of the inhabitants of these two villages to some chromoclastic agent, whose effects had disappeared with time. It is tempting to relate this finding to the relatively high estimates of mutation rate described earlier.

Response to epidemic measles and measles vaccine

Serological studies conducted on 606 blood specimens obtained from Yanomama in 1966 and 1967 revealed them to be essentially a 'virgin soil' population for measles. Accordingly, measles vaccine was obtained for administration during the expedition of 1968. Measles itself was introduced to the Yanomama just before the team reached them. It was possible to observe the response of the Amerindians to the Edmonston B vaccine administered both with and without measles immune gammaglobulin as well as the response to the actual disease (Neel, Centerwall, Chagnon & Casey, 1970). It was also possible to obtain follow-up studies a year later of antibody titers in those exposed both to the vaccine and measles itself. It was concluded that the primary response of these Amerindians to measles did not differ greatly from that of Caucasoids. The relatively high mortality that usually accompanies such epidemics can be largely ascribed to the secondary features of the epidemic: complete collapse of village life, lack of medical care, simultaneous illness of mothers and their nursing infants, etc. If this point of view is correct and can be generalized to other epidemic diseases, then there is little reason to be satisfied with poor results in the treatment of epidemic diseases in newly contacted groups.

132

negoaYannomama:negoaYanomama:negoaYanomamanegoayanomama:negoaYannegoaYannegoaYanegoaYannegoayannegoaYanegoaYannegoaYannegoaYannegoaYannegoaYannegoaYannegoaYannegoaYanomama:

Congenital defects

The etiology of the great majority of congenital defects remains obscure. Fears are frequently expressed that their frequency may increase as our exposure to environmental contaminants of one kind or another increases. It would be desirable to have baselines, derived from the experience of primitive peoples, which would create the perspective from which to view present frequencies. Such baselines are difficult to establish, because of the unsupervised nature (by representatives of our culture) of births among primitive peoples, and the almost universal practice among primitives of killing deformed newborn infants. We have attempted to assemble all our available data, based upon the observations of missionaries and of ourselves on some 800 Amerindians examined systematically (Neel, 1974). It is a most inadequate material. Nevertheless, taken at face value, it provides no evidence for a recent increase in the frequency of congenital defects among civilized peoples. The observations have been advanced in part to encourage others to come forth with comparable data, so that a significant series can be assembled.

Blood pressure, salt intake, and renin and aldosterone levels

Hypertension is commonly considered a disease of civilization. A particular effort has been directed at understanding the regulation of blood pressure among the Yanomama (Oliver, Cohen & Neel, 1975). Unlike the situation in civilized man, mean blood pressure did not continue to rise in the Yanomama after the second decade (sample of 252 adults); the same is true for the Cayapo (Ayres & Salzano, 1972). Because of the recurrent interest in the permissive role of high salt intakes in the etiology of hypertension, 24-h urine samples were obtained from 26 adult male Yanomama. Twenty-four hour values for Na^+ ($M \pm \sigma$) were 1.02 ± 1.51 mequiv., of K^+, 152.16 ± 74.51 mequiv., and of Cl^-, 13.70 ± 7.16 mequiv. Plasma samples were obtained from 11 of these subjects for renin determinations; the mean renin activity was 13.10 ± 14.17 ng/ml/h. In these same subjects, mean urine aldosterone excretion was 74.52 ± 44.94 μg/24 h. Comparable values for Caucasoids on chronic low-sodium diets of this degree are not available, but for Caucasoids on 10 mequiv. sodium diets for approximately a month, comparable aldosterone and renin values are observed; thus the limited data suggest that the normal adjustments to these chronic, very low levels of sodium intake are not very different from the adjustment to acute salt restriction, but the interpretation is complicated by the high potassium intakes of the Amerindians. These studies are continuing, directed especially towards women, who because of pregnancy and lactation are subjected to salt losses not experienced by men; it is difficult to perceive how they maintain sodium balance. It is clear that selection against any disease resulting in a failure to conserve sodium, such as a salt-losing nephropathy, would be severe in such people.

Exposure to heavy metals

As for the chromosome studies, the motive in undertaking studies of heavy metal concentrations in blood serum was to establish (low) baselines – again the results were unexpected (Hecker, Allen, Dinman & Neel, 1976). Controls consisted of 100 specimens collected in Ann Arbor, Michigan; 90 Yanomama specimens were available for comparison. Copper levels were similar in the two groups; lead and cadmium levels were much lower in the Yanomama. The unexpected finding was higher mean mercury levels in the Yanomama than in the Ann Arbor sample, especially in the results from three quite remote villages. The elevated levels are thought to be caused by alkyl mercury. There is no ready explanation for these findings, which suggest that sporadic unusual exposures to some heavy metals may antedate civilization.

Escherichia coli *types*

In an effort to understand the epidemiology of diarrhea under these conditions, an effort was made to isolate Salmonella, Shigella, and pathogenic *E. coli* from Indian stools (Eveland, Oliver & Neel, 1971). Salmonella and Shigella could not be isolated. With respect to *E. coli*, 432 isolates were obtained from 72 individuals. Only 204 of these strains were typable with a standard panel of 147 O antisera; these included 8 enteropathogenic strains. From the untypable strains, 13 were selected for the production of antisera. This resulted in 13 different antisera which typed 50 additional strains. It is clear these Amerindians constitute a reservoir of new *E. coli* strains. Whether these strains are somehow related to the milieu of the Amerindian gastrointestinal tract (high bulk diet with rapid transit time) or whether this is a consequence of mutation in an isolated population is unknown. Studies are still in progress, with particular reference to the isolation of the phage strains present in both the previously defined and the new strains.

Disease pressures, as determined by studies of serum antibodies and stool parasites

An extensive material has been accumulating during the years of the IBP on presence and titers of a wide variety of serum antibodies and intestinal parasites. It is anticipated that when summarized, this material will provide a rather comprehensive picture of the major disease pressures to which rain forest populations are subjected. Although many of the findings to date are much as might be expected, some are notable for occurrence or extent. For instance, Australia antigen was detected in 7.4% of 1179 persons tested by the immuodiffusion technique, and in 7.0% of 487 tested by the complement fixation technique (Layrisse, M., unpublished). Individual villages varied from 0.0 to 31.0%. Such high frequencies have been encountered in other populations living under poor sanitary conditions, but the apparent village differences are puzzling in view of the uniform life style of these Amerindians.

Where next?

Hopefully, the above-enumerated studies and analyses, and others still in progress, have contributed to our understanding of the organization and dynamics of the basic demes ('villages') and higher units (tribes) of man during the long years of his evolution. We do not regard the present contribution as the final synthesis of the studies to date. Major papers are yet to be written on such topics as the true degree of inbreeding in Amerindian tribes (defined as frequency of identity by descent of corresponding alleles in an individual) and the extent of zygotic disequilibrium and some of its implications. Furthermore, it is anticipated that the end of the IBP does not mean the termination of these studies on Amerindian populations. Three chief lines of investigation lie before us.

(1) *The incorporation of these observations into population models, either through formal mathematical treatments or Monte Carlo-type population simulation.* In particular we are moved to explore the implications of the apparent departures from genetic equilibrium in the demes of populations of this nature.

(2) *Further accumulation of data for the direct and indirect estimation of mutation rates.* There are no data on civilized *communities* concerning the frequency of rare protein variants similar to those now available for these Amerindian villages. However, cautious comparison of our data with those which seems most nearly comparable (cf. Harris, Hopkinson & Robson, 1973), plus surmises concerning t_0 in civilized populations, raise the possibility that mutation rates are at least as high in these Indian populations as in civilized groups, and may even be higher. Whether the appropriate contrast is between 'civilized' and 'non-civilized' or, given the residence of the caucasoids, 'tropical' versus 'non-tropical', is not clear. The theoretical and practical implications of this possibility being what they are, this question should be pursued vigorously.

(3) *Further biomedical observations.* The high prevalence among formerly primitive groups undergoing acculturation of such 'diseases of civilization' as hypertension, diabetes mellitus, and gout is extensively documented. The proper extensive studies of groups in rapid transition, as the Yanomama soon will be, cannot fail to yield insights of value to all human beings.

James V. Neel and others

The term 'multi-disciplinary research' is much used – and abused – these days. To the extent that this represents a successful endeavor along these lines, it is not because the principals, meeting in comfortable quarters, then dispatched their students to the field to work things out. We've all been there together – with our students. In closing, we dedicate this summary to those humid, breathless, equatorial nights when weary, itching, and usually less than satisfied with the day's events, we tried around the campfire to get our respective viewpoints across to each other until – sometimes succeeding, often failing – we collapsed into our hammocks to gather our strength for the next day.

The financial support of the US Atomic Energy Commission and National Science Foundation is gratefully acknowledged.

References

Arends, T., Brewer, G., Chagnon, N., Gallango, M., Gershowitz, H., Layrisse, M., Neel, J., Shreffler, D., Tashian R. & Weitkamp, L. (1967). Intratribal genetic differentiation among the Yanomama Indians of Southern Venezuela. *Proceedings of the National Academy of Sciences*, USA, **57**, 1252–9.

Ayres, M. & Salzano, F. M. (1972). Health status of Brazilian Cayapo Indians. *Tropical and Geographical Medicine*, **24**, 178–85.

Bloom, A. D., Neel, J. V., Tsuchimoto, T. & Meilinger, K. (1973). Chromosomal breakage in leukocytes of South American Indians. *Cytogenetics and Cell Genetics*, **12**, 175–86.

Bodmer, W. F. (1973). Population studies and the measurement of natural selection, with special reference to the HL-A system. *Israel Journal of Medical Science*, **9**, 1503–18.

Booth, P. B., Jenkins, W. J. & Marsh, W. L. (1966). Anti-I^T: a new antibody of the I blood group system occurring in certain Melanesian sera. *British Journal of Haematology*, **12**, 341–4.

Brewer-Carias, C., LeBlanc, S. & Neel, J. V. (1976) Genetic structure of a tribal population, the Yanomama Indians. XIII. Dental microdifferentiation. *American Journal of Physical Anthropology*, **44**, 5–14.

Cavalli-Sforza, L. L. & Bodmer, W. F. (1971). *The Genetics of Human Populations*. San Francisco: Freeman & Co.

Cavalli-Sforza, L. L. & Edwards, A. W. F. (1965). Analysis of human evolution. In *Genetics Today*, ed. S. J. Geerts, vol. 3, Proceedings of the 11th International Congress of Genetics, The Hague, The Netherlands, pp. 923–33. New York: Pergamon Press.

Cavalli-Sforza, L. L. & Edwards, A. W. F. (1967). Phylogenetic analysis: models and estimation procedures. *American Journal of Human Genetics*, **19**, 233–7.

Chagnon, N. A. (1968). *Yąnamamö: The Fierce People*. New York: Holt, Rinehart & Winston.

Chagnon, N. A. (1974). *Studying the Yanomamö*. New York: Holt, Rinehart & Winston.

Chagnon, N. A., Neel, J. V., Weitkamp, L., Gershowitz, H. & Ayres, M. (1970). The influence of cultural factors on the demography and pattern of gene flow from the Makiritare to the Yanomama Indians. *American Journal of Physical Anthropology*, **32**, 339–50.

136

Comas, J. (1971). Anthropometric studies in Latin American Indian populations. In *The Ongoing Evolution of Latin American Populations*, ed. F. M. Salzano, pp. 333–94. Springfield, Ill.: C. Thomas.

Crow, J. F. (1958). Some possibilities for measuring selection intensities in man. *American Anthropologist*, **60**, 1–13.

Crow, J. F. & Kimura, M. (1970). *An Introduction to Population Genetics Theory.* New York: Harper & Row.

Crow, J. F. & Kimura, M. (1972). The effective number of a population with overlapping generations: a correction and further discussion. *American Journal of Human Genetics*, **24**, 1–10.

Da Rocha, F. J. & Salzano, F. M. (1972). Anthropometric studies in Brazilian Cayapo Indians. *American Journal of Physical Anthropology*, **36**, 95–102.

Edwards, A. W. F. (1971). Distances between populations on the basis of gene frequencies. *Biometrics*, **27**, 873–81.

Edwards, A. W. F. & Cavalli-Sforza, L. L. (1963). The reconstruction of evolution. *Annals of Human Genetics, London*, **27**, 104–5.

Edwards, A. W. F. & Cavalli-Sforza, L. L. (1964). Reconstruction of evolutionary trees. In *Systematics Association Publication*, no. 6, pp. 67–76. London: British Museum (Natural History).

Edwards, A. W. F. & Cavalli-Sforza, L. L. (1965). A method for cluster analysis. *Biometrics*, **21**, 362–75.

Eveland, W. C., Oliver, W. J. & Neel, J. V. (1971). Characteristics of *Escherichia coli* serotypes in the Yanomama, a primitive Indian tribe of South America. *Infection and Immunity*, **4**, 753–6.

Fisher, R. A. (1929). *The Genetical Theory of Natural Selection.* New York: Dover.

Gershowitz, H., Layrisse, M., Layrisse, Z., Neel, J. V., Chagnon, N. & Ayres, M. (1972). The genetic structure of a tribal population, the Yanomama Indians. ii. Eleven blood-group systems and the ABH-Le secretor traits. *Annals of Human Genetics, London*, **35**, 261–9.

Guttman, L. (1968). A general nonmetric technique for finding the smallest coordinate space for a configuration of points. *Psychometrika*, **33**, 469–506.

Harris, H., Hopkinson, D. A. & Robson, E. B. (1973). The incidence of rare alleles determining electrophoretic variants: data on 43 enzyme loci in man. *Annals of Human Genetics, London*, **37**, 237–53.

Hecker, L. H., Allen, H. E., Dinman, B. D. & Neel, J. V. (1976) Mercury, cadmium, lead and copper levels in acculturated and unacculturated populations. *Archives of Environmental Health*, **29**, 181–5.

Jain, S. K. & Workman, P. (1967). Generalized F statistics and the theory of inbreeding and selection. *Nature, London*, **214**, 674–8.

Kendall, M. G. (1962). *Rank Correlation Methods*, 3rd edn. London: Griffin.

Keyfitz, N. & Flieger, W. (1968). *World Population: An Analysis of Vital Data.* Chicago: University of Chicago Press.

Kimura, M. & Crow, J. F. (1964). The number of alleles that can be maintained in a finite population. *Genetics*, **49**, 725–38.

Kimura, M. & Ohta, T. (1969). The average number of generations until extinction of an individual mutant gene in a finite population. *Genetics*, **63**, 701–9.

Kimura, M. & Ohta, T. (1971). *Theoretical Aspects of Population Genetics.* Princeton: Princeton University Press.

Koch-Grünberg, T. (1917). *Vom Roroima zum Orinoco, Ergebnisse einer Reise in Nord-Brasilien und Venezuela in den Jahren 1911–1913.* Berlin: O. Reimer.

Lance, G. N. & Williams, W. T. (1967). A general theory of classificatory sorting strategies. i. Hierarchical systems. *Computer Journal*, **9**, 373–80.

James V. Neel and others

Layrisse, M. (1971). Blood group polymorphisms in Venezuelan Indians. In *The Ongoing Evolution of Latin American Populations*, ed. F. M. Salzano, pp. 617–30. Springfield, Ill.: C. Thomas.

Layrisse, M. & Wilbert, J. (1966). *Indian Societies of Venezuela*. Caracas: Monograph no. 13, Fundación la Salle de Ciencias Naturales.

Layrisse, Z. & Layrisse, M. (1968). High incidence cold agglutinins of anti-1^T specificity in Yanomama Indians of Venezuela. *Vox Sanguinis*, 14, 369–82.

Layrisse, Z., Layrisse, M., Malavé, I., Terasaki, P., Ward, R. H. & Neel, J. V. (1973). Histocompatibility antigens in a genetically isolated American Indian tribe. *American Journal of Human Genetics*, 25, 493–509.

Lewontin, R. C. & Krakauer, J. (1973). Distribution of gene frequency as a test of the theory of the selective neutrality of polymorphisms. *Genetics*, 74, 175–95.

Li, F. H. F. & Neel, J. V. (1974). Simulation of the fate of a mutant gene of neutral selective value in a primitive population. In *Computer Simulation in Human Population Studies*, ed. B. Dyke & J. W. MacCluer, pp. 221–40. New York: Seminar Press.

Lingoes, J. C. (1965). An IBM 7090 program for Guttman–Lingoes smallest space analysis-I. *Behavioral Science*, 10, 183–4.

Lingoes, J. C. (1971). Some boundary conditions for a monotone analysis of symmetric matrices. *Psychometrika*, 36, 195–203.

Lingoes, J. C. (1973). A statistical note on mean *tau*. *Michigan Mathematical Psychology Program*, 3, 1–20.

Lingoes, J. C. & Roskam, E. (1971). A mathematical and empirical study of two multidimensional scaling algorithms. *Michigan Mathematical Psychology Program*, 1, 1–169.

Lingoes, J. C. & Schönemann, P. H. (1976). Alternative measures of fit for the Schönemann-Carroll matrix fitting algorithm. *Psychometrika* (in press).

MacCluer, J. W., Neel, J. V. & Chagnon, N. A. (1971). Demographic structure of a primitive population: a simulation. *American Journal of Physical Anthropology*, 35, 193–207.

Mahalanobis, P. C., Majumdar, D. N. & Rao, C. R. (1949). An anthropometric survey of the United Provinces, 1941: a statistical study. *Sankhyā*, 9, 89–324.

Migliazza, E. C. (1972). Yanomama grammar and intelligibility. Ph.D. thesis, University of Indiana, Bloomington.

Morton, N. E. (1974). Controversial issues in human population genetics. *American Journal of Human Genetics*, 26, 259–62.

Neel, J. V. (1957). Some problems in the estimation of spontaneous mutation rates in animals and man. In *Effects of Radiation on Human Heredity*, pp. 139–49. Geneva: World Health Organization.

Neel, J. V. (1962). Mutations in the human population. In *Methodology in Human Genetics*, ed. W. J. Burdette, pp. 203–24. San Francisco: Holden-Day, Inc.

Neel, J. V. (1972). The genetic structure of a tribal population, the Yanomama Indians. I. Introduction. *Annals of Human Genetics, London*, 35, 255–9.

Neel, J. V. (1974). A note on congenital defects in two unacculturated Indian tribes. In *Symposium on Epidemiology of Congenital Defects*, ed. D. T. Janerich, R. G. Skalko & I. H. Porter, pp. 3–15. New York: Academic Press.

Neel, J. V. (1975). The study of 'natural' selection in man: last chance. In *The Role of Natural Selection in Human Evolution*, Wenner-Gren Symposium, ed. F. M. Salzano, pp. 355–68. Amsterdam: North-Holland.

Neel, J., Arends, T., Brewer, C., Chagnon, N., Gershowitz, H., Layrisse, M., Layrisse, Z., MacCluer, J., Migliazza, E., Oliver, W., Salzano, F., Spielman, R., Ward, R. & Weitkamp, L. (1972). Studies on the Yanomama Indians. In

138

Human Genetics, Proceedings of the 4th International Congress of Human Genetics, Paris, September, 1971, pp. 96–111. Amsterdam: Excerpta Medica.

Neel, J. V., Centerwall, W. R., Chagnon, N. A. & Casey, H. L. (1970). Notes on the effect of measles and measles vaccine in a virgin-soil population of South American Indians. *American Journal of Epidemiology*, **91**, 418–29.

Neel, J. V., Rothhammer, F. & Lingoes, J. C. (1974). The genetic structure of a tribal population, the Yanomama Indians. x. Agreement between representations of village distances based on different sets of characteristics. *American Journal of Human Genetics*, **26**, 281–303.

Neel, J. V. & Salzano, F. M. (1964). A prospectus for genetic studies of the American Indian. *Cold Spring Harbor Symposia on Quantitative Biology*, **29**, 85–98.

Neel, J. V., Salzano, F. M., Junqueira, P. C., Keiter, F. & Maybury-Lewis, D. (1964). Studies on the Xavante Indians of the Brazilian Mato Grosso. *American Journal of Human Genetics*, **16**, 52–140,

Neel, J. V. & Schull, W. J. (1968). On some trends in understanding the genetics of man. *Perspectives in Biology and Medicine*, **11**, 565–602.

Neel, J. V. & Ward, R. H. (1970). Village and tribal genetic distances among American Indians, and the possible implications for human evolution. *Proceedings of the National Academy of Sciences, USA*, **65**, 323–30.

Neel, J. V. & Ward, R. H. (1972). Genetic structure of a tribal population, the Yanomama Indians. vi. Analysis bv F-statistics, including a comparison with the Makiritare and Xavante. *Genetics*, **72**, 639–66.

Neel, J. V. & Weiss, K. (1975). The genetic structure of a tribal population, the Yanomama Indians. xii. Biodemographic studies. *American Journal of Physical Anthropology*, **42**, 25–52.

Nei, M. (1971). Total number of individuals affected by a single deleterious mutation in large populations. *Theoretical Population Biology*, **2**, 426–30.

Oliver, W. J., Cohen, E. L. & Neel, J. V. (1975). Blood pressure, sodium intake, and sodium related hormones in the Yanomama Indians, a 'No-salt' culture. *Circulation*, **52**, 146–51.

Pearson, K. (1926). On the coefficient of racial likeness. *Biometrika*, **18**, 105–17.

Peña, H. F., Salzano, F. M. & Da Rocha, F. J. (1972). Dermatoglyphics of Brazilian Cayapo Indians. *Human Biology*, **44**, 225–41.

Post, R. H., Neel, J. V. & Schull, W. J. (1968). Tabulations of phenotype and gene frequencies for 11 different genetic systems studied in the American Indian. In *Biomedical Challenges Presented by the American Indian*, Scientific Publication no. 165, pp. 141–85. Washington: Pan American Health Organization.

Roskam, E. & Lingoes, J. C. (1970). MINISSA-I: A FORTRAN IV program for the smallest space analysis of square symmetric matrices. *Behavioral Science*, **15**, 204–5.

Rothhammer, F., Neel, J. V. Da Rocha, F. & Sundling, G. Y. (1973). The genetic structure of a tribal population, the Yanomama Indians. viii. Dermatoglyphic differences among villages. *American Journal of Human Genetics*, **25**, 152–66.

Salzano, F. M. (1968). Intra- and inter-tribal genetic variability in South American Indians. *American Journal of Physical Anthropology*, **28**, 183–90.

Salzano, F. M. (1972). Genetic aspects of the demography of American Indians and Eskimos. In *The Structure of Human Populations*, ed. G. A. Harrison & A. J. Boyce, pp. 234–51. Oxford: Clarendon Press.

Salzano, F. M. (1975). Interpopulation variability in polymorphic systems. In *The Role of Natural Selection in Human Evolution*, ed. F. M. Salzano, pp. 217–29. Amsterdam: North-Holland.

139

James V. Neel and others

Salzano, F. M. (1976). Multidisciplinary studies in tribal societies and man's evolution. In *World Anthropology*, Proceedings of the 9th International Congress of Anthropological and Ethnological Sciences, The Hague, Mouton (in press).

Salzano, F. M., Neel, J. V. & Maybury-Lewis, D. (1967). Further studies on the Xavante Indians. I. Demographic data on two additional villages; genetic structure of the tribe. *American Journal of Human Genetics*, 19, 463–89.

Salzano, F. M., Neel, J. V., Weitkamp, L. R. & Woodall, J. P. (1972). Serum proteins, hemoglobins and erythrocyte enzymes of Brazilian Cayapo Indians. *Human Biology*, 44, 443–58.

Salzano, F. M., Steinberg, A. G. & Tepfenhart, M. A. (1973). Gm and Inv allotypes of Brazilian Cayapo Indians. *American Journal of Human Genetics*, 25, 167–77.

Schönemann, P. H. & Carroll, R. M. (1970). Fitting one matrix to another under choice of a central dilation and a rigid motion. *Psychometrika*, 35, 245–55.

Smouse, P. E. (1974). Likelihood analysis of geographic variation in allelic frequencies. II. The logit model and an extension to multiple loci. *Theoretical and Applied Genetics*, 45, 52–8.

Smouse, P. E. & Kojima, K. (1972). Maximum likelihood analysis of population differences in allelic frequencies. *Genetics*, 72, 709–19.

Spielman, R. S. (1973). Differences among Yanomama Indian villages: Do the patterns of allele frequencies, anthropometrics and map locations correspond? *American Journal of Physical Anthropology*, 39, 461–79.

Spielman, R. S., Da Rocha, F. J., Weitkamp, L. R., Ward, R. H., Neel, J. V. & Chagnon, N. A. (1972). The genetic structure of a tribal population, the Yanomama Indians. VII. Anthropometric differences among Yanomama villages. *American Journal of Physical Anthropology*, 37, 345–56.

Spielman, R. S., Migliazza, E. C. & Neel, J. V. (1974). Regional linguistic and genetic differences among Yanomama Indians. *Science, Washington*, 184, 637–44.

Stevenson, A. C. & Kerr, L. B. (1967). On the distributions of frequencies of mutation to genes determining harmful traits in man. *Mutation Research*, 4, 339–52.

Tanis, R., Ferrell, R. E., Neel, J. V. & Morrow, M. (1974). Albumin Yanomama-2, a 'private' polymorphism of serum albumin. *Annals of Human Genetics, London*, 38, 179–90.

Tanis, R. J., Neel, J. V., Dovey, H. & Morrow, M. (1973). The genetic structure of a tribal population, the Yanomama Indians. IX. Gene frequencies for 18 serum protein and erythrocyte enzyme systems in the Yanomama and five neighboring tribes; nine new variants. *American Journal of Human Genetics*, 25, 655–76.

Ward, R. H. (1972). The genetic structure of a tribal population, the Yanomama Indians. V. Comparison of a series of networks. *Annals of Human Genetics, London*, 36, 21–43.

Ward, R. H., Gershowitz, H., Layrisse, M. & Neel, J. V. (1975). The genetic structure of a tribal population, the Yanomama Indians. XI. Gene frequencies for 7 blood groups in the Yanomama; the uniqueness of the tribe. *American Journal of Human Genetics*, 27, 1–30.

Ward, R. H. & Neel, J. V. (1970). Gene frequencies and microdifferentiation among the Makiritare Indians. IV. Comparison of a genetic network with ethnohistory and migration matrices; a new index of genetic isolation. *American Journal of Human Genetics*, 22, 538–61.

Man in the tropics: the Yanomama Indians

Ward, R. H. & Neel, J. V. (1976). The genetic structure of a tribal population, the Yanomama Indians. XIV. Clines and their interpretation. *Genetics*, **82**, 103–21.

Weiss, K. M. (1972). A general measure of human population growth regulation. *American Journal of Physical Anthropology*, **37**, 337–44.

Weiss, K. M. (1973). Demographic models for anthropology. (Memoirs of the Society for American Archaeology, no. 27.) *American Antiquity*, **38**, 1–186.

Weitkamp, L. R., Arends, T., Gallango, M. L., Neel, J. V., Schultz, J. & Shreffler, D. C. (1972). The genetic structure of a tribal population, the Yanomama Indians. III. Seven serum protein systems. *Annals of Human Genetics, London*, **35**, 271–9.

Weitkamp, L. R. & Neel, J. V. (1972). The genetic structure of a tribal population, the Yanomama Indians. IV. Eleven erythrocyte enzymes and summary of protein variants. *Annals of Human Genetics, London*, **35**, 433–44.

Wright, S. (1943). An analysis of local variability of flower color in *Linanthus Parryae*. *Genetics*, **28**, 139–56.

Wright, S. (1965). The interpretation of population structure by F statistics with special regard to systems of mating. *Evolution*, **19**, 395–420.

Appendix

Participants in the collaborative Integrated Research Project on the Population Genetics of the American Indian.

Dr Tulio Arends, Departamento de Medicina Experimental, Instituto Venezolano de Investigaciones Cientificas, Apartado 1827, Caracas, Venezuela

Dr Nellie Arvello de Jimenez, Departamento de Antropologia, Instituto Venezolano de Investigaciones Cientificas, Apartado 1827, Caracas, Venezuela

Dr Manuel Ayres, Laboratório de Genetica, Centro de Ciências Biológicas. Universidade Federal do Pará, 66000 Belém PA, Brazil

Dr Arthur D. Bloom, Division of Genetics, Department of Pediatrics, College of Physicians and Surgeons of Columbia University, 630 West 168th Street, New York, New York 10032, USA

Dr Charles A. Brewer-Carias, Jr, Edificio Galipan, Avenida Francisco Miranda, Caracas, Venezuela

Dr Napoleon A. Chagnon, Department of Anthropology, 409 Social Science Building, Pennsylvania State University, University Park, Pennsylvania 16802, USA

Dr Warren C. Eveland, Department of Epidemiology, School of Public Health, University of Michigan, Ann Arbor, Michigan 48104, USA

Dr Robert Ferrell, Center for Demographic and Population Genetics, 141 The University of Texas at Houston, Houston, Texas 77030, USA

Dr M. L. Gallango, Hematologia Experimental, Instituto Venezolano de Investigaciones Cientificas, Apartado 1827, Caracas, Venezuela

Dr Henry Gershowitz, Department of Human Genetics, University of Michigan Medical School, Ann Arbor, Michigan 48109, USA

Ms Zulay Layrisse, Instituto Venezolano de Investigaciones Cientificas, Apartado 1827, Caracas, Venezuela

Dr Frances H. F. Li, Department of Human Genetics, University of Michigan Medical School, Ann Arbor, Michigan 48109, USA

Dr James C. Lingoes, Department of Psychology, University of Michigan, Ann Arbor, Michigan 48109, USA

141

James V. Neel and others

Dr Jean W. MacCluer, Department of Biology, The Pennsylvania State University, 208 Life Sciences I, University Park, Pennsylvania 16802, USA

Ms Karen Meilinger, Department of Human Genetics, University of Michigan Medical School, Ann Arbor, Michigan 48109, USA

Dr Ernest C. Migliazza, Department of Anthropology, University of Maryland, College Park, Maryland 20742, USA

Ms Marianne Morrow, Department of Human Genetics, University of Michigan Medical School, Ann Arbor, Michigan 48109, USA

Dr William Oliver, Department of Pediatrics and Communicable Diseases, University of Michigan Medical School, Ann Arbor, Michigan 48109, USA

Dr Fernando Jose da Rocha, Departmento de Genética, Instituto de Biociências, Universidade Federal do Rio Grande do Sul, Caixa Postal 1953, 90000 Porto Algeré RS, Brazil

Dr Francisco Rothhammer Chief, Unit of Human Genetics and Evolution, Universidad de Chile, Departamento de Génetica y Biología, Sede Norte, Casilla 6607, Correo 4, Santiago, Chile

Dr Richard Spielman, Department of Human Genetics & University of Pennsylvania Human Genetics Center, Richards Building/G4, School of Medicine, Philadelphia, Pennsylvania 19174, USA

Ms Grace Y. Sundling, Department of Human Genetics, University of Michigan Medical School, Ann Arbor, Michigan 48109, USA

Dr Robert Tanis, Department of Biochemistry, Biochemistry Building, Michigan State University, East Lansin, Michigan 48823, USA

Dr Richard H. Ward, Child Development and Mental Retardation Centre, WJ-10, University of Washington, Seattle, Washington, 95195, USA

Dr Kenneth Weiss, Center for Demographic and Population Genetics, University of Texas, Houston, Prudential Building, Room 1109, 1100 Holcombe Blvd., Houston, Texas 77025, USA

Dr Lowell R. Weitkamp, Division of Genetics University of Rochester, School of Medicine, 260 Crittenden Blvd., Rochester, New York, 14620

142

5. Culture, human biology and disease in the Solomon Islands

LOT B. PAGE, JONATHAN FRIEDLAENDER &
ROBERT C. MOELLERING, JR

The Harvard Solomon Islands project was conceived and executed as a multi-disciplinary approach to the study of human ecology by Dr Albert Damon, physician and anthropologist, who directed and coordinated all aspects of the project from its inception in 1964 until his death in 1973. Analysis of the data obtained during four field studies, and plans for further follow-up observations in the field have been continued, since 1973, by Damon's anthropological and medical colleagues associated with the Peabody Museum, Harvard University. The cooperation and support of Governmental and public health officials of the Territory of Papua New Guinea and the British Solomon Islands Protectorate have been given unstintingly throughout the study.

Purpose and design

The object of the Solomon Islands project has been to combine the techniques of cultural and physical anthropology and a variety of medical specialties for the purpose of relating culture, habitat, biological variation, and disease patterns. This approach has been applied to a spectrum of defined populations in the Solomon Islands which vary in genetic background, environmental setting, mode of subsistence, and contact with Western cultural influences. The eight different population groups in the study (see Fig. 5.1) include inland mountain and hill dwellers, coastal fishermen, and inhabitants of small islands and atolls. All were living in rural tribal areas where traditional ways of life had been largely preserved, although important differences in acculturation occurred among them at the time of the study.

Each population was a residentially defined sample of a separate Solomon Islands tribal society. A cultural anthropologist (or in some cases a husband and wife team) resided with each society for 18 to 24 months in order to obtain detailed ethnographic data. This included demographic, cultural, economic and dietary details, the establishment of biological kinship, and estimation of ages. Where no written birth records existed, age was estimated by 'triangulation' against events of known dates, together with relative (older–younger) age relationships among individuals. This information is deemed to be accurate to within two years for children and young adults and to within five years for older individuals. The ethnographers usually

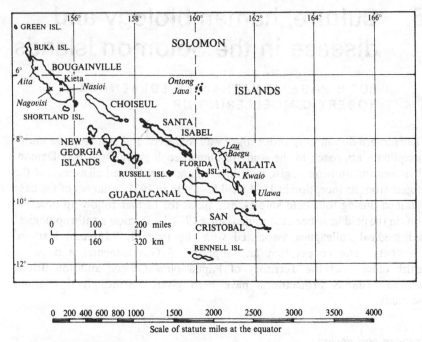

Fig. 5.1. Map of the Solomon Islands region. The location of eight tribal groups is indicated. (From Page *et al.*, 1974, by permission of the American Heart Association, Inc.)

became fluent in the local language, and were able to prepare the population for the visit later by the biomedical team.

Following completion of the ethnographic phase, the biomedical team, accompanied by the ethnographer, spent four to six weeks examining each population. This team included five physical anthropologists and six physicians, each with a specific protocol of examinations to perform. Using buildings of native construction, designed and built for the purpose under the ethnographer's guidance and centrally located in the tribal area, it was practicable to examine 25 to 35 human subjects each day. Subjects were paid, in goods or cash, for time lost from their ordinary activities, and effort was made to enlist cooperation of as close as possible to 100% of the defined sample. Medical care was offered to all, regardless of cooperation in the study.

Field trips by the biomedical team occurred every second year from 1966 through 1972. With few exceptions the same physical measurements were made on all subjects by the same examiners. Protocols of the physical anthropologist included detailed anthropometric measurements, grip strength, somatotype photographs and dermatoglyphics. In addition, observations were made on such qualitative traits as earwax texture, mid-phalangeal hair, color vision, tongue curling and folding, arm folding, hand clasping, and

144

ability to taste phenylthiocarbamide (PTC). Medical examinations included physical examination on all subjects aged 15 years and over by an internist and by all subjects aged 14 years and under by a pediatrician. All individuals except for small infants were subject to a dental examination, including the taking of dental impressions and also an ophthalmologic examination including visual acuity, cycloplegic retinoscopy, ophthalmoscopy, and tonometry. Posterior–anterior chest X-rays were made on children and adults, and bone-age films on infants and children. A standard 12-lead electrocardiogram was recorded on adults. Blood samples were obtained, on all but the infants, for blood typing, malaria smears, cholesterol, uric acid, protein polymorphisms, and hepatitis-associated antigen. During 1972, when work was carried out on the laboratory ship *Alpha Helix*, certain other procedures, not possible under field conditions of the earlier expeditions, were also included. These included determination of glucose-6-phosphate dehydrogenase, collection of frozen plasma for determination of plasma renin activity, and glycerolization of red cells for enzyme studies.

Except where tribal taboos prevented it, samples of urine for sodium, potassium and creatinine were obtained, and stool samples were collected for ova and parasites. In some subjects, samples of hair and finger nails and skin biopsies were obtained. After other examinations were completed, intradermal tests for *Histoplasma capsulatum*, *Mycobacterium tuberculosis* and four atypical mycobacteria were implanted and skin reactions were read 48 hours later.

Due in large measure to the efforts of the ethnographers, more than 90% of subjects in the defined populations participated in the study in seven of the eight societies. A lower figure of 78% participation among the Kwaio resulted from a severe outbreak of acute respiratory disease during the study, later identified from convalescent sera as Influenza A, Asian strain.

In all, 3631 persons, including 2020 adults (15 years of age and older) in 8 different populations were studied. Fig. 5.1 shows the geographical locations of these societies. Three are on the island of Bougainville, Papua New Guinea, and three are on the island of Malaita, British Solomon Islands Protectorate. The remaining two populations are on the small islands of Ulawa and Ontong Java (a coral atoll) both in the British Solomon Islands.

Both Bougainville and Malaita lie from 5° to 7° S of the equator, and are volcanic in origin and tropical in climate, with annual temperatures 21–29 °C and high annual rainfall. Experience with outside influences has been markedly different between the two islands. Parts of the coastal plain of Bougainville have been settled by European planters and traders since 1885. During World War II the island was occupied at various times by both Allied and Japanese military forces and was the site of heavy fighting. After the war, the European census declined, but remained above the pre-war level. The discovery of extensive copper ore deposits in the mid 1960s resulted in a rapid influx of

Lot B. Page and others

Europeans which was just beginning at the time of the first field study in 1966 (Oliver, 1973). By 1970 the rapid development of mining activities had already had a major impact on the economy, activity, and attitudes of the people of Bougainville, including those physically remote from the mines. The 1966 population estimate was 70000 (*Annual Report*, Department of Public Health, Territory of Papua New Guinea, 1967).

In Malaita, European settlements have been small and scattered, and mostly limited to the town of Auki on the southwest coast and mission stations in coastal locations. The island was never occupied during World War II. The indigenous population was estimated at 50000 in 1970 (*BSIP News Sheet* no. 9, May 1970. BSIP Information Service, Honiara, Guadalcanal, BSIP).

Ethnographic descriptions of the study populations have, in most instances, been published by the cultural anthropologist who worked with the biomedical team. These publications should be consulted for authoritative details (Keesing, 1967; Maranda, 1970; Maranda & Maranda, 1970; Mitchell, D. P., 1971; Mitchell, J. N., 1971; Ogan, 1972; Ross, 1973).

Biological variation related to geographical and historical relationships

Although Melanesia is renowned for its cultural and biological heterogeneity from one village or language group to the next (Giles, Wyber & Walsh, 1970; Littlewood, 1972; Friedlaender, 1975), nevertheless the groups on different islands tend to share similar overall patterns of blood gene frequencies, skin color, body and head sizes and shapes, hair form, tooth dimensions, finger-ridge counts, and various other aspects of the phenotype. Papers on these topics will be appearing in the literature in the near future, and here we will only indicate the general coherence of these different patterns of variations, all of which, to a large degree, reflect genetic differences.

The three societies studied on Bougainville Island (the Nasioi, Aita and Nagovisi) speak Papuan languages, and are in many other cultural respects quite distinct. Their biological heterogeneity also attests to a long period of isolation on the island, estimated by some (Terrel, personal communication) to have been inhabited at least 3000 years ago, and possibly a considerable time before that. Yet taken as a group, there are a number of common features which set them and other Bougainville groups apart from the other Solomon Islanders, or from the peoples of the Bismarck Archipelago to the northwest. Their skin color (along with some Islanders immediately to the southeast) is darker than any other group native to the other islands of the Pacific. Their hair, with the typically Melanesian irregular curl (Hrdy, 1973) is quite dark. They are generally broad-faced and narrow across the cranium. Their

146

teeth are especially large for this area of the world. In terms of blood genetics, these people have very low L^{Ms} frequencies, low I^B frequencies, and very high *Inv* frequencies.

Malaita, a population concentration almost at the other end of the Solomons chain, presents a contrast to the Bougainville cluster. The Kwaio and Baegu are inland Malaitan populations, while the Ulawans and Lau people are Malaitans who have settled on off-shore islands (there may be some Polynesian intermixture in the Lau, however). The languages these people speak are part of the large Austronesian phylum, which also includes Polynesian and Malay, as well as most of the other languages of Melanesian islands. In appearance they are brown-skinned, with considerably lighter hair than the Bougainville groups (the hair form is the same). The shortest faces belong to the Baegu and Kwaio, with the Lau and Ulawa values all being substantially longer. As for blood genetics, the trio of Kwaio, Baegu, and Lau samples are especially high in L^{Ms}, with Ulawa slightly lower.

The Ontong Java people are substantially different from the other populations included in the survey. They are highly variable in physical appearance, reflecting a long period of intermixture. They most often resemble Micronesians, with straighter hair, and lighter skin color, yet their culture is largely Polynesian. They are the largest people included in the survey in almost every bodily dimension, and yet their teeth are small. They have the highest L^{NS} frequencies of any group and evidently significant R^o frequencies.

In summary, preliminary findings suggest that in a wide variety of physical and hematological characters, all in some way reflecting genetic variation, the three Bougainville population samples tend to form a cluster in contrast to a more genetically dispersed Malaitan group. Ontong Java appears to be quite distinct from the other populations.

Variation in acculturation

All eight of the Solomon Islands' populations studied were living in traditional tribal groups in rural areas where technological conveniences such as roads, vehicles, electricity etc., were totally absent. A few battery-powered transistor radios were present in most of the areas (excepting Kwaio and Baegu) and doubtless served to increase the horizons of their awareness. In all groups many young men had travelled to towns such as Kieta in Bougainville, and Honiara on Guadalcanal, or had been employed for a period of time on a European-owned plantation, and then had returned to their home territory to resume traditional modes of life. Generally speaking, women did the bulk of agricultural labor, tended to the home and children and travelled little if at all outside the local tribal area.

The eight different groups thus were all at low levels of Western acculturation. Nevertheless, they differed considerably in the intensity and duration of

Lot B. Page and others

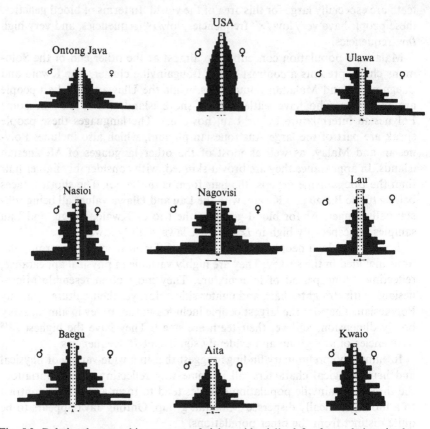

Fig. 5.2. Relative demographic structure of eight residentially defined populations in the Solomon Islands and the USA (not to same scale). Horizontal bars show number of males (*left*) and females (*right*) for each quintile of age.

exposure to European cultural influences, and in the extent to which these influences had altered their traditional life-patterns.

In order to assess the affect of acculturating influence on such things as growth, fertility, nutrition, infectious disease, blood pressure, and cardiovascular disease, a relative scale of acculturation was developed based on cultural, demographic and physical data obtained in the study.

The factors considered in ranking the societies with respect to acculturation were (1) evidence of demographic change within the defined populations, (2) secular increase in adult height, (3) length and intensity of contact with Western cultural influences, (4) religious change from pagan to christian belief, (5) education, (6) availability of medical care, (7) extent of entry into cash economy, and (8) adoption of European-style dietary items.

148

Demography

The demographic structure of six of the eight population samples is shown, in comparison with the United States in Fig. 5.2. An increase in proportionate numbers of young persons connotes an expanding society. Such a pattern is characteristic of 'young' societies in which the effects of reduced infant mortality and increased birth rate are seen, as a result of medical care and improving nutritional status. This pattern is present in most of the Solomon Islands groups, but is least evident in the Kwaio where a balance between moderate fertility and early mortality, characteristic of a 'primitive' society obtains. The demographic structure of the United States, shown for comparison in Fig. 5.2, is that of a society no longer expanding rapidly, and characterized by low rates of birth and early death.

Trends in adult height

A secular trend toward increasing height in successive generations is thought to indicate improving health and nutrition during growth (Tobias, 1962). The presence or absence of this change may, therefore, be considered as one index of the effects of acculturation. Heights by decile of age are shown in Table 5.1. Among the Solomon Islanders, secular increase in height appears to be present in both males and females in the Nasioi, and is beginning in males in the Nagovisi, Lau, Ulawa and Ontong Java groups and in females of the Lau and Beagu and possibly in Ontong Java.

The small numbers represented in certain age groups, and occasional reverse tendencies make interpretation of secular trends uncertain in several of these populations.

Cultural factors

Based on ethnological data accumulated by the cultural anthropologists, comparisons of a wide spectrum of variables has been reduced to a series of ordinal scales to cross-compare the Solomon Islands' societies.

Length of significant contact with Western cultural influences was a matter of historical record in each instance. *Intensity of contact* is harder to define. Estimations were based on proximity to European-dominated towns or missions, and frequency of interpersonal contacts with European officials, missionaries and tradesmen. Negative attitudes toward Christianity and European culture (as in the Lau population sample) were considered to reduce intensity of contact. The abandonment of pagan belief in favor of one or another of the Christian sects is a highly important event, which may influence diet, mate selection, education, life goals and birth spacing. It may reduce certain stresses within a society and introduce others. In the Solo-

149

Table 5.1. *Secular trend in adult height (mm) for eight Solomon Islands' populations and the United States*

Males

Age (years)	Nasioi		Nagovisi		Lau		Baegu		Aita		Kwaio		Ulawa		Ontong Java		USA
	(N)	(61)	(N)	(114)	(N)	(31)	(N)	(137)	(N)	(89)	(N)	(130)	(N)	(110)	(N)	(156)	(3511)
18–24	(11)	1639	(17)	1615	(12)	1657	(21)	1599	(20)	1591	(23)	1612	11	1626	37	1666	1745
25–34	(19)	1633	(18)	1593	(14)	1633	(30)	1627	(28)	1599	(25)	1613	30	1627	50	1659	1755
35–44	(22)	1618	(27)	1589	(14)	1617	(13)	1606	(25)	1618	(26)	1616	28	1608	23	1619	1740
45–54	(7)	1602	(29)	1602	(21)	1637	(34)	1609	(13)	1610	(34)	1596	10	1600	24	1604	1732
55–64	—	a	(12)	1590	(14)	1591	(9)	1630	—	—	(13)	1608	18	1621	11	1649	1712
65+	—	a	(11)	1574	(7)	1623	(11)	1592	—	—	(8)	1554	13	1607	11	1629	1699[b]
Secular increase?	Yes		Youngest only		Youngest only		No		No		No		Yes		Yes		Yes

Females

Age (years)	Nasioi		Nagovisi		Lau		Baegu		Aita		Kwaio		Ulawa		Ontong Java		USA
	(N)	(66)	(N)	(112)	(N)	(113)	(N)	(116)	(N)	(97)	(N)	(131)	(N)	(85)	(N)	(222)	(3511)
18–24	(15)	1541	(27)	1512	(36)	1538	(13)	1535	(23)	1489	(32)	1486	25	1508	65	1549	1621
25–34	(12)	1515	(21)	1524	(18)	1516	(35)	1495	(35)	1503	(34)	1500	29	1513	79	1567	1618
35–44	(21)	1515	(29)	1493	(20)	1520	(34)	1512	(24)	1518	(26)	1502	15	1515	31	1530	1613
45–54	(9)	1458	(20)	1516	—	1516	(16)	1497	(11)	1539	(26)	1477	9	1476	28	1549	1598
55–64	(7)	1470	(11)	1483	(7)	1508	(10)	1475	(4)	—	(11)	1468	—	—	15	1514	1585
65+		a	(11)		(13)	1458	(8)	1491									1562[b]
Secular increase?	Yes		No		Youngest only		Youngest only		No		No		No		Probably yes		Yes

a Fewer than 7 persons. b Means for age group 65–74 years.

mons, the impact of conversion to Christianity appeared also to depend, to some degree, on whether religious activities were presided over by a European missionary, or by a native catechist. In the latter case, there was often little visible change in tribal custom, even though (nominal) Christianity might be widespread. *Medical care* included public health programs chiefly for eradication of malaria, and maternal and infant health services and aid stations, provided either by the government or by missions. The health clinics were usually staffed by native 'dressers' who have been trained in the recognition and treatment of common illnesses. The dressers were in turn in contact with more extensive government or mission-operated health facilities. With the exception of the malaria control programs, the services offered had to be actively sought by the patients. *Education* in these Solomon Islands tribal groups was provided almost entirely by missions. Participation by children in the educational experience depended on the attitude of parents. Most missions provided primary education of three or four years' duration, with heaviest emphasis on Bible study and basic reading skills. Higher education, when available at all, was limited to a small number of the more promising students who could obtain support from relatives or from mission sources.

The extent of use of *Western dietary items* depended on the availability of the foodstuffs and of cash with which to purchase them. The items purchased were chiefly tinned meat and fish, rice and flour. In the Nagovisi, where co-operatively owned stores were present, it was possible to keep records of the quantities of foodstuffs sold and the number of families supplied. The Nasioi purchased these items in Kieta town and trade stores, and the Lau, Ontong Java and Ulawa people from trading vessels. Estimation of the use of these items was thus chiefly dependent on the observations of the cultural anthropologists during their period of residence.

Entry into the *cash economy* varied in mode and degree among the different groups. Enterprises yielding cash flow were strongest where they were encouraged and coordinated by local leaders, as in the Nagovisi and to lesser degrees in the Lau and Ontong Java groups. Proximity to a town (Nasioi) also provided individuals with opportunities for employment, although continuous employment was traditionally seldom sought or desired. Some type of cash-generating activity was sought at least temporarily by some individuals in all groups.

Table 5.2 gives results of the acculturation analysis for six of the eight societies. Analysis of data on Ulawa and Ontong Java is still incomplete, although preliminary analysis suggests that both will fall (in order) between the Lau and Baegu. Habitat, subsistence, and salt intake are included in the Table for comparison but were not used in ranking acculturation.

No attempt has been made, to our knowledge, to assign acculturation rank in this way to any other set of population groups. The process is inevitably somewhat subjective, and the results are no more than semi-quantitative at

Table 5.2. *Indices of acculturation in six Solomon Islands Societies* (*Page et al., 1974*)

Habitat, subsistence and salt intake are included for reference, but were not used in ranking acculturation.

	Nasioi	Nagovisi	Lau	Baegu	Aita	Kwaio
Length of significant contact (yr)	85	40	50	25	6	25
Intensity of contact	+ + +	+ + +	+ +	+	+ +	+
Christianity	+ + +	+ + +	0	+	+ +	0
Medical care	+ +	+ +	+	+	+	0
Literacy/Education	+ + +	+ + +	+	+	0	0
Western diet	+ + +	+ + +	+ +	+	+	0
Cash economy	+ + +	+ + +	+ +	+	+	0
Secular trend in height	Yes	Youngest males only	Youngest only	Youngest females only	No	No
Demographic change under 15/over 45	4.4	2.9	2.6	2.4	7.9	1.7
Habitat	Coastal hills near town	Inland plain no town	Small islands coastal lagoon	Inland hills	Mountains	Inland hills
Subsistence	Agriculture	Agriculture	Fishing	Swidden/ Agriculture	Swidden	Swidden
Salt intake (mequiv./ 24 h)	50–130	50–130	150–230	10–30	10–30	< 20
Acculturation order	1	2	3	4	5	6
Acculturation degree	+ + +	+ + +	+ +	+	+	0

best. Once ordinal scales have been assigned the question arises as to how much weight should be given to each of the eight factors considered. Three different weighting systems were tested and scores calculated. These weightings proved to change rank of acculturation only among the three less acculturated groups. For practical purposes, the exact rank order seems less vital than the ability to distinguish clustering toward greater or lesser degrees of acculturation. Therefore, in further comparisons while data are presented separately, the Nasioi, Nagovisi and Lau (more acculturated) were compared against the Baegu, Aita and Kwaio (less acculturated). Except in general terms, Ulawa and Ontong Java are not yet included in these comparisons.

General health and nutrition

Average values for selected physical measurements and for hemoglobin levels, prevalence of splenomegaly and enteric parasites in the eight societies are given in Table 5.3 for males and in Table 5.4 for females. There is close resemblance among all groups in most of these measurements, although the

Lau and Ontong Java groups are higher in measurements reflecting body fat. Although protein intake is low, physical examinations failed to reveal evidence of malnutrition or protein starvation in any of the eight populations. Furthermore, hemoglobin levels were essentially the same in populations with high rates of splenomegaly (resulting from endemic malaria) as in those with low rates. Heavy infestation of enteric parasites was present in all groups tested. Unfortunately, tribal taboos prevented collection of stool samples among the Kwaio, and females of the Lau and Baegu populations.

No formal tests of physical fitness were performed. However, in the six large island populations large numbers of both males and females carried expedition equipment over kilometres of hilly terrain, and conveyed an impression of excellent physical strength and stamina.

Radiological survey

Radiological surveys revealed a variable number of pulmonary abnormalities among six of the Solomon Islands populations. The prevalence of findings consistent with active granulomatous disease (presumptive active tuberculosis), inactive granulomatous disease, and pulmonary fibrosis in these groups is given in Table 5.5. The highest prevalence of tuberculosis-like disease was found among the Lau and Baegu, closely followed by the Ontong Java and Ulawa populations. Much lower prevalences were noted among the Nagovisi and especially the Aita.

In order better to define the etiology of the pulmonary abnormalities, skin testing was carried out using antigens to *M. tuberculosis* (PPD–S), a number of atypical mycobacteria (PPD–B, PPD–G, PPD–Y, and PPD–platy) and histoplasmin. There were no positive reactors to histoplasmin in any of the populations, suggesting that histoplasmosis is not endemic in these Solomon Islands populations. Among the Lau and Ulawa populations, PPD–S elicited the largest number of positive reactions, making it likely that much of the pulmonary disease found in these populations is indeed tuberculosis. Significant numbers of positive reactors to PPD–S were also noted among the Baegu, Ontong Java, and to a lesser extent, the Nagovisi populations. The Aita, which are the most isolated of the six groups tested had a very low prevalence of radiographic evidence of pulmonary disease and of positive reactors to PPD–S. Among all six groups, a large proportion of the population were positive reactors to one or more of the antigens derived from atypical mycobacteria. PPD–G elicited the largest number of positive reactions among those groups. The distribution and patterns of reactions of positive skin tests and positive findings on chest radiography suggest the *M. tuberculosis* has been introduced into the Solomon Islands from exogenous sources. Moreover, our findings also suggest that a high rate of endogenous infection (most of it

153

Table 5.3. *Physical characteristics of*

Tribe ...	Nasioi		Nagovisi		Lau	
Island ...	Bougainville		Bougainville		Malaita	
Number ...	59		109		77	
	Mean	±s.e.	Mean	±s.e.	Mean	±s.e.
Height (mm)	1622	6.85	1596	4.67	1626	6.66
Weight (kg)	57.64	0.89	57.16	0.65	64.30	0.89
Ponderal index, ht/$^3\sqrt{wt}$[a]	12.72	0.05	12.56	0.04	12.31	0.05
Arm circumference (mm)	267	2.50	276	2.02	295	2.59
Skinfold-triceps (mm)	5.4	0.22	5.9	0.18	6.2	0.38
subscapular (mm)	8.3	0.27	9.9	0.30	9.3	0.46
Chest breadth (mm)	258	1.66	255	1.29	276	1.49
Chest depth (mm)	214	1.59	214	1.16	222	1.38
Chest depth/breadth (%)	83.0	0.67	84.0	0.48	80.3	0.54
Hemoglobin (g/100 ml)	14.4	0.20	13.8	0.19	14.3	0.21
Enlarged spleen (per cent of both sexes, ages 2–9 years)	7.5	—	4.1	—	6.0	—
Parasites, stool (per cent of stools examined, all ages)	35.0	—	87.0	—	94.0	—
No. of parasites/positive stool (all ages)	1.8	—	1.6	—	2.6	—

[a] For purposes of this measurement, inches and pounds were used.
[b] Includes 18- and 19-year olds.

Table 5.4. *Physical characteristics of*

Tribe ...	Nasioi		Nagovisi		Lau	
Island ...	Bougainville		Bougainville		Malaita	
Number ...	63		101		101	
	Mean	±s.e.	Mean	±s.e.	Mean	±s.e.
Height (mm)	1504	6.42	1503	4.60	1519	5.44
Weight (kg)	46.62	0.72	47.67	0.66	54.32	0.93
Ponderal index, ht/$^3\sqrt{wt}$[a]	12.64	0.06	12.60	0.05	12.17	0.06
Arm circumference (mm)	233	2.47	242	2.21	263	2.96
Skinfold-triceps (mm)	7.9	0.39	10.1	0.34	10.9	0.52
subscapular (mm)	9.8	0.49	12.6	0.53	12.7	0.78
Chest breadth (mm)	235	1.39	231	1.36	256	1.57
Chest depth (mm)	194	1.27	195	1.32	198	1.48
Chest depth/breadth (%)	82.6	0.56	84.2	0.50	77.4	0.48
Hemoglobin (g/100 ml)	13.3	0.18	11.6	0.18	12.6	0.13
Enlarged spleen (per cent of both sexes, ages 2–9 years)	7.5	—	4.1	—	6.0	—
Parasites, stool (per cent of stools examined, all ages)	83.0	—	86.0	—	—	—
No. of parasites/positive stool (all ages)	1.6	—	2.2	—	—	—

[a] For purposes of this measurement, inches and pounds were used.
[b] Includes 18- and 19-year olds.

eight societies (males 20 and over)

Baegu		Aita		Kwaio		Ulawa Small volcanic island		Ontong Java Coral atoll		USA[b]
Malaita 126		Bougainville 81		Malaita 127		106		145		3091
Mean	±s.e.	Mean	±s.e.	Mean	±s.e.	Mean	±s.e.	Mean	±s.e.	Mean
1613	4.16	1604	5.29	1604	4.77	1617	5.60	1642	4.61	1732
57.06	0.50	59.77	0.71	55.76	0.53	59.50	0.70	64.40	0.93	76.20
12.68	0.03	12.42	0.04	12.72	0.04	12.55	0.03	12.47	0.04	12.40
264	1.95	272	3.70	260	1.82	276	1.83	284	1.97	307
5.2	0.10	5.6	0.14	5.4	0.10	7.3	0.20	7.0	0.31	13.0
8.3	0.14	10.5	0.31	8.8	0.20	10.3	0.32	9.4	0.33	15.0
258	0.97	262	1.13	259	0.86	268	1.79	278	1.19	300
212	0.86	218	1.17	214	1.08	202	1.52	217	1.14	226
82.2	0.41	83.0	0.42	82.5	0.39	75.5	0.50	78.3	0.40	75.3
13.1	0.18	16.1	0.17	14.0	0.18	12.8	0.27	14.5	0.15	15.5
92.0	—	5.2	—	52.2	—	13.2	—	15.3	—	—
25.0	—	97.0	—	—	—	—	—	—	—	—
1.3	—	2.6	—	—	—	—	—	—	—	—

eight societies (females 20 and over)

Baegu		Aita		Kwaio		Ulawa Small volcanic island		Ontong Java Coral atoll		USA[b]
Malaita 111		Bougainville 88		Malaita 114		83		197		3581
Mean	±s.e.	Mean	±s.e.	Mean	±s.e.	Mean	±s.e.	Mean	±s.e.	Mean
1503	4.0	1507	4.85	1491	4.74	1506	5.28	1550	3.23	1600
47.17	0.46	52.45	0.76	46.11	0.58	48.80	0.87	58.50	0.66	64.41
12.60	0.04	12.21	0.06	12.61	0.04	12.55	0.06	12.12	0.04	12.15
228	1.77	260	2.22	228	208	248	2.74	282	2.29	284
8.4	0.22	10.5	0.36	7.9	3.34	11.9	0.56	17.3	0.43	22.0
10.2	0.32	13.2	0.57	11.8	0.61	17.9	1.05	21.4	0.63	18.0
237	0.96	245	1.37	233	1.16	240	1.69	260	0.98	—
195	1.0	205	1.10	199	1.05	181	1.62	204	1.02	—
82.5	0.48	83.8	0.46	83.5	0.36	75.5	0.50	78.3	0.30	—
11.6	0.13	13.5	0.16	12.7	0.18	11.1	0.26	12.5	0.11	13.7
92.0	—	5.2	—	52.2	—	13.2	—	15.3	—	—
—	—	89.0	—	—	—	—	—	—	—	—
—	—	2.5	—	—	—	—	—	—	—	—

Table 5.5. *Radiological findings in six Solomon Islands societies*

	Aita	Baegu	Lau	Nagovisi	Ontong Java	Ulawa
No. of subjects examined radiographically	467	485	438	493	635	410
Per cent of subjects with active granulomatous disease	0	1.2	1.8	0.4	0.6	1.2
Per cent of subjects with inactive granulomatous disease	0.4	16.5	16.2	5.9	11.3	13.2
Per cent of subjects with pulmonary fibrosis	1.1	4.9	3.9	0.4	3.1	2.0

subclinical) with atypical mycobacteria serves to provide some, but not complete, protection against infection with *M. tuberculosis* among the Solomon Islanders.

Ophthalmological survey

Ophthalmologically, interesting and significant findings included the fact that 95% of the examined eyes had visual acuity of 20/20 or better, and that significant refractive errors, particularly myopia and astigmatism, were virtually non-existent. Further, ophthalmoscopic examination of the retinal vessels demonstrated a remarkable absence of arteriosclerotic changes, although this is not a measure of systemic arteriosclerosis. In addition, the absence of elevated intraocular pressures strongly suggested the absence of chronic simple (open angle) glaucoma.

Cardiovascular disease

Public health records indicate that clinical coronary heart disease and atherosclerosis are very rare in the Solomon Islands. An analysis of electrocardiograms in six of the eight societies (Page, Damon & Moellering, 1974) revealed a striking absence of most abnormalities, and particularly of those abnormalities commonly associated with coronary heart disease in all six groups.

Analysis of cardiovascular 'risk factors' in six societies has recently been published (Page *et al.*, 1974). Fig. 5.3 shows serum cholesterol values by ages for males and females, in comparison with a United States population sample. The six Solomon Islands groups did not show the age-related trends

156

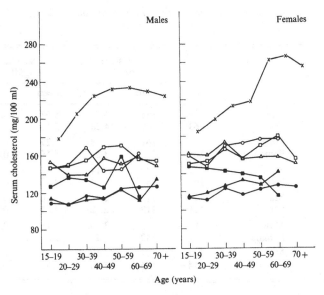

Fig. 5.3. Serum cholesterol levels by age for males and females in six Solomon Island societies. Values for a USA population sample are included for comparison. ×, USA; ○, Nasioi; □, Nagovisi, △, Lau; ● Baegu; ■, Aita; ▲, Kwaio. (From Page *et al.*, 1974, by permission of the American Heart Association, Inc.)

usually seen in Western societies. However, the three more acculturated groups, at almost all ages and in both sexes showed higher cholesterol levels than the three less acculturated groups. The differences could not be correlated with use of coconut, which although used by all, was not an important component of the diet in any of the six. It was concluded that the difference was best correlated with the use of canned meat and fish in these societies since the traditional diet is extremely low in fat.

Mean values for cholesterol and uric acid in six societies are shown in Table 5.6. In contrast to the findings with respect to cholesterol, uric acid values were higher in the three less acculturated societies than in the three more acculturated. High serum uric acid levels have been reported among Pacific Islands people, especially Polynesians (Prior, Rose & Davidson, 1964; Burch, O'Brien, Need & Kurland, 1966). Most previous studies have shown uric acid rising, together with cholesterol, as acculturation proceeds (Prior & Rose, 1966; Blahŏs & Reisenauer, 1956; Dreyfuss, Yoran & Balogh, 1965). The higher values found in the less acculturated Solomon Islanders is, therefore, surprising. Nevertheless, a similar result was reported by Jeremy & Rhodes (1971) in New Guinea. In their study, villagers living primarily on sweet potato showed higher serum uric acid values due to the habitually low salt intake of the villagers. They showed that addition of salt to the

157

Lot B. Page and others

Table 5.6. *Means and standard errors for cholesterol and uric acid for six societies*

		N	Cholesterol mg/dl		Uric acid mg/dl	
			Mean	±S.E.	Mean	±S.E.
Nasioi	Male	59	155	4.56	5.97	0.12
	Female	63	165	4.28	5.16	0.11
Nagovisi	Male	109	161	3.38	5.27	0.08
	Female	101	162	3.52	4.56	0.09
Lau	Male	77	149	3.14	5.47	0.11
	Female	101	160	3.10	4.83	0.10
Baegu	Male	126	115	2.41	6.16	0.14
	Female	111	119	2.72	5.81	0.13
Aita	Male	81	135	3.30	6.11	0.12
	Female	88	142	3.24	5.89	0.11
Kwaio	Male	127	114	2.59	6.47	0.15
	Female	114	125	3.03	5.70	0.13

diet produced a fall in uric acid levels. A similar explanation is likely for the Solomon Islanders, whose diet was very similar to the New Guinea group.

Salt intake

Because of the suspected relationship of salt intake to hypertension, data on habitual salt intake was recorded in each of the societies studied. Information was obtained from several sources. Observations on dietary habits, including salt use were recorded over an 18–24-month period by the ethnographers. Interviews based on recall were conducted. Times urine samples proved impossible to obtain. Urine samples were, therefore, obtained for determination of sodium, potassium and creatinine.

Based on all data available, average daily salt intake was 50–130 mequiv. in the Nasioi and Nagovisi, 150–250 mequiv. in the Lau, 10–30 mequiv. in the Aita and Baegu, and less than 20 mequiv. in the Kwaio (see Table 5.2).

Blood pressure

Blood pressure measurements were made on all subjects of 15 years and older by a single observer. Two or more independent readings were made, using a mercury manometer, and with the subject seated comfortably on a bench. Results were analyzed using the mean of the systolic and of diastolic (phase IV) values. A full analysis of these data has been published elsewhere (Page

158

Table 5.7. *Coefficients of correlation for age, body weight and blood pressure*[a]

	Nasioi	Nagovisi	Lau	Baegu	Aita	Kwaio
Males						
N ...	59	109	77	126	81	127
Age and weight	—	−0.18	−0.29*	−0.36*	—	−0.40*
Age and systolic blood pressure	—	—	—	—	−0.18	—
Age and diastolic blood pressure	—	−0.19	—	−0.24*	−0.29*	−0.20
Weight and height	0.67*	0.60*	0.58*	0.61*	0.63*	0.58*
Weight and systolic blood pressure	—	—	—	—	—	0.24*
Weight and diastolic blood pressure	—	0.22	—	—	—	0.27*
Females						
N ...	63	101	101	111	88	114
Age and weight	−0.40*	−0.26*	−0.38*	−0.49*	−0.48*	−0.45*
Age and systolic blood pressure	0.36*	0.30*	0.36*	—	—	—
Age and diastolic blood pressure	—	—	—	—	—	−0.16
Weight and height	0.64*	0.61*	0.55*	0.50*	0.46*	0.58*
Weight and systolic blood pressure	—	—	—	—	—	—
Weight and diastolic blood pressure	—	0.26*	0.30*	—	—	—

[a] Only those coefficients of correlation which reach statistical significance ($P < 0.05$) are included in the table. An asterisk denotes $P < 0.01$.

et al., 1974). The statistically significant correlations of blood pressure on age for six societies are given in Table 5.7. Most interesting was the finding that in females of the three more acculturated groups, systolic blood pressure increased significantly with age. No such age-related rise was found in females of the three less acculturated groups, nor was it seen among men.

Diastolic blood pressure was found to decrease with age in men of the three less acculturated societies and also in one of the more acculturated, the Nagovisi.

In all groups, weight was found to decline with age, a tendency that was especially strong among females. In view of this, the rising systolic blood pressure seen among more acculturated women clearly cannot be attributed to weight gain. Such a discrepancy between weight and blood pressure trends with age is rarely if ever seen in advanced societies.

Mean values for systolic and diastolic blood pressure in six societies are shown for males in Fig. 5.4 and for females in Fig. 5.5. Highest systolic and diastolic blood pressures for both sexes and nearly all ages were found in the Lau. Blood pressures greater than 140 mm Hg systolic and 90 mm Hg

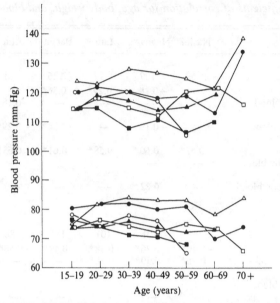

Fig. 5.4. Mean values for systolic and diastolic blood pressure by age in adult males of six Solomon Islands societies. ○, Nasioi; □, Nagovisi; △, Lau; ●, Baegu; ■, Aita; ▲, Kwaio. (From Page *et al.*, 1974, by permission of the American Heart Association, Inc.)

diastolic were found in 7.8% of Lau men and 9.9% of Lau women. Pressures in this range were rare in the other societies.

Whereas blood pressure tends to rise with age in populations in all industrialized societies (Epstein & Eckoff, 1967), a number of non-industrialized people have been described from various parts of the world in which no age-related increase occurs (Page, 1976). The Baegu, Aita and Kwaio appear to be further examples of this phenomenon. Especially provocative is the finding of an age-related rise in blood pressure among the more acculturated Nasioi, Nagovisi and Lau, occurring only among the females. This suggests that the factor or factors which induce such a trend during acculturation must play a role at a very early stage in the acculturation process. The selective effect on women has not been previously commented upon, but is noted in data from several other partially acculturated populations (Abbie & Schroder, 1960; Maddocks, 1967; Prior *et al.*, 1968; Glanville & Geerdink, 1972).

Consideration of the various factors which might or might not influence the blood pressure trends noted in these Solomon Islanders has been discussed elsewhere (Page *et al.*, 1974). Many factors, commonly present in acculturating societies were wholly absent in the Solomon Islands groups and could be effectively eliminated from consideration. The factor which correlates best with blood pressure changes in these peoples is salt intake. This was

160

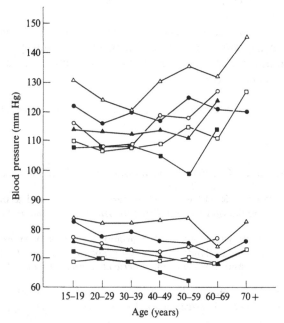

Fig. 5.5. Mean values for systolic and diastolic blood pressure by age in adult females of six Solomon Islands societies. ○, Nasioi; □, Nagovisi; △, Lau; ●, Baegu; ■, Aita; ▲, Kwaio. (From Page *et al.*, 1974, by permission of the American Heart Association, Inc.)

substantially greater in all the more acculturated groups, and especially in the Lau, where the long-established practice of boiling vegetables in sea water results in extremely high salt intake, correlating with substantially higher blood pressures than in any of the other populations.

Preliminary analysis of the data from Ulawa and Ontong Java shows no tendency of blood pressure to rise with age. Furthermore, in spite of their proximity to the ocean, peoples of these small islands appear to have a very low intake of salt.

The hypothesis that salt may induce elevated blood pressure in genetically susceptible individuals is not new. Evidence from both animal experiments (Dahl, 1972) and epidemiologic studies (Page, 1976) give strong support to such an hypothesis. Data from the Solomon Islands populations give further support, and show, in addition, that other acculturating influences need not be present.

Future of the Harvard Solomon Islands project

The Harvard Solomon Islands project has been unusually broad in scope. Many of the data are still being processed, and will be published in the next

161

Lot B. Page and others

several years. The unique scope of this project reflects the unusual talent and training of its late director, Albert Damon.

The eight societies which have been documented are already yielding in various degrees to the influence of the dominant Western European culture. Longitudinal studies to document the effects of this cultural impact will be pursued over the coming decade.

The work was supported by Grant no. GM-1342. National Institutes of Health.

References

Abbie, A. L. & Schroder, J. (1960). Blood pressure in Arnhem Land Aborigines. *Medical Journal of Australia*, 2, 493–6.

Blahŏs, J. & Reisenauer, I. R. (1956). Levels of serum uric acid and serum cholesterol in various population groups in Ethiopia. *American Journal of the Medical Sciences*, 250, 308–14.

Burch, T. A., O'Brien, W. M., Need, R. & Kurland, L. T. (1966). Hyperuricemia and gout in the Mariana Islands. *Annals of Rheumatic Diseases*, 25, 114–16.

Dahl, L. K. (1972). Salt and hypertension. *American Journal of Clinical Nutrition*, 25, 231–44.

Dreyfuss, F., Yoran, E. & Balogh, M. (1965). Blood uric acid levels in various ethnic groups in Israel. *American Journal of the Medical Sciences*, 247, 438–44.

Epstein, R. H. & Eckoff, R. D. (1967). The epidemiology of high blood pressure–geographic distributions and etiological factors. In *The Epidemiology of Hypertension*, ed. J., Stamler, R. Stamler & T. N. Pullman, pp. 155–66. New York: Grune & Stratton.

Friedlaender, J. S. (1975). *Patterns of Biologic Variation: The Demography, Genetics, and Phenetics of Bougainville Islanders*. Cambridge, Mass.: Harvard University Press.

Giles, E., Wyber, S. & Walsh, R. J. (1970). Microevolution in New Guinea: additional evidence for genetic drift. *Archaeology and Physical Anthropology in Oceania*, 5, 60–72.

Glanville, E. V. & Geerdink, R. A. (1972). Blood pressure of Amerindians from Surinam. *American Journal of Physical Anthropology*, 37, 251–4.

Hrdy, D. (1973). Quantitative hair form variation in seven populations. *American Journal of Physical Anthropology*, 39, 7–18.

Jeremy, R. & Rhodes, F. A. (1971). Studies of serum urate levels in New Guineans living in different environments. *Medical Journal of Australia*, 1, 897–9.

Keesing, R. M. (1967). Christians and pagans in Kwaio, Malaita. *Journal of the Polynesian Society*, 76, 82–100.

Littlewood, R. A. (1972). *Physical Anthropology of the Eastern Highlands of New Guinea*. Seattle & London: University of Washington Press.

Maddocks, I. (1967). Blood pressure in Melanesians. *Medical Journal of Australia*, 1, 1123–6.

Maranda, E. (1970). Les femmes dans l'espace socialisé. *Journal de la Societé des Oceanistes*, 26, 155–62.

Maranda, P. & Maranda, E. K. (1970). Le crâne et l'utérus: deux theorêmes nordmalaitain. In *Exchange et Communications*, vol. II, ed. P. Maranda, & J. M. Pouillon, pp. 829–61. Paris: The Hague.

162

Mitchell, D. P. (1971). Gardening for money; land and agriculture in Nagovisi. Ph.D. thesis, Harvard University, Cambridge, Mass.

Mitchell, J. N. (1971). Aspects of matriliny in Nagovisi society. Ph.D. thesis, Harvard University, Cambridge, Mass.

Ogan, E. (1972). Business and cargo: socio-economic change among Nasioi. *New Guinea Research Unit Bulletin*, **44**.

Oliver, D. (1973). *Bougainville, A Personal History*. University of Hawaii Press; Honolulu.

Page, L. B. (1976). Epidemiologic evidence on the etiology of human hypertension and its possible prevention. *American Heart Journal*, **91**, 527–34.

Page, L. B., Damon, A. & Moellering, R. C., Jr (1974). Antecedents of cardiovascular disease in six Solomon Islands societies. *Circulation*, **49**, 1132–46.

Prior, I. A. M., Grimley-Evans, J., Harvey, H. P. B., Davidson, F. & Lindsey, M. (1968). Sodium intake and blood pressure in two Polynesian societies. *New England Journal of Medicine*, **279**, 515–20.

Prior, I. A. M. & Rose, B. S. (1966). Hyperuricemia, gout, and diabetic abnormalities in Polynesian people. *Lancet*, **1**, 333–8.

Prior, I. A. M., Rose, B. S. & Davidson, F. (1964). Metabolic maladies in New Zealand Maoris. *British Medical Journal*, **1**, 1064–9.

Ross, H. (1973). *Baegu's Social and Ecological Organization in Malaita, Solomon Islands*. Illinois Monographs in Anthropology no. 8. Urbana and Chicago: University of Illinois Press.

Tobias, P. V. (1962). On the increasing stature of the Bushmen. *Anthropos*, **57**, 801–10.

6. The Tokelau Island migrant study

I. A. M. PRIOR, ANTONY HOOPER,
JUDITH W. HUNTSMAN, J. M. STANHOPE &
CLARE E. SALMOND

A severe hurricane in the Tokelau Islands in January 1966 focussed the attention of the New Zealand Government on this New Zealand dependency made up of three atolls (Fakaofo, Nukunonu and Atafu), lying some 480 km north of Western Samoa. At that time the total population was 1950 and there were a further 500 Tokelauans in New Zealand. An expanding population and extremely limited opportunity for individual and community development led to the establishment of the Tokelau Resettlement Programme, aiming to resettle 1000 in New Zealand over a period of five or six years.

Since 1966, a major multi-disciplinary study has been developed to examine hypotheses relating to the way in which the physical, social and cultural changes associated with the process of migration and adaptation can influence health and disease. The study emphasises cardiovascular disease and related abnormalities such as diabetes, obesity, hyperuricaemia and gout.

The study population in 1974 comprises a total of 1600 adults and children who are still living in the Tokelau Islands, and the 2000 adults and children who are now living in New Zealand.

The fact that the majority of those migrating to New Zealand were examined in the Tokelaus prior to migration, and in many cases before they had made the decision to migrate, allows the pre-migration characteristics of the migrants to be defined and compared with those who have chosen to remain in the Tokelaus. This is a particular advantage of the study.

The continuing comparison over a period of years of Tokelauans remaining in the Tokelaus with those in the very different physical, social and cultural environment of urban New Zealand is a key part of the study, enabling measurements to be made of the extent, rate and direction of change in a number of variables, both in individuals and in the groups in the two environments. The study would have considerable merit as a long-term epidemiological exercise even if we looked only at the medical variables such as blood pressure, weight, changes in coronary risk factors, development of diabetes and overall morbidity and mortality related to the process of migration. We believe, however, that a much greater contribution will come from the efforts that are being made to measure changes occurring in the social and ideological areas of the individual migrant's life in New Zealand and to test to what extent stressful situations may be contributing to changes in the medical variables.

The dynamics of these changes will be examined, looking for differential prevalences of disorders such as hypertension related to varying degrees of stress. The aim is to seek specific clues to the dynamic relationship between migration, stress and disease.

The present report deals with the general background of the study, the research design and strategy, the hypotheses to be tested, the comparisons to be made, the results from the baseline studies of the isolated populations, the characteristics of pre-migrants compared with non-migrants, and the initial results of the examination of the Tokelauans in New Zealand after varying periods in their new environment.

The Wellington Hospital Medical Unit, which in 1970 became the Wellington Hospital Epidemiology Unit, had been involved since 1962 in epidemiological studies among samples of New Zealand Maoris and Europeans. In 1964 surveys were carried out in the Cook Islands, in Avarua the main town on the island of Rarotonga, and on Pukapuka an atoll in the Northern Cooks only 460 km from the Tokelaus. The results of these studies showed the New Zealand Maoris to be at high risk from a range of disorders including hypertension, coronary heart disease, obesity, diabetes, hyperuricaemia and clinical gout.

New Zealand mortality statistics had also revealed a higher mortality rate for these disorders in Maoris than in Europeans (Rose, 1960).

In 1963 a health survey in the Tokelaus had been sponsored by the New Zealand Government to assess health needs and advise on treatment programmes. Epidemiological data relating to blood pressure, height and weight in adults, and nutritional status and presence of infections in infants and children were obtained on the population. Blood pressures in the Tokelauans were notably lower than in the Rarotongans and New Zealand Maoris but somewhat higher than in the Pukapukans. Tokelauans were heavier than the Pukapukans with weights nearer to those of the New Zealand Maoris.

It was against this background that the opportunities presented by the Tokelau migrant study were viewed. Would the very major shift in physical environment and dietary habits, social and living patterns produce changes in the direction of the New Zealand Maoris? Would they occur to a sufficient extent and to a sufficient degree to be measured over a reasonable period of time? Though the total study population was probably too small to provide sufficient cases of coronary heart disease, manifest by myocardial infarction or sudden death, to make valid analyses of mortality, would the process of migration result in changes in coronary risk factors and blood pressure levels sufficient to allow specific hypotheses about morbidity to be tested?

The good liaison developed with the medical staff on each of the home islands, Fakaofo, Atafu and Nukunonu, gave the opportunity for collection of data relating to pregnancies, births, deaths, and hospital admissions which would allow comparison with similar data being collected in New Zealand.

The Tokelau Islands

The three Tokelau atolls lie in the central Pacific area bounded by the parallels of 8° and 10° S and the meridians of 171° and 173° W. Fakaofo is about 480 km north of Western Samoa, separated from the more northerly islands of Nukunonu and Atafu by distances of 65 and 160 km, respectively. The people are Polynesian, with a distinct and unitary cultural tradition. The language has its closest affinities with those spoken in the neighbouring Polynesian groups.

The total land area of the group is 1043 ha (Atafu 205 ha, Fakaofo 243 ha, Nukunonu 547 ha) and is made up of numerous low-lying islets varying in size from some with an area of only a few square metres to one which is 200 m wide and some 6.5 km in length. None of the islets rise to more than 5 m above sea level. The vegetation consists of forest, dominated by coconut palms, with edible pandanus and a number of valuable timber trees. The mean daily temperature is 29.4 °C and there is an average annual rainfall of 287 mm which is spread throughout the year, with the fall from April through September approximately half of that for the remaining months.

There are four significant food crops, of which coconut is by far the most important. Breadfruit is grown extensively in the village areas and provide three good crops each year. Numerous edible forms of pandanus grow on all the islets and are cultivated to some extent. Pulaka (*Cytosperma chamissonis*) is the only other significant food crop and is cultivated on both Fakaofo and Atafu in artificial humus pits dug down to the freshwater lens. The main sources of protein are fish, pigs, and chickens, the last two introduced since European contact.

Although the islands lie in close proximity to one another, their discovery by Europeans was spread over the period from 1765 to 1835. The first detailed account of the people was made by members of the US Exploring Expedition, who visited both Atafu and Fakaofo briefly in 1841. At that time the population of the two islands was estimated to be between 600 and 700, and Fakaofo gave the appearance of dense settlement, in spite of a shortage of fresh water and a meagre subsistence economy based entirely on coconut, pandanus and fish. In the early 1850s, after an extended period of famine caused by drought, several hundred people were taken from Fakaofo by a Roman Catholic mission ship to the island of Uvea, where they were given both sustenance and religious instruction, but the islands remained under pagan control until 1863. In that year the group's relative isolation from the outside world was finally broken, with disastrous results. The ship which brought Protestant mission teachers to Fakaofo also carried dysentery, and 64 out of that island's population of 260 died in the subsequent epidemic. Shortly afterwards, all three islands were raided by ships engaged in the Peruvian slave trade, who reduced the total population of the group to about

167

200 people, most of whom were women and children. Only three of the men taken by the slavers ever returned to their home islands.

The following years saw the establishment of some half dozen foreign traders in the atolls (three of whom were of Portuguese or mixed Portuguese–African descent), who brought with them about 10 workmen from various other Polynesian groups. All of these immigrants had children by local women and their descendants are now numerous. The demographic history of the islands since 1863 is one of steady growth – very rapid until the early years of this century and then more gradual as developing contacts allowed emigration to Samoa. The islands became British Protectorates in 1889 and were annexed by Great Britain in 1916 and included in the Gilbert and Ellice Islands Colony. New Zealand assumed responsibility for the administration of the group in 1925, and the islands were included within the boundaries of New Zealand in 1948. Since then, Tokelauans have been citizens with unrestricted right of entry into New Zealand.

During the following 14 years, however, most Tokelau emigration continued to be to Western Samoa, where the migrants formed a self-conscious ethnic minority of over 500 during the late 1950s and early 1960s. Then in January 1962 Western Samoa became an independent state and the Tokelauans there were placed at a severe disadvantage as aliens in a situation of intense competition for paid employment. This stimulated the migration of Tokelauans from both the atolls and Western Samoa to New Zealand.

From the outset, this migration has been encouraged and to a large extent sponsored by the New Zealand Government. Between 1963 and 1965, 70 single men and women were assisted to migrate from the atolls, and the programme gained impetus after the destructive hurricane of 1966, becoming known as the Tokelau Islands Resettlement Scheme. Since 1967, whole families have been sponsored in preference to single persons, and the total number of Government-sponsored migrants up to the end of March 1974 was 456. The assisted migrants have sponsored others to come from both the home atolls and from Western Samoa, and the resulting pattern of chain migration has meant a rapid increase of the migrant population. In late 1973 the register of the Tokelau Island Migrant Study recorded some 2000 Tokelauans and part-Tokelauans in New Zealand, and a population of approximately 1600 in the atolls. In addition there are known to be over 100 Tokelauans in Western Samoa and American Samoa, and smaller numbers established in Hawaii and the west coast of the United States.

Conceptual basis of the study

Owing to their relative isolation, Tokelauans have preserved an autonomous and distinctively Polynesian culture to a far greater extent than other societies of the South Pacific which have been more accessible to commercial and colonial influences. The kinship system, patterns of land tenure, social and economic organisation and the demographic history of the islands have already been analysed and described (Hooper, 1970a, b; Huntsman, 1971; Hooper & Huntsman, 1973) and further articles are in press.

Even in 1967, however, when the two social anthropologists began their field studies, there were signs that Tokelau culture and ideology were beginning to change under the influence of an increasing flow of information and money from the emigrants established overseas. Since 1971, the anthropological research has been concentrated on the pattern of socio-cultural changes in the atolls and the adaptation of the migrants to the monetised urban milieu of New Zealand. As well as attempting to answer such basic questions as 'who migrates, and why?', 'what happens to the migrants?' and 'what effect does migration have on the home society?', the research has been designed to test a hypothesis about the social and psychic stresses involved in migration and rapid social and cultural change (Hooper & Huntsman, 1974). Briefly, this hypothesis states that those individuals who experience a disjunction between their personal values and ideology on the one hand, and those governing the social system in which they find themselves on the other, will experience greater stress than those who do not have this experience. The effects of stress and other aspects of migration will be examined by means of the following subhypotheses.

(1) The process of migration from the Tokelaus to New Zealand will be associated with a changing pattern of disease, leading to increased morbidity and mortality in migrants as compared to those persons who remain on their home islands. It is also anticipated that the pattern will approach that shown by New Zealand Maoris and Europeans.

(2) Those individuals who experience a conflict between their personal values and ideology and those governing the social system and interaction patterns in which they find themselves, will experience a greater elevation of blood pressure and greater morbidity than those who do not experience such conflict.

(3) A higher rate of obesity and diabetes will develop in New Zealand migrants, possibly as the result of the changes in type of food and availability of money, as compared with those remaining on their home islands.

(4) The pattern of coronary risk factors, including development of higher levels of cholesterol, triglycerides, blood pressure, and smoking rates, will change at a faster rate among those in New Zealand than in those remaining on their home islands.

(5) High-cholesterol dairy products and sugar intake in New Zealand will probably be associated with the development of higher cholesterol and triglyceride levels in those in New Zealand than in those in the Tokelau Islands.

169

The long-term plan of the study includes comparisons of characteristics of the following groups:

(1) Within individuals, persons examined before migration on their home islands and at various times after their migration to New Zealand.
(2) Pre-migrants and non-migrants.
(3) Migrants in New Zealand and non-migrants still in the islands.
(4) Tokelau migrants and New Zealand Maoris.
(5) Tokelau migrants and New Zealand Europeans.
(6) Migrants living in areas where other Tokelauans live and migrants living in relative isolation from other Tokelauans.
(7) Migrants who have spent significant periods in Samoa (primarily in the port town of Apia) before coming to New Zealand and migrants coming directly from the Tokelaus.

Development of the study: 1966–73

The study has evolved considerably since the first young Tokelauan male and female migrants were examined on arrival in New Zealand in 1966. In 1967 two of the authors (A. H. and J. W. H.) commenced sociological and ethnographic studies in the islands, gaining a firm basic knowledge of the way of life on the three islands. In 1968 the medical team visited the group in order to obtain base-line medical data. The logistics were such that detailed examinations were carried out only on Fakaofo. Limited data were collected on Atafu and Nukunonu.

In 1970, the medical and anthropological workers were brought together and their studies integrated with the help of the Division of Research in Epidemiology and Communications Science of the World Health Organisation.

Sociological and ideological questionnaires were developed for administration to all those aged 15 years and over, on the home islands and in New Zealand, in order to test hypotheses relating to the process of adaptation to migration. The team revisited the group in 1971 for eight weeks to undertake detailed examinations of the populations on Atafu and Nukunonu and a limited second examination of those on Fakaofo. The behavioural scientists remained in the atolls for nine months in 1971 administering the questionnaires and collecting data on the economic resources and organisation of the different *kaiga* (land-holding units).

In 1972 and 1973 medical and socio-anthropological examinations of all those aged 15 years and over living in New Zealand were carried out. Infants and children up to 14 years of age in New Zealand have been examined medically.

Genealogical data back to eight or nine generations ago have been transferred to computer cards to allow genetic factors to be related to a range of

170

Table 6.1. *Combined Tokelau Atolls adult population, 1968*

Age (years)	Males				Females				
	Pre-migrant	Non-migrant	Not ex-amined	Total	Non-pregnant				
					Pre-migrant	Non-migrant	Pregnant	Not ex-amined	Total
15–19	41	31	2	74	38	48	0	2	88
20–24	9	18	0	27	14	28	3	1	46
25–34	21	40	2	63	19	57	24	6	106
35–44	32	47	6	86[a]	22	65	14	3	104
45–54	31	36	7	74	25	41	0	1	67
55–64	18	23	2	43	19	36	0	1	56
65–74	11	21	1	33	9	36	0	0	45
75 and over	1	9	0	10	0	26	0	0	26
Total	164	225	20	410	146	337	41	14	538

[a] Includes one examined but migrant status undetermined.

physiological and biochemical variables, particularly risk factors for coronary heart disease, hypertension and diabetes.

In 1974 the medical team continued to examine new migrants soon after arrival in New Zealand and in this way has renewed contact with them.

Study populations

A census of the islands was made in May 1968 and the combined population was 1792: 948 aged 15 years and over and 844 under 15 years. The combined population by age and sex is set out in Table 6.1. Of the 34 aged 15 years and over who were not seen, 20 were attending a church celebration in Western Samoa, 7 were visiting relatives out of the Tokelaus, 1 was in hospital in Western Samoa, and 4 were on one of the other two atolls, being missed by the survey team. The reason why the remaining two were not seen is uncertain. This gives an effective response rate of 99.3% (914/920).

Seven of the 34 not seen subsequently migrated to New Zealand and have been examined, and 27 remained in the Tokelaus and were examined during the second visit in 1971.

In 1971 when the team returned, the total population had declined to 1640 made up of 872 aged 15 and over and 768 below 15 years.

In 1973 a further census was carried out and there was a total of 1567 made up of 649 on Fakaofo, 572 on Atafu and 366 on Nukunonu.

The total numbers in New Zealand have increased from around 500 in

171

1966 to 1920 in 1974. This population is made up of 1042 aged 15 and over, and 878 below 15 years.

A definitive demographic study of the growth of the Tokelau population in New Zealand is currently being undertaken by A. H. and J. W. H.

Methods

The Tokelau register

All subjects are given a discrete identity number including a designation of their island of origin. The register now includes all those on their home islands and all those in New Zealand.

Official notification of migration is received through the Island Affairs Department of all persons coming to New Zealand under the Resettlement Programme. Local medical officers on the three islands also examine persons leaving for New Zealand, and a copy of their report is sent to the Unit. A system has also been developed whereby the medical officers send a statement of those arriving and those leaving the islands, together with births and deaths, every six months.

Contact is maintained with Tokelau communities in New Zealand through liaison staff. Changes of address and other demographic data are recorded. A register of all those dying in the Tokelaus and in New Zealand is kept.

Anthropological methods

The research methods were designed to elucidate both the social and ideological dimensions of change, among the migrants as well as those remaining on their home islands. During 1971, detailed studies were made of households and the descent groupings which form the basic land-holding and economic units of Tokelauan society. Those at the total population of the atolls aged 15 years and above were interviewed in Tokelauan for between one and two hours per person, using two separate interview schedules. The first of these covered the subjects of educational attainment, economic resources, social position, achieved status and complete residential and migration history. The second schedule was designed to measure each person's degree of commitment to the key values of Tokelauan culture – cooperation, sharing, obedience to elders and the obligations to the natal family.

During 1972 and early 1973 two comparable schedules were administered to all Tokelauans in New Zealand who were aged 15 years and above.

Biomedical methods

The methods followed in these studies have been detailed in a previous publication (Prior, Stanhope, Evans & Salmond, 1974) and are summarised

172

here. Essentially, they follow the World Health Organisation recommendations. Since 1968, blood pressures have been taken using a zero muddler (Evans & Prior, 1970) and the fourth phase has been used in the analysis of the diastolic pressures. Hypertension is defined by accepted WHO criteria, classed as definite if the systolic pressure is equal to or greater than 160 mm Hg. or diastolic pressure is equal to or greater than 95 mm Hg. Cases not meeting the criterion for definite hypertension are classified as borderline if the systolic pressure is equal to or greater than 140 mm Hg or if the diastolic pressure is equal to or greater than 90 mm Hg. Anthropometric examination has included height, weight, arm circumference, triceps and subscapular skinfolds. Harpenden skinfold calipers have been used. The medical history has been taken by trained Tokelauan-speaking interviewers on a precoded field sheet. A physician's assessment of angina was made using WHO criteria (WHO, 1959). The London School of Hygiene and Tropical Medicine chest pain questionnaire has been used together with the British Medical Research Council cough and sputum questionnaire (Rose & Blackburn, 1968).

Frequency and duration of migration from and back to the Tokelaus has been documented as well as information concerning previous medical history and symptoms relating to joint disorders. The medical examination includes auscultation of the heart and lungs, examination of the nervous system and recording of the presence or absence of arcus senilis, thyroid enlargement, varicose veins and skin problems.

Since 1968, subjects aged 20 and over have reported fasting and have had their electrocardiograph (ECG) recorded prior to a 100 g glucose load, with blood sampling one hour later. This has allowed estimation of plasma glucose, serum cholesterol, triglycerides, uric acid and urea. In Fakaofo in 1968 plasma proteins were measured and lipoprotein electrophoresis was carried out. The ECG tracings were classified by the Minnesota code.

The maintenance of the laboratory in the Tokelaus involved taking petrol generators to supply power for the laboratory equipment and ECG machines. Since 1973 a Beckman glucose analyser has been used so that the one-hour plasma glucose levels can be estimated in the field, and an immediate assessment made of the diabetic status of individuals. Persons classed as definite diabetics had one-hour plasma glucose levels of 250 mg/dl or more, or were known diabetics being treated. Persons classed as probable diabetics had one-hour plasma glucose levels of 200–249 mg/dl.

Since 1962 the Unit has taken part in the lipid standardisation programme based on the Communicable Disease Center of the National Institutes of Health, Atlanta, Georgia, USA. The performance of the laboratory has been very good. In addition, pooled sera and commercial standards are used in the Unit's quality control programme. Urinary electrolytes have been carried out by flame photometry using commercial standards as a means of quality control. Immunoglobulins have been estimated in sera from Tokelau infants and children in the Tokelaus and in New Zealand, and anti-HL-A anti-bodies have also been measured in a subsample of Tokelauans in New Zealand.

Special whole-blood smears were prepared for examination for micro-filaria in the 1968 Tokelau survey and these have been examined by Dr Tin Maung Maung, WHO filaria consultant in Western Samoa.

Results – medical

These are dealt with in five sections, (*a*) comparison of findings in pre-migrants and non-migrants based on the 1968 examinations, (*b*) baseline data for the pooled population on the three islands in 1968, (*c*) prevalence data on clinical conditions from detailed medical examinations in 1968 and 1971, (*d*) a preliminary report of findings in migrants living in New Zealand who were examined in 1972 and 1973, and (*e*) aspects of the health of infants and children in the Tokelaus and in New Zealand.

Pre-migrants and non-migrants: 1968 status

Table 6.1. includes details by age, sex and pregnancy status of those classed as pre-migrants and non-migrants, combining data for the three atolls.

Pre-migrants and non-migrants among males and non-pregnant females aged 15–64 years have been compared by three methods: directly, within standard age groups; across age groups using *z*-transformations; and across age groups using matched pairs. Comparisons within age groups highlighted very few statistically significant differences.

Standardising for age by employing sex-specific *z*-transformations within age groups, mean scores for pre-migrants and non-migrants of both sexes were obtained for height, weight, casual urinary sodium and potassium concentration, and serum cholesterol, triglycerides and uric acid levels. No important pre-migrant/non-migrant differences are discernible.

An analysis of those pre-migrant and non-migrant individuals who could be matched for sex and year of age (266 pairs) allowed comparisons to be statistically examined across the total group of pairs. This overcame the problem of small samples within age groups. It also enabled comparisons across age groups to be made for blood pressures, where possible trends with age in the two migrant groups made pooled *z*-scores inapplicable. By pairing, we found that both systolic and diastolic pressures were statistically higher in younger male pre-migrants than in their non-migrant peers, and that this was reversed in the older males. Twenty-four-hour urinary sodium concentrations in males were lower in pre-migrants than non-migrants.

Tokelau baseline data: 1968 survey

Comparison of the findings between the three islands shows small inconsistent statistical differences in some variables and the results have therefore been pooled on the assumption that they represent one population. Again, analysis is limited to males and non-pregnant females aged 15–64 years.

The basic physiological, anthropometric and biochemical data are presented in some detail as a means of characterising the Tokelauan population on their home islands in 1968 (Table 6.2). These data will be used for comparisons with other Polynesian groups and with the patterns that develop in Tokelauan migrants in New Zealand.

Blood pressures, both systolic and diastolic, rose slowly in males and females, the female levels exceeding the male levels from age 35–44 on. Age-standardised rates of definite and borderline hypertension per 1000 are set out in Table 6.3, and can be compared with those in the New Zealand Maori, Carterton European, Rarotongan and Pukapukan groups studied by the Unit. The Pukapukans have the lowest rates while the Tokelauans are next and there is then a considerable gap to the New Zealand Maoris and Europeans. The Rarotongan females have the highest rates.

There is a tendency for weight to increase with age in both sexes to age 45–54 and then decline (Table 6.2). The extent of fatness, particularly in the females, can be seen from the skinfold results. There is a steady increase with age in both sexes, the peak for triceps and subscapular being reached in the 55–64 years age group in the males and in the 45–54 years age group for the females.

Serum cholesterol levels have mean values of 184.5 mg/dl and 198.2 mg/dl in young adult males and females respectively, and show some increase with age, reaching a peak of 220.4 in males aged 45–54 years and 245.4 in females aged 55–64. These levels are considerably higher than those recorded in the atoll-dwelling Pukapukans in the Northern Cook Islands, while the males have lower and the females higher levels than those recorded in New Zealand Maori samples, and those in turn were lower than those recorded in the New Zealand European study.

The triglyceride levels are the lowest recorded among Polynesian groups studied by the Unit, and show an increase with age in both sexes, from 0.35 mmole/l to 0.68 mmole/l in males, and from 0.42 mmole/l to 0.64 mmole/l in females.

Serum uric acid peak levels of 7.65 mg/dl for males and 6.71 mg/dl for females are found in the 55–64 years age class. These findings confirm the presence of hyperuricaemia with levels similar to those in New Zealand Maori samples. Male levels exceed female by 0.66 mg/dl to 1.21 mg/dl at various ages. Five cases of clinical gout were found in men on Fakaofo.

Table 6.2. *Tokelau 1968: physiological, anthropometric and biochemical data*

Variable	Sex		Age in years					
			15–19	20–24	25–34	35–44	45–54	55–64
Systolic blood	M	mean	115.1	121.5	122.5	126.1	131.0	134.4
pressure		±S.E.	1.6	2.4	2.2	1.5	3.4	3.8
(mm Hg)		N	69	26	57	76	64	35
	F	mean	121.2	115.7	118.7	129.1	141.2	145.3
		±S.E.	1.4	1.8	1.5	2.0	3.2	3.4
		N	81	41	74	86	63	52
Diastolic blood	M	mean	65.7	68.1	71.1	76.7	77.8	78.4
		±S.E.	1.5	2.3	1.6	1.1	2.2	2.5
		N	69	26	57	76	64	35
	F	mean	74.6	75.3	74.8	79.3	85.5	84.7
		±S.E.	1.2	1.2	1.2	1.5	1.8	2.1
		N	81	41	74	86	63	52
Height (cm)	M	mean	165.0	167.4	170.8	169.2	168.4	166.7
		±S.E.	0.6	0.8	0.7	0.7	0.8	1.0
		N	72	26	60	79	63	38
	F	mean	160.4	161.0	159.7	159.8	158.0	159.4
		±S.E.	0.6	0.8	0.6	0.5	0.8	0.5
		N	85	41	71	87	65	55
Weight (kg)	M	mean	60.4	69.7	76.1	77.9	78.9	74.1
		S.E.	0.9	1.6	1.2	1.5	1.4	1.7
		N	73	26	60	79	63	38
	F	mean	66.0	70.6	71.8	77.7	78.1	74.2
		S.E.	1.1	1.8	1.5	1.3	1.8	1.7
		N	86	41	71	87	65	55
Triceps skin-	M	median	9.2	8.2	9.1	10.0	12.8	13.0
folds (mm)		N	32	5	22	27	25	13
	F	median	19.5	21.5	25.8	27.2	29.5	27.5
		N	35	13	30	32	31	25
Subscapular	M	median	9.4	13.5	12.1	14.5	17.5	18.5
skinfolds (mm)		N	32	5	22	27	25	11
	F	median	24.5	23.5	31.2	36.5	36.7	36.5
		N	35	13	30	32	31	25
Cholesterol	M	mean	184.5	195.9	209.5	215.7	220.4	216.9
(mg/dl)		±S.E.	6.4	12.0	6.2	6.1	5.5	4.9
		N	40	13	35	48	45	25
	F	mean	198.2	176.1	213.3	222.5	220.6	245.4
		±S.E.	4.1	4.0	8.1	6.0	5.1	7.2
		N	43	18	43	53	50	38
Triglycerides	M	mean	0.35	0.44	0.53	0.68	0.62	0.63
(mmole/l)		±S.E.	0.03	0.07	0.05	0.05	0.05	0.07
		N	38	12	33	47	43	24
	F	mean	0.42	0.44	0.47	0.57	0.57	0.64
		±S.E.	0.03	0.06	0.04	0.04	0.04	0.05
		N	40	19	44	51	50	34

Table 6.2. *cont.*

Variable	Sex		Age in years					
			15–19	20–24	25–34	35–44	45–54	55–64
Serum uric	M	mean	6.95	6.52	7.05	6.96	7.09	7.65
acid (mg/dl)		±S.E.	0.16	0.45	0.17	0.15	0.20	0.30
		N	40	13	33	48	54	23
	F	mean	5.99	5.78	5.84	6.18	6.43	6.71
		±S.E.	0.13	0.27	0.18	0.16	0.43	0.19
		N	40	19	42	49	49	37
Urine volume	M	mean	733	1037	1075	1176	937	927
(ml/day)		±S.E.	104	151	70	92	81	60
		N	10	12	32	42	31	16
	F	mean	705	607	694	966	789	752
		±S.E.	76	85	63	79	71	67
		N	8	19	37	40	40	29
Urine sodium	M	mean	30.1	39.0	51.5	42.3	45.2	34.6
(mequiv./day)		±S.E.	6.9	7.6	7.7	5.4	6.7	5.7
		N	9	11	31	40	30	17
	F	mean	28.6	33.6	32.4	44.0	32.2	28.5
		±S.E.	2.8	5.7	4.1	4.6	3.6	4.1
		N	8	18	35	39	39	28
Urine potassium	M	mean	66.4	72.3	98.1	84.8	77.0	67.6
(mequiv./day)		±S.E.	14.4	11.5	9.3	6.4	11.6	5.4
		N	9	11	30	40	30	17
	F	mean	55.0	52.8	49.2	56.9	56.5	46.2
		±S.E.	10.6	18.0	3.5	5.5	5.8	4.6
		N	8	6.6	35	39	39	28
Urine uric acid	M	mean	548	601	539	511	411	353
(mg/day)		±S.E.	120	55	49	42	47	89
		N	5	11	28	37	29	13
	F	mean	361	349	324	398	301	270
		±S.E.	45	109	29	38	27	50
		N	8	17	34	33	34	26
Early morning	M	mean	38.3	27.5	33.5	42.3	51.0	37.8
urinary sodium		±S.E.	4.4	7.9	4.3	3.5	6.9	3.9
(mequiv./l)		N	49	11	42	56	46	29
	F	mean	47.9	44.0	42.9	51.1	43.0	30.4
		±S.E.	4.8	7.4	4.6	4.8	5.8	4.2
		N	53	25	54	64	44	37
1-hr plasma	M	mean	—	131.6	132.9	138.6	151.1	157.4
glucose (mg/dl)		±.S.E.	—	3.4	2.0	1.1	1.7	1.8
		N	—	5	22	27	24	13
	F	mean	—	135.8	153.4	158.4	161.6	182.7
		±S.E.	—	2.8	1.4	1.9	1.5	3.0
		N	—	13	29	32	30	24

M = male; F = female.

Table 6.3. *Blood pressure categories age-standardised rates per 1000*

	Males		Females	
	Borderline hypertension	Definite hypertension	Borderline hypertension	Definite hypertension
New Zealand Maoris	243	288	202	318
New Zealand Europeans	242	266	209	225
Rarotongans	242	343	232	366
Pukapukans	74	47	76	82
Tokelauans 1968	102	80	152	151

Twenty-four-hour urine samples were obtained from a total of 143 males and 173 females in the 15–64 years age group and the combined results from the three islands are presented in Table 6.2 for volume, sodium, potassium, and uric acid.

The volumes and sodium outputs were higher in males than females, the peak sodium outputs being 51.5 mequiv. in males aged 25–34 and 44.0 mequiv. in females aged 35–44.

The potassium outputs were generally lower in females than in males despite their overall lower sodium output.

Early morning urinary sodium concentrations were determined for almost all participants. Female levels were higher than male and no age trend was apparent.

Plasma glucose levels were determined one hour after a 100 g glucose load. Female levels were consistently higher than male levels and rose with age.

Tokelauan prevalence data: 1968 and 1971

Differences between 1968 and 1971 clinical, dietary and tobacco data are minor. Therefore these data are pooled for the whole island group. Alcohol consumption data are also presented.

The prevalence of diabetes was found to be high (Table 6.4) in both sexes but showed no consistent rise with age after 35 years. The rates of definite diabetes, for age 35–74, were 8.3% in males and 12.7% in females, higher than rates seen in New Zealand Europeans but not as high as in Maoris studied by the same screening method (18.6% males and 23.0% females) and standardised to the pooled age distribution. An association between body mass as measured by Quetelet's Index and age-standardised diabetes prevalence was apparent (Table 6.5, 1968 data only). In Tokelauans and Maoris, both male and female, diabetes rates were two to six times higher in the highest tertiles of body mass than in the lowest.

Table 6.4. *Tokelau Islands: prevalence rates for 1968 and 1971*

	Age in years						
	15–19	20–24	25–34	35–44	45–54	55–64	65–74
Diabetes[a] × 10³	—	0	67	74	224	79	222
Male							
Female	—	40	85	193	250	176	104
COLD[b] × 10³							
Male	0	0	0	14	0	125	108
Female	0	19	14	45	15	38	113
Gout × 10³							
Male	0	0	0	0	59	25	216
(no females)							

[a] Diabetes includes definite, probable and known categories.
[b] COLD: Chronic obstructive lung disease, including asthma and bronchitis.

Table 6.5. *Age-standardised prevalence (%) of diabetes (definite)[a] by quetelet index (QI) Maori and Tokelauan males and non-pregnant females ages 35–74*

	QI tertiles	A ≤2.68	2.69–3.08	B ≥3.09	Ratio B:A 1
Maoris	Males	4.6	21.7	29.6	6.4
	Females	12.0	26.3	29.6	2.5
Tokelauans	Males	5.7	5.0	27.7	4.9
	Females	3.2	9.4	20.8	6.5

[a] Plasma glucose at 1 h 250 mg/dl and greater.

Angina on efforts, as elicited by the standard questionnaire, occurred in 1.8 % (7/389) of males and 3.7 % (18/493) of females. There was poor agreement with the physician's (which we consider to be the more reliable standard) assessment of angina (1968 data only), with only four of nine questionnaire positive cases being confirmed by physician's assessment and four of five physician's assessment positives being also positive on the questionnaire. The six discrepant cases were all female. Chest pain of skeletal origin was thought to account for some of the unconfirmed questionnaire positives.

When compared with New Zealand European and Maori males, Tokelau males showed no significant differences in prevalence of angina by physician's assessment, but Tokelau females had less angina than expected, and Maori females more. The risks for the females were 0.24 Tokelau, 0.90 European and 1.43 Maori, relative to the pooled samples.

Previous myocardial infarction (defined by ECG codes 1.1, 1.2 and 1.3) was detected in two males and one female.

Residual evidence of stroke was present in two males and two females.

Asthma was uncommon, affecting two males and six females. Chronic bronchitis was present in nine males and eleven females. Pooled asthma–bronchitis rates are presented in Table 6.4.

Gout was common in males (Table 6.4) there being 13 cases, all aged 45 or over. No cases were seen in females. Among males, the rate when age standardised, while high by comparison with Europeans, was only 40% of that of the Maoris.

Diet

The diet in the Tokelaus is a traditional one of breadfruit, taro, pulaka, fish and coconut, with chicken and pork added on special occasions. Small amounts of flour, rice and refined sugar are used. The outstanding dietary finding is the very high saturated fat intake, contributing 56% of calories. Coconut supplies three quarters of this saturated fat, whereas cholesterol intake is very low. Dairy products are rarely consumed. In 1968 the average daily consumption of coconuts was four to five per head (Wodzicki, 1972). An analysis of pork fat has shown a very high content of short chain saturated fatty acids, as was previously found in specimens obtained in Pukapuka (Shorland, Czochanska & Prior, 1969). In Pukapuka the overall calorie intake is lower, while fat intake is moderate, contributing around 37% of calories and again, 80% of these are from coconut. The daily cholesterol intake is also low, around 100 mg.

The Tokelau diet can be contrasted also with the New Zealand Maori diet, where around 44% of calories are from fat with high meat and dairy consumption and high cholesterol intake.

Tobacco

Most islanders use a type of Derby plug which is shredded from a block and then rolled in a pandanus leaf. Men with a regular income smoke either hand-rolled cigarettes made from imported tobacco or manufactured cigarettes. The percentage of men who smoked was 62%, and varied little after the age of 20. Only 26% of women smoked. Most smokers had less than five cigarette equivalents per day.

Alcohol

No females were current drinkers of alcohol. On Fakaofo, 21% of males aged 15–74 years classed themselves as current drinkers, but in the other two atolls a significantly higher proportion of men were current drinkers, 33% on Atafu and 38% on Nukunonu. The age-standardised risks of being a current drinker were, relative to the pooled population, 0.75 for Fakaofo, 1.16 for Atafu and 1.29 for Nukunonu. The most likely explanation for this

180

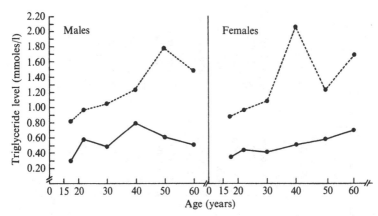

Fig. 6.1. 1968 Tokelau Island pre-migrant (●—●) and 1972–73 migrant (● - - - ●) mean triglyceride levels, by age and sex.

finding is that the importation of beer and spirits into the islands was still increasing between 1968 and 1971, following the removal of an official ban on alcohol importation in 1966 and the introduction of home brewing of coconut liquor in 1964. There was no significant difference in drinking patterns between the wholly Protestant island Atafu and the wholly Catholic island Nukunonu.

Examination of Tokelau migrants: 1972–73

Only limited reference can be made to post-migration characteristics as they are still being analysed and related to the results of the ideological and socio-logical questionnaires administered in New Zealand over the same period. The total migrant group contains some subjects who were not examined prior to arrival in New Zealand because they left the atolls for New Zealand or Samoa before 1968.

The mean blood pressures, both systolic and diastolic, of migrants were not significantly higher than those observed in the atolls. The mean weights of migrants were also not significantly higher than those of atoll dwellers. Serum cholesterol and uric acid and one-hour plasma glucose levels showed no consistent change.

Serum triglyceride levels were significantly higher in migrants than in the Tokelau 1968 subjects, at all ages in both sexes (Fig. 6.1).

Early morning urine sodium concentrations were three to four times higher in the migrants, ranging from 131.5 mequiv. in the 15–19-year to 83.3 mequiv. in the 55–64-year males, and 114.0 mequiv. in the 15–19-year to 79.9 mequiv. in the 65–74-year females. The levels in the atoll subjects were in the 27–51 mequiv. range (Table 6.2).

Table 6.6. *Tokelaun children aged under five years*

N	Fakaofo 90	Atafu 104	Nukunonu 46	Auckland 120	Hutt Valley 58	Porirua 147	Probability that Porirua is not different from pooled others	Probability that Auckland+Hutt is not different from pooled atolls
Ear infection only	0	2	2	6	4	32	$\chi_1^2 = 46.904$, $p < 0.0001$	$\chi_1^2 = 1.918$, $p = 0.1661$
Chest infection only	0	0	1	2	2	11	$\chi_1^2 = 13.420$, $p = 0.002$	Fisher's test, $p = 0.1070$
Chest+ear infection	0	0	0	0	0	7	Fisher's test, $p = 0.0001$	No cases in either
Asthma	0	0	0	2	5	21	$\chi_1^2 = 34.092$, $p < 0.0001$	Fisher's test, $p = 0.0024$
Pyoderma	0	2	0	5	2	10	$\chi_1^2 = 8.736$, $p = 0.0031$	Fisher's test, $p = 0.0637$
Eczema	0	0	0	9	3	16	$\chi_1^2 = 13.175$, $p = 0.0003$	Fisher's test, $p < 0.001$
Scabies	3	15	3	22	7	48	$\chi_1^2 = 0.060$, $p = 0.8065$	$\chi_1^2 = 30.538$, $p < 0.0001$

Tokelau children

Major health differences have emerged between atoll and migrant children aged under five years (Table 6.6), when children seen in the 1971 island survey are compared with migrants examined in New Zealand 1972–73. Tokelauan children resident in Porirua East, an outer lower-socio-economic suburb of Wellington, had high prevalences of ear and chest infections, asthma, purulent skin lesions and eczema.

Tokelauan children in middle-socio-economic areas in Auckland and the well-established Hutt Valley suburban area of Wellington had intermediate rates of asthma and eczema, and did not differ significantly from atoll children in ear, chest and skin infections.

A major change in infant feeding occurred, with the proportion of children breast-fed for more than three months falling from 99 % in the atolls to 49 % in New Zealand, and the proportion of children never breast-fed rising from virtually zero (1/240) to 30 % in New Zealand (90/297). Of the 90 artificially fed children, 19 had eczema, while 26 had asthma. The prevalences of eczema and asthma were both significantly related to lack of breast feeding, and the asthma relationship was stronger than the eczema relationship (testing differences of contingency coefficients, $p = 0.016$).

There was no significant difference in anaemia prevalence, defined as the proportion of children having haemoglobin levels below 10 g/dl between atoll (11 %) and New Zealand (7 %) groups.

Results – anthropological

So far, only the data from the 1971 interviews in the atolls have been available for analysis. The responses to the items in the 'values' schedule have been tested by factor analysis, confirming the validity of the four factors *a priori*. The pattern of responses to the 'Values' items taken by sex, age group and island has confirmed many of the impressions which the anthropologists had previously formed on the basis of more qualitative field research, and the pattern of inter-island differences in values responses corresponds neatly with subtle differences of social, political and economic organisation in the three atolls. A paper has been prepared for publication on this topic. In general, the degree of commitment to Tokelaun values is not associated with either age or achieved status, as was expected, but is highly associated with income from employment and experience outside the atolls. The association of any of these factors with the medical data is being undertaken.

Discussion

An important advantage of the present study is the opportunity to examine all Tokelauans on their home islands and in New Zealand, thereby encompassing almost the entire Tokelauan population apart from a small group living in Samoa and in other parts of the world. The people who remain on their home islands are regarded as a control group in whom changes are expected to occur at a much slower rate than in migrants to New Zealand.

We have compared the characteristics of those who were to migrate following the 1968 examination with those who were to remain on the islands. The only differences to reach statistical significance were systolic and diastolic blood pressures in males. The younger male pre-migrants had higher systolic and diastolic blood pressures than their non-migrant peers. This trend is reversed in the older males. This could suggest a selective process. The younger males are known to initiate migration. Their higher blood pressures possibly indicate either a greater degree of prior Westernisation and/or personality differences from the non-migrants. The females and the older males who migrate tend to do so in response to the decision made by the younger males, and can be regarded as passive migrants not differing socially or biologically from non-migrants.

The anthropometric, physiological, biochemical and clinical documentation of the island population is a necessary step in the Tokelau Island Migrant Study in order to characterise the baseline from which migrants move in to the New Zealand environment.

The Tokelauans have higher blood pressures than the Pukapukans (also atoll dwellers), in the Northern Cook Islands, but notably lower levels than the New Zealand Maoris. The weight patterns and cholesterol levels are similar to the New Zealand Maoris, but triglycerides are much lower. The Tokelauan diet pattern is very different from the New Zealand Maoris with 56% of calories coming from fat, of which most is derived from coconut, as compared to 44% of calories from fat in the New Zealand Maoris with a high cholesterol and dairy product intake (Prior, 1970). The cholesterol intake in the Tokelauans, at less than 100 mg, is very low. The sodium intake is very much lower as measured by the casual urinary sodium concentration and the 24-hour outputs, and has been shown to be changing considerably in their new environment (Prior, 1970; Prior & Evans, 1970).

Tokelauans share, with other Polynesian groups, high serum uric acid levels and a high male rate of clinical gout (Evans, Prior & Cooke 1969; Evans, Prior & Morrison, 1969; Prior, Rose, Harvey & Davidson, 1966). Hyperuricaemia may contribute to the development of hypertension. Gouty men have higher blood pressure than non-gouty. There is also a significant correlation of body weight and body mass with uric acid levels in the New Zealand and Cook Island Maoris (Evans, Prior & Harvey, 1968).

The initial analysis of the New Zealand 1972–73 examinations indicate that some changes are occurring in the New Zealand migrants. The casual urinary sodium concentrations are higher, indicating a much freer use of salt. The lack of significant change of the serum cholesterols despite the considerable changes in diet pattern with a much greater use of eggs and dairy products is of considerable interest. More detailed interpretation will be possible when more New Zealand dietary data are available. The changes in serum triglycerides are considerable and may be the result of a higher carbohydrate intake and refined sugar use. The levels are moving in the direction of those found in the New Zealand Maoris.

Migrant children under five years of age show a higher morbidity rate in the New Zealand environment. The higher rates of asthma, allergic and pyogenic skin disorders and ear infections are in contrast to the relative freedom from those disorders in infants and young children in the home islands. Analyses of blood pressure, sodium output and coronary risk factors in children are being undertaken and should allow comparison between migrant and atoll groups.

The file of genealogies will allow estimates to be made of the genetic contribution to blood pressure, weight, cholesterol, triglycerides and uric acid, as well as indicating the effect of different environments on members of families.

The continuing updating of demographic data in the Tokelaus and in New Zealand relating to pregnancies, births, stillbirths and deaths will allow life tables to be calculated to compare the migrant and the non-migrant populations.

It is planned to carry out a further examination of Tokelauan migrants in New Zealand in 1975–76, three years after their last examination. Both medical and sociological data will be collected. In 1976 a return visit is to be made to the Tokelau Islands when the total population on the three atolls are to be again examined. This will be five years after the last survey.

Conclusion

The acute change experienced by Tokelau Islanders in moving from their traditional environment and complex society to urban New Zealand offers favourable opportunities to test hypotheses relating to the process of adaptation and the way in which health and disease may be influenced by stresses and changes accompanying migration involving social, biological, medical and genetic factors. Anthropological insights may be gained into small populations and how they cope with living in a new environment. The effects on the home atoll communities of losing emigrants should also be evident.

The authors wish to acknowledge the support and guidance received from the Tokelau Project Committee of the South Pacific Research Committee of the Medical Research Council of New Zealand, the Medical Research Council of New Zealand itself, the Cardiovascular Disease Unit and Division of Research in Epidemiology and Communications Science of the World Health Organization, the Wellington Hospital Board, and the Tokelau communities in New Zealand and in Tokelau.

I. A. M. Prior and others

References

Evans, J. G. & Prior, I. A. M. (1970). Experience with the random-zero sphygmomanometer. *British Journal of Preventive and Social Medicine*, **24**, 10–15.

Evans, J. G., Prior, I. A. M. & Cooke, N. J. (1969). The Carterton study. 4. Serum cholesterol levels of a sample of New Zealand European adults. *New Zealand Medical Journal*, **70**, 346–50.

Evans, J. G., Prior, I. A. M. & Harvey, H. P. B. (1968). Relation of serum uric acid to body bulk, haemoglobin and alcohol intake in two South Pacific Polynesian populations. *Annals of Rheumatic Diseases*, **27**, 319–25.

Evans, J. G., Prior, I. A. M. & Morrison, R. B. I. (1969). The Carterton study. 3. Blood pressure of a sample of New Zealand European adults. *New Zealand Medical Journal*, **69**, 146–52.

Hooper, A. B. (1970a). Land tenure in the Tokelau Islands. In *Proceedings of the South Pacific Commission Symposium on Land Tenure*. Noumea, New Caledonia: South Pacific Commission.

Hooper, A. B. (1970b). Socio-economic organisation of the Tokelau Islands. In *Proceedings of the 8th International Congress of Anthropological and Ethnological Sciences, Tokyo*, pp. 238–40.

Hooper, A. & Huntsman, J. (1973). A demographic history of the Tokelau Islands. *Journal of the Polynesian Society*, **82**, 366–411.

Hooper, A. & Huntsman, J. (1974). The Tokelau Island migrant study: behavioural studies. In *Migration and Related Social and Health Problems in New Zealand and the Pacific*, ed. J. M. Stanhope & J. S. Dodge, pp. 97–102. Wellington: Epidemiology Unit, Wellington Hospital.

Huntsman, J. (1971). Concepts of kinship and categories of kinsmen in the Tokelau Islands. *Journal of the Polynesian Society*, **80**, 317–54.

Prior, I. A. M. (1970). Population studies in New Zealand and the South Pacific. *World Health Organization Report on Cardiovascular Epidemiology in the Pacific*, ed. Z. Fejfar, pp. 28–42. Geneva: WHO.

Prior, I. A. M. & Evans, J. G. (1970). Current developments in the Pacific in atherosclerosis. *Proceedings of the 2nd International Symposium on Atherosclerosis*, ed. R. J. Jones, pp. 335–42. New York, Heidelberg, Berlin: Springer-Verlag.

Prior, I. A. M., Rose, B. S., Harvey, H. P. B. & Davidson, F. (1966). Hyperuricaemia, gout and diabetic abnormality in Polynesian people. *Lancet*, **i**, 333–8.

Prior, I. A. M., Stanhope, J. M. Evans, J. G., Salmond, C. E. (1974). The Tokelau Island migrant study. *International Journal of Epidemiology*, **3**, 225–32.

Rose, R. F. (1960). *Maori European Standards of Health*. Department of Health Special Report Series No. 1. Wellington, New Zealand.

Rose, G. A. & Blackburn, H. (1968). *Cardiovascular Survey Methods*, pp. 172–78. Geneva: WHO

Shorland, F. B., Czochanska, Z. & Prior, I. A. M. (1969). Studies on fatty acid composition of adipose tissue and blood lipids of Polynesians. *American Journal of Clinical Nutrition*, **22**, 594–605.

Wodzicki, K. (1972). The Tokelau Islands: men and introduced animals in an atoll ecosystem. *South Pacific Bulletin*, **22**, 37–41.

World Health Organization (1959). Hypertension and coronary heart disease: classification and criteria for epidemiological studies. *Technical Report Series*, **168**, 15. Geneva.

7. Long-term biological effects of human migration from the African savanna to the equatorial forest: a case study of human adaptation to a hot and wet climate

JEAN HIERNAUX

The African rain forest covers two areas divided by an extension of the savanna down to the west coast in Dahomey, Togo and eastern Ghana: west of this interruption, a smaller area of forest runs along the coast to Sierra Leone (the Guinea rain forest); east of it, lies a much larger area which covers most of the Congo basin. The latter is the domain of the equatorial forest (see Fig. 7.1).

As yet, no human fossil remains have been discovered in this area; owing to the nature of the soil, it is doubtful if any ever will be found, except possibly in caves. The earliest archaeological remains from the equatorial forest are very late compared with other parts of Africa; it would seem that it is not until about 20000 years ago that men were living permanently within the forest (Clark, 1970). It seems, therefore, that the first wave of migration into the equatorial forest consisted of communities of hunter–gatherers of *Homo sapiens*. They could have come only from the surrounding forest–savanna mosaic, where they had originated if not in a drier savanna then at least further away from the equator.

Equatorial climate is characterised by heavy rainfall and nearly constant high temperature and humidity throughout the year. The more one moves away from the equatorial climatic zone into the zone of tropical climate, the more two seasons are marked: the hotter rainy season and the cooler dry season. Mean annual temperatures are higher in the northern tropical zone than in the forest, and also than in the southern tropical zone which experiences a strong cooling influence from the oceans.

It is evident then that human migration into the equatorial forest entailed exposure to a change of climate, i.e. towards moister and more uniform climatic conditions. Moreover, air movement is much reduced under the dense canopy of the evergreen forest. Also reduced there, by comparison with the conditions in the savanna, is the level of incident ultraviolet radiation.

Fig. 7.1. Localisation of the populations analysed. Pygmies and Pygmoids in italics; equatorial forest shaded.

The IBP HA project synthesised here was designed to investigate the question of how man adapts biologically to the change of biome when he migrates from the savanna into the African equatorial forest.

Sociological and historical background

Today, two socio-economic classes of populations inhabit the equatorial forest: a class of populations living, at least predominantly, on hunting and gathering, and a class of populations living predominantly on agriculture, with fishing as another important source of subsistance for many of the latter. The populations of the second class cover all the forest, while those of the first class live in more localised areas. All the hunter–gatherer groups are regarded as vassals by the agricultural populations with which they co-habit. Vassals and suzerains are associated in a symbiotic relationship, the vassals exchanging game and forest products for agricultural produce and iron artifacts (and pottery in most groups).

188

In the anthropological literature, the populations of the first class are called Pygmies (when their mean male adult stature is lower than 150 cm, which is the case only for the Mbuti of the Ituri forest) or Pygmoids (when their stature is greater). Mean stature does not exceed 160 cm in any Pygmoid population, but some populations of agriculturalists are shorter statured than some Pygmoid ones: primarily, the Pygmoids are defined in socio-economic, not biological, terms (Hiernaux, 1974).

The history of the migration of the agricultural populations into the equatorial region has been fairly well reconstructed. Nearly all of them speak a Bantu language. This linguistic class now encompasses approximately 700 languages spoken by some 70 million people spread over nearly one-half of subsaharan Africa. Nevertheless, it merely forms part of the Benue–Congo branch of Niger–Congo languages; it has no higher rank than the other members of this branch – the individual Bantoid languages spoken in eastern Nigeria and central Cameroon. This implies that the speakers of Bantu languages must have expanded very rapidly and recently to have achieved such a wide geographic dispersion along with such a small degree of linguistic divergence (Greenberg, 1963).

All linguists are now agreed in locating the ultimate source of Bantu origins north of the rain forest. According to Guthrie (1962), a first small band of Bantu speakers migrated across the forest and founded a secondary Bantu cradle in the moist savanna which lies to the south of it, probably in the present Luba territory. From there, Bantu expansion proceeded by stages, of which only the latest occupied the equatorial forest. A recent analysis suggests that while this may be true for some Bantu linguistic classes, some other groups of Bantu speakers more probably came directly from the north to occupy their present territory in the forest (Henrici, 1973).

The date of the occupation of the forest by the Bantu agriculturalists is not precisely known. If Henrici is right, it may differ with the linguistic class. However, according to linguistic data, it may not be older than the first millenium B.C.; Guthrie's views put it well into the Christian era.

There is every reason to believe that the Pygmies and Pygmoids inhabited the forest before the arrival of the Bantu agriculturalists, whose language they adopted. By their way of life based on hunting and gathering, they are the direct descendants of the Stone Age people who settled in the forest some 20000 years ago. They now use iron, but still do not practise metallurgy. Some of them may well be the descendants of the first Stone Age wave of migrants, but there is no reason for postulating a common ancestral forest population for all of them. It might be that several waves of hunter–gatherers penetrated the immense equatorial forest at different times, and in different areas. If this has been the case, it offers an opportunity to study adaptive convergence: the resulting populations will have evolved independently under similar selective pressures released by the common change of environment. They will have tended to resemble each other more and more in those hereditary features which are adaptive to the rain forest, whereas that part of the genome which is not involved in this adaptation will not manifest such a convergence.

If all Pygmy and Pygmoid populations have a common, forest-adapted ancestry, possible factors causing differences are gene admixture and random processes like genetic drift, founder effect, and bottleneck effect. If not, differences in the savanna-adapted ancestral populations and number of generations since migration into the forest are additional possible factors.

For identifying the right historical schema as well as for studying the possible adaptive convergence, we need data on a number of Pgymy and Pygmoid populations from different parts of the forest. We also need data on the neighbouring agriculturalist populations so that we can investigate possible gene admixture and have reference populations who have recently arrived in the forest.

In this connection, in 1965 when the present study was programmed, the only pairs of hunter versus agriculturalist populations on which a fair body of data was available were the Mbuti versus Bira pair from the Ituri forest in northeastern Zaïre, and the Kuba Cwa versus Bushong pair from the southern forest boundary in the Kasai region of Zaïre (if the Twa of Rwanda and Burundi, who live in part in the open country and in part in the mountain forest, are excluded because of their peculiar environment). It seemed necessary, therefore, to add one more pair of populations to the record, from a different area of the equatorial forest. The two castes, Twa and Oto, of the Konda ethnic group were selected for this purpose. They live side-by-side, east of Lake Tumba, in the heart of the forest near to the equator, far away from the other studied pairs.

As a reference, data on a population of Bantu agriculturalists who lived outside the forest were needed. The already studied Bushong suited this need. They live in the belt of moist savanna where Guthrie locates the secondary heartland of Bantu expansion, not far from the territory of the Luba–Kasai from whom they differ only moderately (Hiernaux, 1966*a*).

Moreover, it was necessary to collect comparable data on an African population descended from a stock, also ancestral to the Bantu but living further away from the equator than the Bushong, in a tropical environment more contrasted with the rain forest. The Sara, who live in the savanna of southern Chad, were selected for this purpose.

Previous state of the problem

In subsaharan Africa, many anthropometric characters are associated with the biome; they present a cline of values from drier to moister zones. Table 7.1 demonstrates such clines in average stature and nasal, facial and

Table 7.1. *Means of four characters in the African arid zone, savanna, and wet forest (excluding the Pygmies), and in the Mbuti Pygmies (from Hiernaux, 1966b)*

	Stature (cm)	Nasal index	Facial index	Cephalic index
Arid zone	171	85.0	89.7	74.8
Savanna	169	85.9	85.8	75.3
Wet forest	164	94.0	81.9	76.1
Mbuti Pygmies	144	103.8	78.3	77.0

cephalic indices, from the arid zone through the savanna to the wet forest. With increasing moisture of the biome, the stature becomes shorter, and the nose, face and head relatively broader. Each of the four clines is extended by the Mbuti value.

Table 7.2 reproduces some of the total or partial correlation coefficients between anthropobiological attributes and six climatic variables in subsaharan Africa, computed by Hiernaux (1968): those between a number of population means or frequencies and three climatic variables (mean annual rainfall, mean humidity mixing ratio of the driest month, and mean daily maximum temperature of the hottest month).

Some variables show no correlation with climate: this is the case for the frequency of I^A, I^O and Hb^S, fingerprints, and the mean of weight, head length, total face height, and nose height. Stature tends to increase with maximum ambient temperature and dryness. When stature is held constant, trunk length tends to increase with air moisture (and accordingly, lower limb length tends to increase with dryness). Shoulder breadth increases, and hip breadth decreases, with air moisture; neither is significantly correlated with ambient temperature. Head, face and nose breadths increase with air moisture and decrease with air temperature. In other words, a higher stature with relatively longer legs, narrower head, face and nose, narrower shoulders and wider hips are associated with a drier climate with higher peaks of heat and of dryness; the inverse morphological tendencies are associated with a moister climate with lower peaks of heat and dryness. The arid zone and the equatorial forest stand at the opposite ends of this climatic scale, to which a cline of human morphology is associated.

How is this association explained? Biological adaptation to climate is the answer generally given. As reviewed by Hiernaux (1974), the physiological advantages of the more common phenotype in the climate in which it prevails have been propounded or verified for body size and linearity, and head and nose shape. It has, however, been argued that the concentration of populations with a tall and slender physique in the arid zone, and of populations with a short and lateral body build in the wet forest, may result merely

191

Jean Hiernaux

Table 7.2. *Correlation between anthropological means or frequencies and three climatic variables in subsaharan Africa (from Hiernaux, 1968)*

Anthropobiological variable(s)	No. of populations	Correlation with		
		Rainfall	Hum. −	Temp. +
Anthropometry				
Stature	312	−0.26**	−0.35**	+0.45**
Weight	63	−0.09	−0.07	+0.19
Weight · stature	63	+0.28*	+0.15	−0.11
Sitting height · stature	87	+0.23*	+0.39*	+0.12
Biacromial	106	+0.55**	+0.07	+0.02
Bicristal	62	−0.35**	+0.01	+0.21
Head length	190	−0.04	+0.04	+0.01
Head breadth	190	+0.24**	+0.28**	−0.33**
Total face height	86	−0.05	+0.06	−0.20
Bizygomatic	160	+0.41**	+0.31**	−0·21*
Nose height	105	−0.16	−0.15	−0.11
Nose breadth	179	+0.49**	+0.10	−0.33**
ABO				
I^A	184	−0.10	−0·03	−0·08
I^B	184	0.00	−0.09	+0.24**
I^O	184	+0.07	+0.07	−0.09
Haemoblobin S (where present)				
Hb^S	169	+0.05	+0.11	+0.06
Dermatoglyphics				
Arches	54	+0.03	+0.02	+0.08
Loops	54	+0.08	+0.03	+0.01
Whorls	54	−0.10	+0.03	−0.04

In the case of partial correlation, the variable held constant is the second one (first column).
* *r* significant at the 0.05 level; ** *r* significant at the 0.01 level.
Rainfall = mean annual rainfall.
Hum. − = mean humidity mixing ratio of the driest month.
Temp. + = mean daily maximum temperature of the hottest month.

from the expansion in each biome of nuclear populations whose morphology had been shaped by random factors, not by selective forces. For example, Gourou (1970), a human-geographer, denies that the adaptation of human morphology to climate in Africa has been convincingly demonstrated. However low the probability of the second interpretation may look to the ecologist, the explanation of the associations between human biology and climate in Africa will rest on a firmer ground if it is restricted to a set of populations descending from a common stock which expanded comparatively lately over several climatic zones.

From an analysis of the biological diversity of the Africans in the present and in the past, I concluded (Hiernaux, 1974) that a number of sets of populations have evolved relatively independently in subsaharan Africa during long periods of time. One line led to the Khoisan; another is ancestral to the bulk of the present populations of West and Central Africa and to the carriers of Bantu expansion; one or several other lines of evolution led to those various populations of East and West Africa which have in common an elongated body build with narrow head, face, and nose (among whom are the Ful of West Africa, the Tutsi and Hima of the Great Lakes area, and the Galla of Ethiopia). These stocks must be seen as expansions of relatively homogeneous groups of related populations out of the large biological diversity of the total human population of subsaharan Africa, not as a cutting up of this total diversity into discrete classes.

All the populations on the study of which this synthesis is based clearly descend from the second stock, the West–Central African. This will leave little support for any explanation other than adaptations of the associations of human biology and climate. It might, however, still be argued, as Gourou (1970) does, that the ancestors of the Pygmies may have developed their peculiar physique in the savanna, before their migration into the forest (by which evolutionary mechanism, he does not discuss). If adaptive convergence of populations which migrated independently into the forest is brought out by the present study, then no other explanation than adaptation may be maintained.

As Table 7.2 shows, the associations between human biology and climate in subsaharan Africa almost exclusively concern morphology: the frequency of fingerprints is completely unrelated to climate, as also is that of haemoglobin S where present; out of nine correlation coefficients of the ABO allele frequencies, only one is significant: a positive correlation of I^B with the peak of temperature. Table 7.2 considers only those biological variables for which the number of populations sampled exceeds 50, but the available allele frequencies of other loci do not suggest that migration from the savanna into the forest causes systematic selective pressures on them either.

The search for adaptive convergence in the equatorial forest must therefore be based on anthropometry. However, data on monogenic traits that are not involved in this adaptive process are also necessary: genetic distances between populations for that part of the genome which has not responded to the ecological challenge under study permit one to assess the phylogenetic affinities more accurately and to estimate the amount of interbreeding.

As this discussion makes clear, the study of the Sara, Oto and Twa was not considered as an end in itself, but was undertaken in order to enlarge the available corpus of data which is relevant to the study of biological adaptation to the equatorial forest of Africa. The present synthesis will present our fresh data in this perspective, as a contribution to a corpus which includes not only older data, but also data published by other teams since the outset of our IBP HA project, like the data now available on one more pair of populations of the equatorial forest: the Pygmoid Binga versus the agriculturalist Mbimu in the Central African Republic.

The three populations studied

The two Konda castes

The Konda live in Zaïre on the equator, in longitude 18° E. They are members of the Mongo larger ethnic entity which is the major group in the equatorial forest inside the Congo bend. According to the 1957 census, they number approximately 80000; their density varies between 5.60 and 11.20 inhabitants per km² (Gourou, 1960). The Twa make approximately one-third of the total Konda population. The Konda villages comprise from 50 to 10000 inhabitants and are built in small clearings of the rain forest. Each of them is subdivided into Oto and Twa quarters, the latter standing peripherally and appearing to be in a poorer state.

Social rules oppose intercaste marriages. In those rare, acknowledged cases of a child being born from intercaste sexual relations, the father always belongs to the suzerain caste, the Oto; the child is reared by his mother's family and is classified as a Twa. If gene flow between the two Konda castes really conforms to this pattern, it follows that it flows in one direction only – from the suzerain Oto to the vassal Twa (unidirectional gene flow in the opposite direction has been inferred by Cavalli-Sforza *et al.* (1969) from the social rules of at least two other pairs of forest-dwelling populations, i.e. from the Mbuti Pygmies and the Binga Pygmoids to their respective suzerains).

The Konda territory is nearly flat, at an altitude of *c.* 350 m. It abounds in rivers and marshes. The climate is equatorial. Annual rainfall exceeds 2 m, and shows no important variation throughout the year. The mean relative humidity during the day is 76% with an amplitude of variation of its monthly average of 15%. Mean annual temperature is about 26 °C with an amplitude of 2 °C. Mean daily maximum temperature varies around 30 °C (from 28 °C in July to 32 °C in March) (data collected at Mabali, on the eastern shore of Lake Tumba, and cited by Pagézy, 1973).

The Twa live predominantly by hunting and gathering, the Oto by practising agriculture (mainly cassava and oil palmtree) and fishing. An economic symbiosis binds the two castes which is based mainly on the exchange of foodstuffs; moreover, the Twa depend on the Oto for iron objects and pottery.

In an attempt to minimise the possible influence of the different ways of life of the Twa and Oto on their anthropometric differences, the samples for somatometry and maximum oxygen intake were taken from a local agricultural labour force. Large numbers of Konda work on the wide local *Hevea* plantations of the Compagnie Equatoriale (CEQUA) (the Twa are in the majority among them: they see in the worker condition a way to escape their bondage to the Oto). Male samples were of 66 Oto and 163 Twa for somatometry, and 27 Oto and 23 Twa for maximum oxygen intake. Their places of origin covered the whole Konda territory.

The possibility of some selection at migration from the villages to the *Hevea* plantation camps must be considered. The only comparative data available on Konda villagers are those of Müller (1964) on stature: both his Oto and Twa samples are taller by a little more than 1 cm than the corresponding CEQUA samples. The difference cannot be tested, since Müller did not publish the standard deviations, but it is most probably insignificant in the Oto, and of small magnitude and in the same direction in any case. From these data, it looks very unlikely that selective migration might have disturbed the pattern of the differences between the Oto and the Twa.

Skin reflectance was measured in the male population of a Konda village, Njalekenga (0° 9′ S – 18° 2′ E). The samples are of 278 Oto and 122 Twa. E. Vincke and I performed the somatometry. I measured skin reflectance; J. Ghesquière worked out the working capacity study; and E. Sulzmann worked on ethnology. H. Pagézy studied the physical fitness of Oto and Twa women and the pattern of their daily activities. She also carried out a quantitative food survey, the results of which have not been fully elaborated yet. It has not been possible to effect the planned collecting of blood specimens.

The Sara Majingay

At the time when the decision to include the Sara in the study was taken, it was already evident that this so-called ethnic group was too large to be considered as a single population unit: It occupies most of the Moyen–Chari Province of Chad, and includes more than 200000 people. I carried out a preliminary study of the whole area in 1965 which showed that the term Sara covers a number of ethnic groups, in each of which the frequency of endogamy (mating within the group) is near to 97.5%. They display some anthropometric heterogeneity, which seemingly is not accounted for by environmental differences (Hiernaux, 1969).

It was then decided to concentrate the study on a village in the territory of the largest Sara group, the Majingay, within which no regional anthropometric variation had been disclosed. Ndila (8° 50′ N – 17° 40′ E) was chosen for that purpose. In this community of 1200 persons who practise virilocal residence, 28% of the wives were born in one of 27 other villages, most of which are within a radius of 40 km of Ndila (Crognier, 1973).

At an altitude roughly similar to that of the Konda territory, Ndila has a tropical climate with two marked seasons. Annual rainfall varies around 1 m; most of it occurs from May to October. The mean relative humidity is 64% with monthly averages varying from 38% in February up to 84% in August.

Mean annual temperature is near to 27 °C with a monthly minimum of 24.2 °C in January and a monthly maximum of 26.6 °C in October. Mean daily maximum temperature varies between 30.1 in August and 39.5 in March, with a yearly average of 34.5 °C.

During the preliminary survey, I interviewed 706 young men and measured them according to the IBP basic list of measurements (Tanner, Hiernaux & Jarman, 1969). They were members of the Mouvement de la Jeunesse Tchadienne in Fort-Archambault (now Sahr); their birth places covered the whole Moyen–Chari Province. At the same time, A. Asnes measured 170 Sara boys and 157 Sara girls, aged from 3 to 10 years, in the kindergarten and schools of the same city according to the IBP full list of measurements. In Ndila, Crognier collected data and measurements on most inhabitants (his samples, including small nearby villages, totalled 1400 individuals). I measured the skin reflectance of 774 individuals and J. Lejeune made casts of the dentition of some 300 subjects. Blood specimens were obtained from 358 individuals and sent for typing to the Laboratoire d'Hémotypologie du CNRS in Toulouse (Professor J. Ruffié). In addition, 177 blood specimens were submitted to Professor J. Frézal for test for abnormal amino acids in the plasma (none was found). Differences in access, local facilities, and availability of staff and equipment, plus some problems of local feasibility, resulted in a lack of total concordance between the sets of data collected in the Chad and Zaïre. This unfortunately handicaps some comparisons but highlights the problems faced in field studies.

A comparison of climate, diet, fertility and child mortality in the Konda and Majingay

Compared to the climate of the Konda territory, that of Ndila is much drier, slightly warmer as an annual average, but with much more pronounced daily peaks of heat, and generally much less uniformity throughout the year. Unfortunately, no quantitative data on air movement have been obtained. It is, however, evident that, on average, it is much slower at ground level in the equatorial forest than in the steppe savanna where the Majingay live.

According to the climatic maps of Landsberg, Lippman, Paffen & Troll (1965), Ndila receives approximately 2500 h sunshine per year whereas Njalekenga gets approximately 1800. The difference between ultraviolet irradiations of the people living in the two places is certainly more pronounced than that between hours of sunshine: the life of a farmer in the savanna around Ndila necessitates working a large part of the day in the sun without shade, whereas the inhabitants of Njalekenga devote much less time to farming, and spend an important part of the day under the shade in the dense rain forest. At least today, the Oto and Twa do not differ much in this respect.

196

The diet of the Majingay is based on millet and ground peas during the rainy season. A period of more or less severe shortage occurs from the end of July to the end of October. Meat is rare, but fish is plentiful. The only fruit eaten is the mango, and it is available in March and April only. Vegetables are consumed in minute amounts only. No quantitative food survey has been carried out in Ndila, but apparently the diet is adequate in calories and in proteins. At least during part of the year, it is deficient in vitamins A, B_2, and C. Goitre affects 6% of the men and 17% of the women (Crognier, 1973).

The still unpublished food survey of H. Pagézy shows that the calorific needs of the Konda are met, without any marked variation throughout the year. Cassava provides one-quarter of the total calories, and palm fruits one-half of them. The protein needs are also met mainly by fish, game and insects, more rarely by domestic animals and poultry. Wild fruits and leaves, cultivated fruits, cassava leaves, and palm fruits provide the Konda with plenty of vitamins and iron. Goitre is absent from the population.

Athough no severe dietary deficiency appears in either the Majingay or the Konda (in particular, kwashiorkor is exceptional in both groups), the former inventories apparently favour the Konda. No data precise enough for comparative purposes are available on the pathology of the populations studied.

The Majingay women are more fertile than those of the Oto and the Twa; the mean number of children born to a woman in the 40–44 year age class is 6.3 in Ndila, 5.7 in the Oto, and 4.7 in the Twa (but the last two figures are based on very small samples). However, from birth until school age, the mortality of the Majingay children is higher, which results in the Majingay women having fewer surviving children of school age than the Konda women (Pagézy, 1973).

From the foregoing discussion, it looks as if the Majingay would be submitted to more acute stresses than the Konda, i.e., higher peaks of temperature, more intense solar irradiation during work, and periods of food shortage. Possibly their higher mortality during the first years of life partly results from such stresses. If climate is stressful to the Konda, it is rather by the constancy of high air moisture, high temperature, and still air.

Results and discussion

On the basis of the data collected in the Majingay and the two Konda castes, and of comparative material on other populations of subsaharan Africa, answers to a number of questions will now be sought.

Genetic affinities of Pygmies, Pygmoids, and other Central African populations for genes supposedly unresponsive to climatic selective pressures

As already stated, I^O, I^A and Hb^S frequencies in subsaharan Africa show no correlation with climate, while I^B shows a low positive correlation with the peak of temperature only. In the present state of our knowledge, as condensed in Table 7.2, it may be supposed that the set of monofactorial hereditary blood traits is but little sensitive to the selective pressures exercised by climate, or at least much less so than the polygenes which control body size and shape. Therefore, comparing the patterns of population diversity for blood traits with those for anthropometry is likely to be highly informative for answering the basic questions of the present study.

The first question is 'are exotic genetic influences a factor of diversity in and around the equatorial forest?' From the frequencies of those gene markers which offer a sharp contrast between Europe, Arabia and North Africa on the one hand, and the rest of Africa on the other, a negative answer has been given for a large area including the forest (Hiernaux, 1974). The allele frequencies of the Sara agree with the preceding statement: 0.53 for R^O, 0.03 for R^1, 0.004 for $Fy(a+)$, 0.998 for K, 0.22 for Gm (6), 1.00 for Gm (1, -2, 5) (Hiernaux, 1976b).

The genetic differentiation of the populations of West and Central Africa therefore appears to have been a process entirely internal to subsaharan Africa. In order to assess this differentiation in the field of blood traits, two matrices of genetic distances have been computed. They constitute two different compromises between two requirements: to include as many popu-lations in the matrix as possible, and to compute the distances from the frequency of as many alleles as possible. The statistic used is the G_c^2 of Balakrishnan & Sanghvi (1968). The matrices obtained have been reduced to a two-dimensional configuration by principal co-ordinates analysis (Gower, 1966) as perfected by Lalouel (1973).

The first matrix has been computed from the frequencies of 25 alleles at nine loci (ABO, MN, Rh, haptoglobins, transferrins, acid phosphatase, 6-phosphogluconate dehydrogenase, adenylate kinase, and phosphogluco-mutase-1). Comparisons are made of the Majingay, the Bedik of eastern Senegal, the Bantu-speaking Shangan, Pedi and Tswana of southern Africa, the Khoisan-speaking !Kung and Khoikhoi of southern Africa, and the Amhara of Ethiopia (Hiernaux, 1976b). In this set, the only population known to have a substantial exotic element is the Amhara, in whom Rh, Duffy and Gm frequencies evidence the gene flow from southern Arabia which has been assessed by history (Harrison *et al.*, 1969).

Fig. 7.2 shows the two-dimensional configuration obtained. Despite their shorter geographic distance from the Amhara, the Majingay are genetically

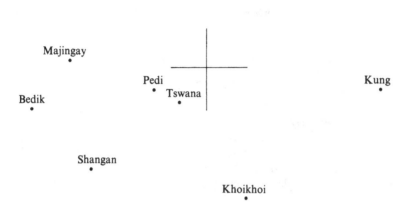

Fig. 7.2. Two-dimensional configuration of the G_c distances between eight populations for nine polymorphisms (from Hiernaux, 1976*b*).

much closer to the Bedik. They are also closer to the three Bantu-speaking groups of Southern Africa, despite the known divergence of the latter from more northerly living Bantu speakers which has resulted from a strong gene admixture from the Khoisan speakers (Jenkins, Zoutendyk & Steinberg, 1970; Jenkins & Corfield, 1972). The latter stand the farthest away from the Majingay. This shows that the Majingay, Bedik, and Bantu speakers (especially when the Khoisan influence is taken into account) are genetically similar to each other in blood genetics, despite the vastness of their geographical scatter and the large variety of their environments. One seems justified in seeing in them the descendants of a single stock.

The second (original) matrix concerns fewer alleles, but more populations. It is based on the frequencies of 13 alleles at four loci (ABO, MN, Rh and Tf). The populations compared are the Majingay, Bedik, Shangan and Amhara of the first matrix and six other ones:

(1) The Amarar of eastern Sudan (Hassan *et al.*, 1968), in whom a strong exotic element (of Arab and, possibly, Egyptian origin) is evidenced by the marker genes.

(2), (3) The Tutsi and Hutu of Rwanda. The Tutsi are descended from the East African stock characterised by its elongated body build, whereas the Hutu are an offshoot of Bantu expansion (see references in Hiernaux, 1968).

199

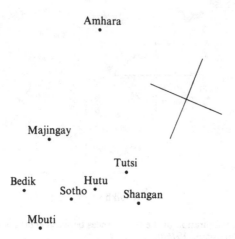

Fig. 7.3. Two-dimensional configuration of the G_c distances between 10 populations for ABO, MN, Rh and Tf polymorphisms (with the direction of the Bedik–Amhara line set as on Fig. 7.2).

    (4) The Sotho, the Bantu-speaking population of Lesotho (Moullec, Mendrez & Nguyen Van Cong, 1966), in whom a strong Khoisan admixture is conspicuous, though weaker than in the Tswana (Jenkins *et al.*, 1970; Jenkins & Corfield, 1972).

    (5) The Mbuti Pygmies (data of Jadin, 1935 for ABO, and of Fraser, Giblett & Motulsky (1966) for the other systems).

    (6) The Binga Pygmoids of the Central African Republic (Cavalli-Sforza *et al.*, 1969).

From the second matrix of G_c distances, Fig. 7.3 has been obtained by the same procedure as Fig. 7.2. The relative positions of the four populations common to the two matrices are approximately the same on the two figures. On Fig. 7.3, Majingay, Bedik and the three Bantu-speaking agriculturalist populations (the Hutu, Shangan & Sotho) are relatively near to each other, which supports the view that all of them are descended from a single stock. The intermediacy of the Amhara between the Amarar and the preceding populations is consistent with the historical data. Far away from the Amhara, the Tutsi are near to the Hutu and Shangan, which confirms that they arose from a purely African differentiation, without any conspicuous exotic admixture (Hiernaux, 1974).

On the basis of Fig. 7.3, the Mbuti Pygmies are also near to the Majingay, Bedik and Bantu speakers, whereas the Binga stand far away. The picture changes considerably when one looks at the matrix of distances computed by Cavalli-Sforza *et al.* (1969) from a larger set of alleles (22 alleles at six loci: A_1A_2BO, Rh, Hp, MNSs, Tf and Gm). They compare, among others, Mbuti, Binga and a mixture of agriculturalist populations of the rain forest and adjacent Rwanda and Burundi, labelled 'Central Africans'. The pooling of a number of populations scattered over an immense area, such as this Central African sample, is highly questionable since it conceals any local diversity. However, Fig. 7.3 suggests that, in blood traits, the agriculturalist populations included in the pool, most of them offshoots of the recent Bantu expansion, are only moderately heterogeneous. Their pooling may therefore be accepted for a general appraisal of the genetic position of the Mbuti and Binga in relation to their suzerains. It must, however, be kept in mind that the Central African point on a two-dimensional configuration is no more than the centre of a cloud of points, the size and shape of which we are ignoring.

On Cavalli-Sforza's configuration (Fig. 3 of Cavalli-Sforza *et al.*, 1969), Mbuti and Binga are relatively far apart; both populations are also distant from the Central Africans, the Mbuti showing the furthest distance from the latter, and the Binga being nearer to the Central Africans than to the Mbuti. From this configuration and an analysis of the possibility of the Binga representing a mixture of Mbuti and Central Africans, Cavalli-Sforza *et al.* conclude that the three groups have a common origin, that there is no clearcut evidence for a mixed origin of the Binga, and that the Mbuti and Binga ancestral lines either diverged independently from the Central African line (the Mbuti line before the Binga one), or diverged from a common line early after the latter's branching from the Central African one.

Since the agriculturalists are recent invaders of the rain forest where they rejoined the anciently settled Pygmies and Pygmoids, the branching of the Mbuti and Binga lines, or of a short common Mbuti–Binga trunk, did not occur in the forest, but most probably resulted from their isolation, through migration into the forest, from their savanna relatives. Whether the ancestors of the Mbuti and those of the Binga started their migration independently or parted soon after it does not much change the picture: in any case, the two lines have lived most of their period of genetic adaptation to the rain forest in genetic isolation from each other.

Jean Hiernaux

The lack of sufficient genetic information prevents us extending the multivariate approach to other Pygmoid populations. At the ABO locus, genetic differences between a number of Pygmy or Pygmoid versus agriculturalists pairs have been analysed by Hiernaux (1962). The Mbuti largely differ from their suzerains, the Bira and Balese, in having much higher *A* and *B* allele frequencies. The Kuba Cwa, the Mongo Twa (among whom the Konda Twa) and the Lia Twa, all differ from their suzerains in the opposite direction. In Zaïre, the Mbuti and the Twa (or Cwa) stand at the opposite ends of the scale of ABO allele frequencies. On this basis, the various Twa groups do not appear at all to represent a Mbuti × local suzerains mixture, no more than the Binga of the Central African Republic on the basis of multivariate genetic distances.

From all this, it seems highly probable that the Mbuti Pygmies have evolved in nearly complete genetic isolation from the Pygmoids of the northwestern and southern parts of the forest. What Pygmies and Pygmoids may have in common is little more than their common descent from the West–Central African stock. Consequently, the affinities in body build that may exist between the Mbuti and the Pygmoids have to be explained in terms of convergence, or at least of parallel evolution; this possibly applies also to the affinities between some groups of Pygmoids, for whom we lack adequate data on blood traits to assess their genetic distance.

The Pygmies or Pygmoids and their suzerains: an anthropometric comparison

In order to assess the differences within and between the four pairs of populations on which adequate comparative data are available, the following anthropometric samples (all of male adults) have been used: 510 Mbuti (Gusinde, 1948) and 178 forest Bira (Sporcq, 1972); 87 Binga and 139 Mbimu (Cresta, 1965); 100 Kuba Cwa and 120 Bushong (Hiernaux, 1966*a*); 66 Konda Oto and 163 Konda Twa (Hiernaux, Vincke & Commelin, 1976).

On Fig. 7.4*a–d*, the graphic profile of each Pygmy or Pygmoid group has been drawn by reference to the corresponding suzerain population. Such an intra-pair comparison presents three advantages: (1) the possible differentiating effect of the environment on the mean phenotypes of the populations compared is minimal; (2) the probable uni- or bidirectional gene flow between members of the same pair can only reduce their differentiation, not change its pattern; (3) in three out of four cases, the two populations have been measured by the same person, which avoids the possible introduction of inter-observer biases in the differences.

In drawing Fig. 7.4*b*, the standard deviations of the Konda Oto were arbitrarily ascribed to the Mbimu, as those for the latter have not been published. Most probably, this could not have changed the pattern of the

202

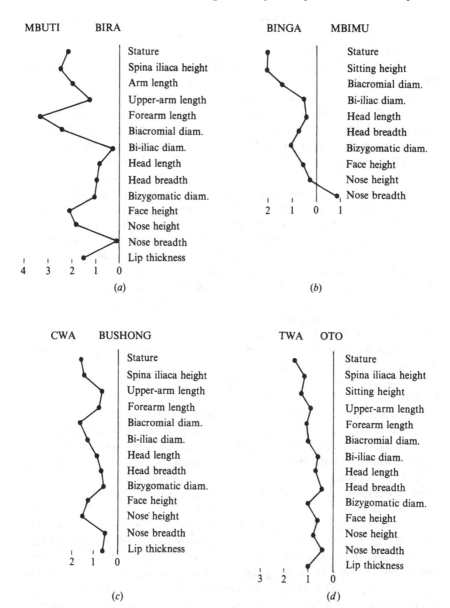

Fig. 7.4. Four graphic profiles comparing a Pygmy or Pygmoid population by reference to the corresponding suzerain population (the scales are in standard deviations of the latter).

differentiation between Binga and Mbimu to a large extent: Oto and Mbimu differ little in size, they live in a similar environment and both originate from the relatively recent Bantu expansion.

On all figures, the smaller size of the Pygmies or Pygmoids is evident. In one case only has an anthropometric variable a higher mean in a Pygmoid population: the Binga have broader noses than the Mbimu. However, size reduction differs with the measurement. In all cases, stature is among the most reduced, together with trunk height or lower-limb length (as measured by the spina iliaca height). In the Mbuti, lower-limb length is more shortened than stature, whereas the reverse is observed in the Twa and Cwa; in the Binga the reduction is equal. Upper-limb length is less shortened than stature in each of the three groups in which it has been measured. In the Mbuti, this reduction concerns the forearm much more than the upper arm; the Cwa and Twa show the same pattern, but much more slightly. Reduction in shoulder breadth does not differ much from that of stature nor between the four groups. In each of them, hip breadth is less reduced than stature, but to a small degree in the Cwa and much more so in the Mbuti, Binga and Twa. In all four groups, head length and breadth show a moderate and roughly similar reduction. Bizygomatic diameter is considerably less reduced than face height in the Mbuti and Cwa, whereas the reverse is observed in the Binga and Twa. Nose height is much more reduced than nose breadth in all groups, but the Binga are unique in having a nose nearly as high, and much broader than that of their suzerains.

As this inventory shows, the four Pygmy or Pygmoid populations are far from presenting the same pattern of differentiation from their suzerains. They have in common a general reduction in body size, with preservation of a big head, wide hips, and a broad nose. The biological requirements underlying these common features look clear: universal selective pressures tend to maintain the volume of the braincase, and consequently the size of the pelvis, whereas the equatorial climate favours a broad nose (Weiner, 1954). In less essential features, the four Pygmy or Pygmoid populations largely vary in their local differentiation. In particular, the profile of the Konda Twa is more rectilinear than that of the three other groups; in other words, the Konda Twa are nearer to a model of uniform reduction of their suzerains' dimensions.

There have thus apparently been different ways for a line of savanna dwellers to become smaller in body size as a result of their migration into the equatorial forest. The analysis of morphological data implies a historical reconstitution similar to that suggested by blood monogenic traits; it suggests several lines descended from a common West–Central African stock being submitted to the same selective pressures in different parts of the equatorial forest, all of which responded by a reduction in body size. The biological differences between these lines are such that, whatever the details of the branching, they must have had a relatively long history of independent evolu-

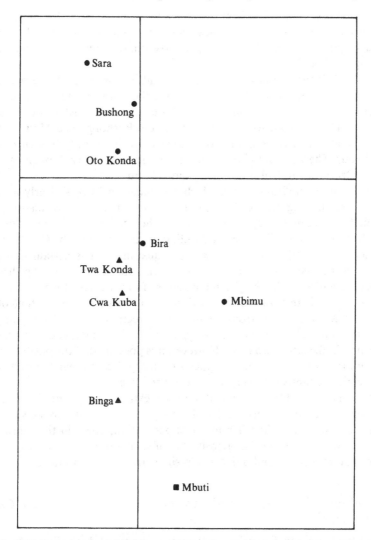

Fig. 7.5. Two-dimensional configuration of the Δg distances between nine populations for ten anthropometric variables.

tion. This evolution kept parallel only in that part of the genome responding to the common environmental change.

The overall morphological differentiation of the four pairs of populations has been assessed by a method similar to that used in Figs. 7.3 and 7.4, i.e., the computation of a matrix of distances, and its two-dimensional projection (Fig. 7.5). Since the standard deviations of the Mbuti and Mbimu are unknown, a statistic which does not require these values has been used: the

Jean Hiernaux

Δg (Hiernaux, 1965). The Sara also have been included in the matrix. This has been computed from 10 measurements: stature, biacromial, bi-iliac, head length and breadth, face height, bizygomatic, nose height and width, and lip thickness.

In Fig. 7.5, the scatter of points is long and narrow, with the Sara and the Mbuti at the opposite ends. Nearest to the Sara are the Bushong, the only other agriculturalist population to live outside the forest (though on its border). The Konda Oto are not far from the Bushong. The Mbuti are the most isolated population, with the Binga as their nearest, though distant, neighbour. The Twa and Cwa are in the middle, near to each other but also not far from the agriculturalist Bira and Mbimu.

The climatic environment in which a population lives is clearly a major factor determining its position in the scatter: the two savanna-dwelling populations are in the upper part of it, with the one living in the driest climate standing at the top; all the forest-dwelling populations but the Oto are much farther down. Although it has not been quantified, it is certain that gene exchange between members of the pairs also influences both the distance between them and their height in the scatter. In particular, it is probable that the relatively high position of the Kondo Oto as compared to that of the Bira and Mbimu results from the different direction of the within-pair gene flow: as aforesaid, it flows from the Pygmoids in the former, and in the reverse direction in the latter two pairs. However, it is possible that the position of the agriculturalist forest-dwelling populations in Fig. 7.5 also partly results from an incipient adaptive convergence toward the Pygmies.

Why have the Mbuti attained such an extreme differentiation? In the absence of any known peculiarity of their environment, it may be suggested that they owe it to a higher number of generations spent in the forest. The Konda Twa could be at the opposite end of such a time scale: they show the lesser size reduction, and differ less in shape from their suzerains.

The relationship between height, weight and biome in the West–Central African stock

From the foregoing discussion, it appears that powerful selective pressures in favour of a smaller body size are at work on African populations once they have migrated into the equatorial forest. That they concern thermoregulation has often been stated, but little documented. As early as 1951, Schreider observed that, in the human species, the ratio of body surface to mass tends to be higher in those climates which put a stress on thermoregulatory mechanisms during at least a part of the year. The physiological basis of this association is clear: heat production tends to be proportional to the metabolic mass of the organism, whereas heat loss occurs at skin level. Peaks of heat are lower in the rain forest than in the savanna, but, in the former biome,

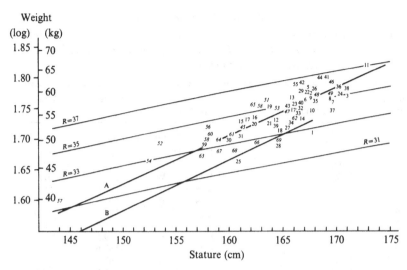

Fig. 7.6. Log weight versus height diagram. The 69 populations are represented by their number in Table 7.3. In italics: populations of the rain forest. Line A: major axis of the log weight versus height simultaneous distribution. Line B: trend line of the log weight versus height relationship during growth in the male Hutu of Rwanda (from Hiernaux, 1964). Lines *R*: lines of equal values of the weight:surface ratio, for the ratios 31, 33, 35 and 37 kg/m². (Reproduced from Hiernaux *et al.*, 1975).

a much higher air moisture content and still air put a fairly low limit to the cooling power of sweating. Conduction–convection, the other main mechanism of heat loss, can be improved per unit of surface only by increasing skin temperature, also a narrowly limited process. Thermoregulation adaptation to the rain forest therefore strongly relies on reduction of the ratio of body mass:surface.

The ratio of body weight to stature is closely related to that of body mass to surface: various functions of height and weight (including the much used du Bois' (1916) formula) are fair means of estimating body surface (Boyd, 1935), and weight is used for mass in practice. The study of body mass/surface ratio reduction as a consequence of migration into the equatorial forest may therefore be approached by comparison of the weight/stature relationship in populations descended from the West–Central African stock, now in the savanna, with those now in the forest.

Data on weight and stature are available in 69 male adult samples of such populations (Table 7.3). They have been plotted on a log weight versus stature diagram (Fig. 7.6). Here also, as on Fig. 7.5, the scatter is long and narrow, with the Sara and Mbuti at the opposite ends. The relationship between mean log weight and mean stature is rectilinear (line A is its major axis). Four curves of equal values of the weight/surface ratio, as estimated by du Bois'

Table 7.3. *Mean stature and body weight in adult male samples of 69 populations descended from the West-Central African stock, grouped in two zones: open country and rain forest (reproduced from Hiernaux, Rudan & Brambati, 1975, where the references can be found)*

Number on figure	Population	Locality				Stature (cm)	Weight (kg)
		A. Open country					
1	Teda	21	N	16	E	167.7	51.5
2	Dogon	15	N	4	W	167.6	59.1
3	Fulse	15	N	1	W	171.1	59.0
4	Matakam	12	N	13	E	165.8	56.2
5	Bedik	12	N	13	W	167.4	60.8
6	Kapsiki	11	N	13	E	167.0	58.0
7	Fali	10	N	13	E	169.5	58.4
8	Nago	8	N	1	E	169.5	59.2
9	Adja	8	N	2	E	167.6	58.9
10	Kotafon	8	N	2	E	167.7	55.6
11	Sara Majingay	8	N	20	E	173.9	66.1
12	Baya	6	N	14	E	164.2	53.9
13	Nzakara	6	N	22	E	165.8	58.2
14	Banda	5	N	16	E	166.6	54.6
15	Savanna Bira	2	N	30	E	161.2	53.2
16	Shu	1	N	29	E	162.3	54.0
17	Swaga	0	S	29	E	161.6	53.6
18	Kikuyu	0	S	36	E	164.6	51.9
19	Hunde	1	S	28	E	164.0	56.5
20	Havu	1	S	28	E	162.3	53.4
21	Shi	2	S	28	E	163.9	53.5
22	Hutu (Rwanda)	3	S	30	E	167.4	59.5
23	Hutu (Burundi)	3	S	30	E	166.1	56.7
24	Bushong	4	S	21	E	170.4	59.5
25	Twa Kuba	4	S	22	E	160.8	46.1
26	Songye	5	S	25	E	167.8	59.8
27	Nyaturu	5	S	34	E	165.4	52.2
28	Sandawi	5	S	35	E	164.6	49.4
29	Luba-Kasai	5	S	23	E	166.9	60.0
30	Hungu	7	S	16	E	160.0	50.3
31	Luango	8	S	14	E	161.0	51.5
32	Luba-Katanga	8	S	25	E	166.3	56.5
33	Makonde	11	S	39	E	166.4	55.1
34	Kakonda	13	S	15	E	166.0	53.5
35	Nyungwe	15	S	33	E	168.0	57.9
36	Mbukuso	16	S	20	E	170.2	61.0
37	Kazama	16	S	20	E	169.7	55.9
38	Kwangar	17	S	19	E	170.8	60.6
39	Chuabo	18	S	37	E	164.3	59.5
40	Venda	22	S	30	E	166.5	56.8
41	Thonga	23	S	35	E	168.8	63.2
42	Chopi	24	S	35	E	166.9	61.6
43	Pedi	25	S	29	E	165.9	56.2
44	Ronga	25	S	32	E	168.5	63.0

Table 7.3. (*cont.*)

Number on figure	Population	Locality			Stature (cm)	Weight (kg)
		B. Rain forest				
45	Anang (Ibibio)	5 N	7 E		161.0	52.6
46	Yambasa	5 N	10 E		169.5	62.5
47	Mangisa	5 N	11 E		165.5	56.0
48	Tanga	4 N	9 E		168.1	59.3
49	Ewondo	4 N	11 E		169.4	59.5
50	Jem	4 N	13 E		162.9	56.6
51	Zimu	4 N	13 E		163.4	57.7
52	Binga (Cameroun)	4 N	14 E		153.4	49.5
53	Mbimu	4 N	15 E		164.2	56.2
54	Binga (Lobaye)	4 N	15 E		152.3	46.4
55	Fang	3 N	12 E		166.4	61.7
56	Forest Bira	2 N	28 E		158.0	52.5
57	Mbuti	2 N	28 E		144.0	39.8
58	Binga (Mekambo)	1 N	13 E		157.9	50.1
59	Humu	1 N	29 E		157.8	49.6
60	Mvuba	1 N	29 E		158.1	50.9
61	Nyanga	0 S	27 E		160.2	51.6
62	Oto Konda	1 S	19 E		166.2	54.4
63	Twa Konda	1 S	19 E		157.5	46.8
64	Tembo	1 S	28 E		159.0	49.9
65	Lega	2 S	27 E		162.2	57.2
66	Yans	3 S	17 E		162.6	49.7
67	Fuliru	3 S	29 E		159.1	47.8
68	Mbala	5 S	18 E		160.5	47.7
69	Mbun	5 S	19 E		164.5	49.9

formula, have also been drawn for the values of 31, 33, 35 and 37. However objectionable the absolute application of this formula may be to non-Euroamerican populations, it may be used for comparative purposes within a single stock.

Line A cuts the lines of equal weight/surface ratio. From the Sara Majingay to the Mbuti, the fall of the ratio comes near to six units. This represents two-thirds of the total range of population means of the ratio in the human species which comes to nine units, from 30.2 to 39.2 (Schreider, 1963). The size reduction has therefore been a very efficient means of reducing the ratio, thence of making heat loss easier. The width of the scatter nowhere exceeds 5 kg on both sides of line A. Most probably, this variation is nutritional for a large part, if not totally.

There is no reason to believe that the reduced size of the forest-dwelling populations results from a poorer nutrition, either as a phenotypic response or as a genetic adaptation: weight is more sensitive to malnutrition than stature; the Mbuti and Binga, the shortest-statured populations, stand well above line A; and more generally, weight is positively correlated with rainfall in subsaharan Africa when stature is held constant (see Table 7.2). There seems to be no alternative to the explanation of size reduction in terms of genetic adaptation to the climate of the equatorial forest.

On Fig. 7.6, line A is parallel to line B, which is the trend line of the log weight versus height relationship during growth in the male Hutu of Rwanda, one of the populations plotted. Such trend lines are known to be parallel in a large variety of human populations (Hiernaux, 1964). Line A therefore appears to be similar to a growth trend. The data suggest that, in body size, all populations descended from the West–Central African stock might differ genetically from each other only by the target that growth has to reach on a common log weight versus height trend line. Most probably, such differences in the target are achieved through differences in speed of growth rather than in duration of growth (Hiernaux *et al.*, 1975).

Nutritional status and work capacity of the Majingay, Oto and Twa

As aforesaid, the diet of the Sara Majingay looks poorer than that of the Konda. However, in Fig. 7.6, the Majingay stand a little above line A, whereas both Konda castes stand below this line, with the Twa nearer to it than the Oto. The discrepancy probably results from the fact that the Konda measured were not living the village life but that of workers in a monetary economy in which the traditional means of obtaining food were partly replaced by buying it in shops, with a high risk of impoverishment of the diet in quantity and variety. In the rural Ntumba, close neighbours of the Konda, whose environment and means of subsistence are the same as those of the latter, Pagézy (1973) observed a mean stature higher by *c.* 3 cm, and a mean weight by *c.* 6 kg in both the Oto and the Twa than the means of the corresponding Konda castes in the plantation camps. In Fig. 7.6, they would stand very near to line A. It may be presumed that the log weight versus height relationship of the Konda villagers is similar, and therefore similar to that of the Majingay.

At the level of the middle of the upper arm, the Majingay have a mean circumference of 277 mm. In the Konda, this measurement comes to 256 in the Oto and 240 in the Twa, and to 283 and 260 in the same castes of Ntumba villagers. In these five populations, the retrotricipital skinfold comes to 6.4, 6.3, 5.6, 5.7 and 5.5 mm respectively. This skinfold has been subtracted from arm diameter (estimated by dividing arm circumference by π) for obtaining the lean arm diameter. Its average is about 82 mm in the Majingay, 75 in the Konda Oto, 71 in the Konda Twa, 84 in the Ntumba Oto, and 77 in the Ntumba Twa.

The lean arm diameter shows a low, but significant positive correlation with arm length in the two samples in which it has been computed, the Majingay and the Konda Twa (Hiernaux *et al.*, 1974). The comparison of the

210

lateral development of the lean tissues of the arm in populations of different body size must take this correlation into account. The joint distribution of the two variables has parallel major axes in the Majingay and the Twa (*ibid.*, Fig. 1), which makes the comparison easier. When the other samples are plotted on this figure, which shows the Konda Twa inferior to the Majingay, the Ntumba Twa are slightly superior to the latter, and the Ntumba Oto greatly so; whereas the Konda Oto are similar to the Konda Twa in their relation to the Majingay. Bone plus muscle lateral development of the arm is thus larger in the traditional rural populations, either Pygmoid or agriculturalist, of the Lake Tumba area in the equatorial forest than in the rural Majingay of the Chadian savanna.

Ghesquière (1972) measured the maximum aerobic power of 27 Konda Oto and 23 Twa out of the larger male anthropometric samples by the direct method using a bicycle ergometer. When the V_{O_2} max is expressed in ml/Kg/min, the Twa are significantly superior to the Oto. The means are 47.5 (s.d. $= \pm 5.3$) in the Twa, 42.7 (s.d. $= \pm 4.1$) in the Oto. In female Ntumba samples of 25 Oto and 20 Twa, Pagézy (1973) found no significant difference between the two castes when estimating the maximum aerobic power by the indirect method using the step test.

Possibly, the work load of the Twa in the *Hevea* plantations is heavier than that of the Oto; this factor is known to influence work capacity strongly. Whatever the case, it may be said that the Twa are at least as physically fit as their suzerains (when using work capacity as a criterium of physical fitness). This most probably applies also to the Binga, who stand high above line A on Fig. 7.6 without having more fat than the Twa, as judged from their subscapular skinfold of 8.5 mm (Cresta, 1965). It would be surprising if the Majingay, with a lean arm diameter inferior to that of the Twa and Oto villagers when arm length is taken into account, and having physical activities of similar type, had a higher maximum aerobic capacity when expressed per unit of body weight. As well as nutritional state, work capacity is at least as high in the equatorial forest as in the Sara of the savanna, however renowned the latter are for their robustness.

Stabilising and directional selection in the Sara

As Table 7.2 shows, human morphology is associated with climate in subsaharan Africa in many of its aspects. These associations suggest that morphological variation between populations is due, in an important part, to adaptation to a variety of climates. That it is so for body size is strongly supported by the present study. Weiner (1954) has explained the physiological basis of the association with climate of another feature; nose shape, as measured by the nasal index. This also shows especially strong correlations (*r*) with climate in subsaharan Africa: $+0.46$ with rainfall, $+0.28$ with humidity of the driest month, and -0.21 with daily maximum temperature of the hottest month (Hiernaux, 1968).

It may therefore be expected that, in a homogeneous African environment, stabilising selection, which selects against both tails of the distribution curve, will be especially active in maintaining the mean of such variables as stature and nasal index at the adaptive optimum value once this has been reached, and that directional selection, which acts on one tail more than on the other one, will be especially active in displacing such means toward their optimum once the population has departed from the latter (in consequence of migration from a different climatic zone, or of gene admixture, or of a stochastic process). Interpopulational variation within the Sara ethnic group conforms to this expectation: the F-test of homogeneity between samples of Majingay, Mbay, Day, Ngama and Kaba amounts to 1.90 for nasal index ($P > 0.05$), 2.84 for stature ($0.05 > P > 0.01$), 6.56 for facial index ($P < 0.001$) and 18.35 for cephalic index ($P < 0.001$) (Hiernaux, 1969).

In this group of populations in which the probability of the matings is largely distance–dependent (Crognier, 1973), multivariate biometric distance is very closely associated with geographic distance when eight villages (four Majingay, two Mbay and two Kaba) are compared: the two-dimensional configuration obtained from a matrix of D^2 distances can be superposed nearly exactly on the geographical map (Ramirez, 1975).

The Sara, who are almost totally isolated genetically from the non-Sara world, live in a homogeneous environment. Most probably, stochastic processes (like the founder effect) are the main source of genetic differentiation between their communities. Isolation by distance appears to be the mechanism of maintenance of this differentiation. The only process which could oppose it is a directional selection tending to bring the diversified allele frequencies or anthropometric means back to the common adaptive equilibrium. As judged by the analysis of variance between the Sara groups, this process seems to act especially strongly on nose shape and body size.

Crognier (1973) searched for demographic evidence of stabilising or directional selection in the Sara Majingay of Ndila. From data on 240 married couples, he observed a significantly higher fertility in the couples whose members were both in the middle of the range than in the couples whose members were both either in the lower or in the upper part of the distribution of the following variables: sitting height, lower-limb length, arm length, biacromial and bi-iliac. 'High' couples for arm length have a significantly lower fertility. The only significant difference in survival rate of the offspring between the same three categories of couples is a lower survival rate of the 'middle' couples for total limb length, for which the 'high' couples show a lower survival rate at the probability level of $0.20 < P < 0.10$.

This rather small sample has thus revealed stabilising selection at work in the Majingay for several longitudinal and transversal body measurements.

The data also suggest that directional selection might favour shorter limbs. Maybe the Majingay are above the optimum in their body size. Possibly their relatively high weight/surface ratio (they stand at the highest position in Fig. 7.6, with a ratio of nearly 37) makes the tallest of them especially less fit in their hot environment. At similar latitudes in Africa, similar or taller statures are found only in the Nilotic populations of Southern Sudan, like the Dinka and Shilluk who display a very slender body build and have a mean weight/surface ratio of no more than 35 despite an average stature of *c.* 180 cm (Roberts & Bainbridge, 1963).

Skin reflectance in the Sara, Oto and Twa

On a world basis, skin pigmentation in human populations is correlated with the amount of ultraviolet radiation to which people are exposed (see reviews by Daniels, 1964, and Loomis, 1967). This correlation has generally been interpreted as reflecting a genetic adaptation of the content of melanin in the skin to the amount of ultraviolet radiation that reaches the skin. However, Daniels (1964) thinks it likely that the dark skin of the African is an adaptation both to ultraviolet and a variety of the other hazards associated with nakedness and tropical environments.

The environments of the Majingay, Oto and Twa apparently do not differ much in such other hazards. It looks reasonable, therefore, to explain the possible differences in skin reflectance between the three populations in terms of adaptation to ultraviolet radiation (as aforesaid, the Majingay are submitted to notably higher doses of this radiation than the Oto and the Twa).

If skin colour is really adaptive to ultraviolet radiation, we may expect the Majingay to be the darkest-skinned of the three populations. In the forest, we may expect the Twa to be at, or at least nearer to, adaptive equilibrium, and therefore to be lighter-skinned than the Oto. Very high selective values of the 'light skin' genotypes, like those of the $Hb^A Hb^S$ genotype in malaria-ridden areas, would have been required to bring the Oto to adaptive equilibrium in the low number of generations which, according to the linguists, they have spent in the forest; they are unlikely. However, we may expect the distance in skin reflectance between the Oto and the Twa to be moderate, since two forces are at work for bringing them nearer to each other: the pressure of a common environment on both genome and phenotype, and gene exchange.

For testing these expectations, the reflectance of the skin was measured with an EEL reflectometer equipped with the 609 (685 nm) filter, at the inner surface of the arm. The choice of the site was determined by the emphasis on genetic variation: the inner surface of the arm is relatively little exposed to variation in suntanning and reflectance measured at this site shows a high heritability (Garn, Selby & Crawford, 1956). Only the 609 filter was used

Jean Hiernaux

because the reflectance with such a filter shows the highest correlation with the concentration of melanin, whose variation this study aims to interpret (Harrison & Owen, 1956/57).

Only male subjects were measured. In the largest sample, that of 415 Sara, there appears no influence of age on skin reflectance although age ranges from primary school age up to old age (Hiernaux, 1972). The age composition of the two Konda samples (278 Oto and 122 Twa) is roughly similar to that of the Sara. It seems therefore very unlikely that age might play a role in the differentiation.

The results fit the two expectations. A Sara–Oto–Twa sequence is observed, with reflectance means of 23.2–28.8–30.0%. All differences are highly significant (Hiernaux, 1976a).

It looks certain that this sequence of phenotypic means reflects, at least partly, a genetic gradient. Nowadays Oto and Twa live side by side. Oto males wear shirts more often than Twa do; since the inner surface of the arm is not entirely deprived of sunlight, this would favor more suntanning of this site in the Twa, and therefore strengthens the truth of the suggested genetic difference between the two Konda castes. Maybe the inner surface of the arm is more irradiated in the Sara than in the Konda, but it seems unlikely that the difference of five units in reflectance between the Sara and Oto results totally from differences in suntanning. The data therefore support the view that genetic adaptation to ultraviolet radiation is an important factor in the variation of skin colour in Central Africa.

Conclusions

The comparison of three populations – Majingay, Oto and Twa – between each other and with other descendants of the West–Central African stock has thrown light on several adaptive changes that migration into the equatorial region has apparently caused: mostly a reduction in body size and a lightening of skin colour. It has shown that the nutritional state of the forest dwellers is, in general, at least as good as that of the savanna dwellers, this being especially the case for the Pygmies and Pygmoids. Size reduction is by no means a response to undernutrition. There seems to be no alternative to an explanation in terms of adaptation to moist heat. It permits a combination of very low weight/surface ratio and strong muscular development, hence a high work capacity.

The finding that different ancestral lines display size reduction while largely differing in some other features (like the set of hereditary blood traits, and a number of head and body shape attributes) supports the conclusion that size reduction results from migration into the forest.

As in any adaptive process, the time factor, in this case in terms of the number of generations spent in the forest, is of importance. It is evidently a

214

major factor in determining the magnitude of the differences between forest agriculturalists and Pygmies or Pygmoids, but it possibly also assumes a part of the morphological differences between populations of the second group. Gene exchange between populations of the two groups is another factor in evolution of the pattern of differences. All three factors – change in adaptive optimum, time, and gene exchange – have been determinants of the differences in skin colour between the three populations. The environmental attribute responsible for the variation of this character appears to be the ultraviolet irradiation.

Selective pressures exerted by climate appear strong in Central Africa, especially those acting on body size and nose shape. They are strong enough to prevent heterogeneity developing in these features in the Sara despite a marked genetic isolation by distance. Stabilising selection has been shown to be at work in the Sara Majingay, and less clearly so directional selection toward a mode which would better reconcile their body size with the requirements of thermoregulation.

By throwing light on deterministic evolutionary mechanisms, such findings have not only an explanatory value, but also a predictive power.

References

Balakrishnan, V. & Sanghvi, L. D. (1968). Distance between populations on the basis of attribute data. *Biometrics*, **24**, 859–65.

Boyd, E. (1935). *The Growth of the Surface Area of the Human Body*. Minnesota: University of Minnesota Press.

Cavalli-Sforza, L. L., Zonta, L. A., Nuzzo, F., Bernini, L., De Jong, W. W. W., Meera Khan, P., Ray, A. K., Went, L. N., Siniscalco, M., Nijenhuis, L. E., van Loghem, E. & Modiano, G. (1969). Studies on African Pygmies. I. A pilot investigation of Babinga Pygmies in the Central African Republic (with an analysis of genetic distances). *American Journal of Human Genetics*, **21**, 252–74.

Clark, J. D. (1970). *The Prehistory of Africa*. London: Thames & Hudson.

Cresta, M. (1965). Contributo alla conoscenza antropologica dei Babinga. *Quaderni de la Ricerca Scientifica*, **28**, 81–102.

Crognier, E. (1973). Adaptation morphologique d'une population africaine au biotope tropical: les Sara du Tchad. *Bulletins et Mémoires de la Société d'Anthropologie de Paris*, **10**, 3–151.

Daniels, F. Jr (1964). Man and radiant energy: solar radiation. In *Handbook of Physiology*, vol. 4, *Adaptation to the environment*, ed. D. B. Dill, pp. 969–87. Washington D.C.: American Physiological Society.

DuBois, D. & Dubois, E. F. (1916). A formula to estimate the approximate surface area if height and weight are known. *Archives of Internal Medicine*, **17**, 863–71.

Fraser, G. R., Giblett, E. R. & Motulsky, A. G. (1966). Population genetic studies in the Congo. III. Blood groups (ABO, MNSs, Rh, Js^a). *American Journal of Human Genetics*, **18**, 546–52.

Garn, S. M., Selby, S. & Crawford, M. R. (1956). Skin reflectance studies in children and adults. *American Journal of Physical Anthropology*, **14**, 101–17.

215

Ghesquière, J. L. A. (1972). Physical development and working capacity of Congolese. In *Human Biology of Environmental Change*, ed. D. J. M. Vorster, pp. 117–20. London: International Biological Programme.

Giblett, E. R., Motulsky, A. G. & Fraser, G. R. (1966). Population genetic studies in the Congo. IV. Haptoglobin and transferrin serum groups in the Congo and in other African population. *American Journal of Human Genetics*, **18**, 553–8.

Gourou, P. (1960). *Atlas général du Congo*. Bruxelles: Académie Royale des Sciences d'Outre-Mer.

Gourou, P. (1970). *L'Afrique*. Paris: Hachette.

Gower, J. C. (1966). Multivariate analysis and multidimensional geometry. *Statistician*, **17**, 13–28.

Greenberg, J. H. (1963). *Languages of Africa*. The Hague: Mouton.

Gusinde, M. (1948). *Urwaldmenschen am Ituri*. Wien: Springer.

Guthrie, M. (1962). Some developments in the pre-history of the Bantu languages. *Journal of African History*, **3**, 273–82.

Harrison, G. A., Küchemann, C. F., Moore, M. A. S., Boyce, A. J., Baju, T., Mourant, A. E., Godber, M. J., Glasgow, B. G., Kopeć, A. C., Tills, D. & Clegg, E. J. (1969). The effects of altitude variation in Ethiopian populations. *Philosophical Transactions of the Royal Society*, B, **256**, 147–82.

Harrison, G. A. & Owen, J. J. I. (1956/57). The application of spectrophotometry to the study of skin color inheritance. *Acta genetica*, **6**, 481–4.

Hassan, A. M. El, Godber, M. G., Kopeć, A. C., Mourant, A. E., Tills, D. & Lehmann, H. (1968). The hereditary biood factors of the Beja of the Sudan. *Man*, **3**, 272–83.

Henrici, A. (1973). Numerical classification of Bantu languages. *African Language Studies*, **14**, 82–104.

Hiernaux, J. (1962). Données génétiques sur six populations de la République du Congo (groupes sanguins ABO et Rh, et taux de sicklémie). *Annales de la Société Belge de Médecine Tropicale*, **2**, 145–74.

Hiernaux, J. (1964). Weight/height relationship during growth in Africans and Europeans. *Human Biology*, **36**, 273–93.

Hiernaux, J. (1965). Une nouvelle mesure de distance anthropologique entre populations, utilisant simultanément des fréquences géniques, des pourcentages de traits descriptifs et des moyennes métriques. *Comptes Rendus de l'Académie des Sciences de Paris*, **260**, 1748–50.

Hiernaux, J. (1966a). Les Bushong et les Cwa du royaume Kuba (Congo-Kishasa): pygmées, pygmoïdes et pygméisation; anthropologie, linguistique et expansion bantoue. *Bulletins et Mémoires de la Société d'Anthropologie de Paris*, **9**, 299–336.

Hiernaux, J. (1966b). Human biological diversity in central Africa. *Man*, **1**, 287–306.

Hiernaux, J. (1968). *La diversité humaine en Afrique subsaharienne. Recherches biologiques*. Bruxelles: Institut de Sociologie de l'Université Libre de Bruxelles.

Hiernaux, J. (1969). Investigations anthropobiologiques au Moyen-Chari (République du Tchad) préliminaires à des recherches multidisciplinaires. *Homo*, **20**, 1–11.

Hiernaux, J. (1972). La réflectance de la peau dans une communauté de Sara Madjingay (République du Tchad). *L'Anthropologie*, **76**, 279–99.

Hiernaux, J. (1973). Numerical taxonomy of man: an application to a set of thirty-two African populations. In *Physical Anthropology and its Extending Horizons*, ed. A. Basu, A. K. Ghosh, S. K. Biswas & R. Ghosh, pp. 151–161. Calcutta: Orient Longman.

Hiernaux, J. (1974). *The People of Africa*. London: Weidenfeld and Nicolson.

Adaptation of the African to the rain forest

Hiernaux, J. (1976a). Skin color and climate in Central Africa: a comparison of three populations. *Journal of Human Ecology*, **4**, 69–73.

Hiernaux, J. (1976b). Blood polymorphism frequencies in the Sara Majingay of Chad. *Annals of Human Biology*, **3**, 127–40.

Hiernaux, J., Crognier, E. & Vincke, A. (1974). A comparison of the development of the upper limb in two African populations living in contrasting environments. *International Journal of Ecology and Environmental Sciences*, **1**, 41–6.

Hiernaux, J., Rudan, P. & Brambati, A. (1975). Climate and the weight/height relationship in sub-Saharan Africa. *Annals of Human Biology*, **2**, 3–12.

Hiernaux, J., Vincke, E. & Commelin, D. (1976). Les Oto et les Twa des Konda. *L'Anthropologie*, **80**, 449–64.

Jadin, J. (1935). *Les Groupes Sanguins des Pygmées*. Bruxelles: Institut Royal Colonial Belge.

Jenkins, T. & Corfield, V. (1972). The red cell acid phosphatase polymorphism in Southern Africa: population data and studies on the R, RA and RB phenotypes. *American Journal of Human Genetics*, **23**, 513–32.

Jenkins, T., Zoutendyk, A. & Steinberg, A. G. (1970). Gammaglobulin groups (Gm and Inv) of various Southern African populations. *American Journal of Physical Anthropology*, **32**, 197–218.

Lalouel, J. M. (1973). Topology of population structure. In *Genetic Structure of Populations*, ed. N. E. Morton, pp. 139–52. Honolulu: University Press of Hawaii.

Landsberg, H. E., Lippmann, H., Paffen, K. & Troll, C. (1965). *World Maps of Climatology*. Berlin: Springer-Verlag.

Loomis, W. F. (1967). Skin-pigment regulation of vitamin D bio-synthesis in man. *Science, Washington*, **157**, 501–6.

Moullec, J., Mendrez, C. & Nguyen Van Cong (1966). Les groupes sanguins au Lessouto. *Bulletins et Mémoires de la Société d'Anthropologie de Paris*, **9**, 363–6.

Müller, E. W. (1964). Die Batwa. Eine kleinwüchsige Jägerkaste bei den Móngo-Ekonda. *Zeitschrift fur Ethnologie*, **89**, 206–15.

Pagézy, H. (1973). Adaptation physique et organisation des activités quotidiennes de femmes pygmoïdes Twa et non pygmoïdes Oto de la forêt équatoriale. Thèse de Doctorat 3è cycle, Université Paris VII.

Ramirez, M. E. (1975). Différenciation morphologique parmi huit villages Sara. Thèse de Doctorat 3è cycle, Université Paris VII.

Roberts, D. F. & Bainbridge, D. R. (1963). Nilotic physique. *American Journal of Physical Anthropology*, **21**, 341–70.

Schreider, E. (1951). Race, constitution, thermolyse. *Revue scientifique, Paris*, **89**, 110–19.

Schreider, E. (1963). Anthropologie physiologique et variations climatiques. In *Physiologie et Psychologie en Milieu Aride. Compte rendu de Recherches*, pp. 39–76. Paris: UNESCO

Sporcq, J. (1972). Les Bira de la savane et les Bira de la forêt. *Bulletins et Mémoires de la Société d'Anthropologie de Paris*, **9**, 97–120.

Tanner, J. M., Hiernaux, J. & Jarman, S. (1969). Growth and physique studies. In *Human Biology. A Guide to Field Methods*, ed. J. S. Weiner & J. A. Lourie, pp. 2–76. Oxford & Edinburgh: Blackwell.

Weiner, J. S. (1954). Nose shape and climate. *American Journal of Physical Anthropology*, **12**, 1–4.

8. The Israel study: the anatomy of a population study

O. G. EDHOLM & S. SAMUELOFF

The study carried out in Israel was based on a plan developed and modified over a number of years. During a period of two months (O.G.E.) spent in Israel in 1958 as a visiting professor in the Department of Physiology, at the Hebrew University, Jerusalem, it became clear to all concerned that there were in this country exceptional opportunities for the study of human biology. A preliminary report to the Medical Research Council was prepared in 1963 in which a research plan was outlined. This report owed much to encouragement from Dr Mark Hollis, at that time Director of the Division of Environmental Health at the World Health Organisation (WHO), and from the discussions which marked the inception of the International Biological Programme. Although many of the basic ideas underlying the Israel study had been developed before the IBP was launched, these were so similar to the ideas inherent in the HA Section of the IBP that from then on the future planning for the Isarel study was carried out in the closest possible way with the development of IBP plans and ideas, and specifically in discussions with Professor J. S. Weiner, Convenor of the HA Section.

A second visit to Israel in May–June 1963 provided the opportunity for specific discussions with all those who might either be associated with the research scheme or could provide information or support. In a detailed report to the Medical Research Council on this visit, the problem to be studied was set out as follows: 'To identify and evaluate the relative role of environmental and genetic factors in determining the physiological status of the individual.'

Reasons for proposing that such a study should be made in Israel were put forward, and may be briefly summarised as follows: the basic requirements for carrying out a study, on a problem like the one set out were to find two groups of people with contrasting genetic characteristics, living in similar environmental conditions. The principal difficulty in such a case is to fulfill the last need, i.e. 'similar environmental conditions'. It is quite inadequate to rely only on an identical climate; physical activity, living and social conditions must also be similar. However, in Israel these requirements could be met. The greater part of the Israeli population consists of immigrants who have entered the state since 1948 from nearly every country in the world. It had been deliberate policy to establish rural settlements occupied by individuals of similar ethnic origin, and in many cases the inhabitants of such a village would have left their village of origin as a group and remained in their new homeland as a group. So a village might consist entirely of Jews from the Yemen, originating from the same village(s) in the Yemen, and the adjacent village would be inhabited by a group of immigrants from a different country. The inhabitants of two neighbouring villages would not only live in the same climate but would also farm similar land, live in similar houses and have similar social conditions. Furthermore, inhabitants of two such villages could have distinctly different genetic characteristics; indeed, studies carried out in Israel showed quite clearly that there were considerable genetic differences between many of the immigrant groups although all of them are Jews (Goldschmidt, 1963).

The original proposals and the final research scheme differed not in principle but in scale and scope. The first plan was to study three contrasting ethnic groups such as:

(e) European Ashkenazi Jews, e.g. Poles, Romanians or Bulgarians;

(b) Oriental Jews, e.g. Yemenite or Iraqui;

(c) North African Jews or groups such as Indian Cochin Jews.

Furthermore, it was proposed that the subjects should be settlers in different climatic zones. Although Israel is a small country, the climate varies considerably, encompassing a coastal strip, a southern semi-arid region (Negev), the highland country round Jerusalem and the semi-tropical Jordan valley. Omitting the coastal zone, similar ethnic groups were to be examined in each of the other three regions, making a total of nine communities. A population of 1000 in each ethnic group was envisaged, making a total of 9000 subjects altogether. The observations to be made were set out (in 1963) as follows.

Physiology

Objectives: the measurement of energy expenditure, water and electrolyte balance, work capacity, tolerance to heat and cold, diurnal rhythms, and studies on peripheral circulation; seasonal factors to be included.

Nutrition

Objectives: to measure the food intake in terms of calories, proportions of fat, protein and carbohydrate, vitamins and minerals, of the whole population. This would be done by measuring household or family group intakes, with a sample of some 10% on an individual basis.

Physical anthropology

Objectives: to record height, weight, skinfold thickness, girth, and limb length in all members of the population studied. The growth rate in children

220

and adolescents to be studied over a five-year period. Characteristics of posture, movement, and the use of equipment at home and work to be assessed.

Genetics and demography

Objectives: (i) Genetics – to study genetic markers with all present methods available, including blood groups, haemoglobin chemistry, serum groups and enzymes, and incidence of genetically determined characteristics such as taste and colour blindness, congenital abnormalities, etc. in order to follow dynamics and consequence of gene flow. (ii) Demography – to obtain, from existing records and by further enquiry, complete demographic data, including occupation, degree of parental consanguinity, mating and reproduction records (pregnancies, still-births, multiple births, etc.). Information to be maintained throughout the survey period.

Epidemiology and clinical medicine

Objectives: to obtain and examine all available medical records, and to conduct detailed clinical examinations of the subjects participating in the survey; to maintain a morbidity study, and in conjunction with demographic enquiry to maintain records of births and deaths, including post-mortem data.

Social psychology

Objectives: to examine details of social structure and organisation, intelligence levels, including educational attainments, income levels, hours of work and type of work, and housing. To examine degrees and levels of psychological tensions and satisfaction.

Meterology

Objectives: to measure dry- and wet-bulb temperature, solar and reflected radiation, air speed, and to record cloudiness, rainfall, dust storms, etc. in the villages of the examined populations. In conjunction with meteorological recording, comfort votes to be obtained from a sample population with observation on clothing worn and activity.

Water and atmosphere

Objectives: to measure, as continuously as possible, the quality of water supply and the degree of atmospheric pollution. Where insecticides are used, to measure pollution of food, crops etc.

Details of staff required were worked out and a provisional budget for the proposed five-year study period was $(US)1 000 000 (£(sterling)400 000). At that time, there appeared to be good reasons to believe that a budget of this magnitude and facilities to carry out this study were possible, since the plan as developed included many leading Israeli research workers as participants, and there was promise of strong financial support from the USA.

In 1964, a group assembled by Professor J. S. Weiner met at Burg Wartenstein and discussed various human adaptability proposals and possibilities. During the two weeks of the conference there was a detailed consideration of the Israel scheme, with thorough criticism, comments and suggestions. A paper on the state of human biology of south-west Asia included an outline of the Israel research proposals and was published in the proceedings of the conference (Edholm, 1966). Starting in 1963 there had been a series of meetings, sponsored by the HA Section of the IBP, of internationally selected specialists to discuss and develop agreed methods for use in IBP studies. The Israel study had also been provisionally accepted as part of the British and Israeli IBP research projects. It soon became clear that the original (1963) proposals were on too large a scale, since adequate financial support was not available. Furthermore, the detailed planning showed that the total size of the teams required to carry out all the proposed work was too big and that the impact on the villages would have been severe. Finally, as the implications of the required field work became clearer a number of colleagues in Israel realised it would be impossible for them to fulfill their obligations. However, the determination to carry out the collaborative study became stronger with the occurrence of each new obstacle.

Drastic pruning or scrapping of the scheme were the only other choices. At this stage, many discussions were held and much encouragement received from colleagues, above all from Professor J. Magnes, Head of the Department of Physiology in Jerusalem, and Professor J. S. Weiner. The decisions reached by the various IBP planning groups were also essential in shaping the final programme. One such decision was of critical importance, that work should be concentrated on the age group of 20–30 years, since in most cases it was evident, apart from Israel, that it would be extremely difficult to examine all members of a population group. It was considered, from the point of view of comparisons to be made between studies in various parts of the world, that as complete work as possible be made in this one age group. Where possible, organisers of regional studies would also include other age groups but not at the expense of those aged 20–30. Figures from the 1961 census in Israel showed that, in the rural villages, approximately 10% of the population were between 20 and 30 years of age. Taking the original estimate of 1000 subjects in each population (all ages) there would be approximately 100 subjects in the age group of 20–30 years. This represented a very large reduction as far as the preliminary budget of the Israel study was concerned, and a tentative figure of £(sterling)50000 was calculated. However, it was realised that even this sum would be difficult to obtain. Therefore it was decided further to reduce the scale of study by examining two ethnic communities instead of three. Furthermore, at this time there were still many uncertainties regarding the effectiveness of the scheme, and the various ways in which it might fail. A proposal was prepared which included only the first stage of the scheme, i.e. a study of two ethnic groups living in the semi-arid Negev region of Israel, and to defer consideration of further work in other climatic zones until the first study had been completed and evaluated. This proposal was accepted by the Medical Research Council in May 1967: in June, war broke out between Israel and Egypt together with Jordan and Syria. In October, during a four-week meeting in Israel, final details were worked out and decisions were taken regarding the villagers to be studied. This last point involved discussions with members of the Bureau of Statistics, who were able to provide, in terms of population and ethnic origin, details of the villages in the Negev. The next stage in choice was advice from one of the chief physicians of the Beersheva Hospital, Dr E. Lehmann, regarding the specific ethnic communities to be studied. Dr Lehmann first began work in Beersheva shortly after the Second World War, and had been closely associated with the development of medical services for the rural community in the Negev so was familiar with the characteristics of the villagers. The organisation of medical services in Israel is complex, but as far as rural villages are concerned medical care is provided by the National Federation of Labour, the Histadrut, which also operates hospital services such as the Beersheva Hospital. Each village has a clinic and a nurse, frequently male, who may look after several villages depending on their size, and sees patients sometimes daily but, in any case, several times a week. A physician can be called on request by the nurse, and usually has one session a week in the village clinic; such a physician is sometimes a member of the staff of the district hospital. After discussions and advice from Dr Lehmann it was decided that the two ethnic communities to be studied would be Kurdish and Yemenite Jews living in villages adjacent to Beersheva, the main town of the Negev. From genetic studies already carried out, it was known that there were substantial differences between the markers genes of Kurdish and Yemenite Jews. Another reason for choosing these two ethnic groups was that they had both emigrated to Israel at the same time (1950–51), whereas the influx of North African Jews had come much later. It was also evident that there were suitable villages inhabited either by Kurdish or Yemenite Jews close to each other and reasonably accessible to Beersheva.

At this stage, detailed planning of the organisation of the study was carried out in close contact with Professor Magnes. Professor Samueloff took the responsibility of coordination of the research project in Israel. Professor

O. G. Edholm and S. Samueloff

Samueloff's main research interest was in human environmental physiology and his close relationship to Professor Edholm stemmed from the time of his postdoctoral training and research work at the Division of Human Physiology at Hampstead. On return to Israel he had been engaged in developing work in environmental physiology at the Negev Institute for Arid Zone Research, situated in Beersheva. As a result, the Director of the Institute and the Secretary of the Israel National Council for Research and Development, which was responsible for the Institute, had generously agreed to make facilities there available for the study.

The final plan as it was worked out in October 1967 differed from the original both in scale and in detail, although the principles remained the same. It is easiest to describe in terms of field work and laboratory studies.

Field work

Clinical examination

All those in the age group 20–30 years were to be examined. Standardised forms were prepared for recording all clinical findings, so the data could be transferred to punch cards or tapes. The information was required in order to be able to compare the state of health of the two ethnic groups, to determine if there were individuals with disease of sufficient severity to warrant exclusion from the study, and to ensure that any apparent differences in the further study between the Kurdish and Yemenite Jews could be correctly interpreted. The clinical examination included routine blood and urine tests, and scrutiny of teeth.

Anthropometry

The intention was to measure all those who had attended the clinical examination. The measurements made and the techniques used were to be as described in the IBP Handbook on methodology (Weiner & Lourie, 1969). Skin colour was examined on the inner surface of the upper arm, colour vision was tested and dermatoglyphs recorded.

Demography

Details of the population of the selected villages, including age and sex distribution, were obtained from the Israel Bureau of Statistics based on the 1961 census. This information was then checked with the village secretary, as well as the local government office in the nearby regional centre. As a preliminary to the clinical examination, a house-to-house enquiry was made to ensure that all those in the 20–30 age group received an invitation to attend at the local clinic, and that so far as possible all were interviewed, the objectives of the

224

study explained, and questions and objections answered. During the clinical examination itself, details were to be obtained of occupation, primary and secondary education, marital status, age, number of children and, in the case of women, number of miscarriages, still-births, etc.

Climatic conditions

Details were available, from a local meterological station, of temperature and humidity (continuous recording), sunshine (continuous recording), wind speed and direction, cloud over (four-hourly) and precipitation. In addition, observations were made in the villages of dry- and wet-bulb temperature, and globe thermometer readings.

Daily energy expenditure and habitual activity

The intention was to observe each individual subject throughout a working day, and to record time spent in each activity. Subsequently, the subject was to be interviewed daily to obtain details of activities throughout the preceding day, noting time of day and periods spent indoors, outdoors or in vehicles. Energy expenditure and heart rates were to be measured during the performance of specific tasks, and estimates of the total daily or nightly energy expenditure were to be obtained by continuous recording of heart rate during the day and night, using the SAMI (Socially Acceptable Monitoring Instrument) heart rate integrator. So far as possible it was intended that all subjects considered to be in good health should be studied.

Food intake

Daily food intake was planned to be assessed from interviews with housewives, using a standardised questionnaire, and checked with an inventory of food supplies in the house. Housewives were interviewed three times during the week of study.

The plan for the field studies envisaged a week as the measurement period for each subject, that is, the assessment of food intake, habitual activity, energy expenditure and the climatic conditions to which he was exposed. The studies were to be carried out in the warmest and coldest months of the year, i.e. June–July and January–February, respectively.

Laboratory studies

These were the critical investigations in which physiological responses of the individual were to be measured. The objective of the laboratory studies was to measure physiological responses to carefully standardised stimuli and so be able to compare the physiological status of the Kurdish and Yemenite Jews. Two tests were chosen, following recommendations of the HA Section of the IBP. These were the gas exchange response to exercise and the sweat response to heat. In both cases, standardised procedures had been tested thoroughly by internal committees of experts and agreement reached by the majority of interested physiologists throughout the world as to the method to be used.

These physiological tests were to be carried out at the Negev Institute for Arid Zone Research in Beersheva. Laboratory space was provided but virtually all the equipment required had to be set up specifically for the trial.

The number of subjects who could be examined physiologically, i.e. heat and exercise response, was determined by the size of the field team and the length of time needed for carrying out the studies. All this was largely a question of cost and it was cost again which brought a substantial reduction in the original scheme of the laboratory investigations.

The work load, size and membership of the various teams were planned by Professors Magnes, Samueloff and Edholm. It was agreed that whenever feasible the participants in the research teams should be recruited in Israel, for obvious reasons of language and saving in travelling costs. However, a substantial number of researchers had to be brought from England as the number of those engaged in human physiology in Israel was small and their university duties made it difficult to devote sufficient time to the project. The final division of labour was as follows.

Laboratory studies

Heat tolerance test

Dr R. H. Fox (MRC) assisted by Mr J. W. Jack and Mr A. J. Hackett (MRC), with general help from two Israeli junior laboratory assistants. Dr Z. Even Paz (Jerusaelm) took part in these tests and eventually took over the main direction from Dr Fox.

Exercise test

Supervised and organised throughout by Professor Samueloff (Israel); in the initial planning phase Dr C. T. M. Davies (MRC) played a consultant role and was present in Israel for the first part of the actual laboratory studies. Other colleagues of Professor Samueloff who were included in the team for all or part of the study were Drs Schwarz and Beer.

Field studies – under the general direction of Dr O. G. Edholm

Clinical examination

Dr E. Lehmann was responsible for all the clinical examinations carried out in the Negev trial. A number of his colleagues and house-staff at the Beersheva Hospital assisted him and Professor Samueloff, who organised the examinations in the different villages, was responsible for part of the clinical examination.

Genetic markers

Venous blood samples were collected from each subject at the clinical examination, and taken immediately, packed in ice, to the airport at Lod (approximately three hours' drive from Beersheva) and thence flown direct to Heathrow from where the samples were collected and taken to Dr A. E. Mourant's MRC Laboratory of Blood Group Studies. Dr Mourant and Mr D. Tills, assisted by the staff of the laboratory, were responsible for the studies of genetic markers. In addition, Dr G. Beaven also examined the blood samples for thalassaemia and abnormal haemoglobins.

Nutrition

Dr Sarah Bavly, Head of the School of Domestic Science at Jerusalem, was responsible for the assessment of food intake by the individual participants, and a number of her assistants carried out the actual field work.

Anthropology

In the Negev study, Dr J. Lourie (MRC) carried out all the measurements made of body dimensions, including skin colour, forced expiratory volume, forced ventilatory capacity, dermatoglyphs and colour vision, assisted principally by S. Humphrey (MRC). In the Jerusalem study Dr Gwen Gerrard (MRC) assisted by Miss Y. Shechter (Israel) carried out the anthropological survey.

Habitual activity and energy expenditure, and environmental conditions

This heading includes the major component of the field studies, and the team responsible changed during the course of the studies as the total time spent was considerably greater than on any other single aspect of the project. Amongst the members were Dr J. Brotherhood, Dr Barbara Tredre, Dr J. Lourie, Mr S. Humphrey, Dr Gwen Garrard, under the leadership of Dr O. G. Edholm; all were members of the MRC Division of Human Physiology.

O. G. Edholm and S. Samueloff

In addition, assistance was obtained from various Israeli students and laboratory technicians who helped with the field studies for varying periods of time.

General organisation and planning

This was essentially Professor Samueloff's responsibility. His problems were to get the full cooperation of the villagers, to maintain their interest, and to achieve this he had to establish contact with local authorities, the health service and the various government departments concerned. When Professor Samueloff began his work there was very little guidance available on the best methods for carrying out such a complex tesk. However, it soon became evident that in the preparation and carrying out of a population study a large proportion of time and energy should be invested in para-scientific activities such as public relations, consideration of psychological characteristics of the potential subjects and, most important of all, careful preparation of the presentation to explain the objectives of the study. There is nothing worse than to underestimate the natural mental abilities of the prospective subjects. In our case, the Kurdish Jews had the reputation of being aloof and suspicious; the Yemenite Jews were understood to have marked reservations about blood samples. However, in both groups obstacles were overcome with patience and human understanding and close friendly relationships were established.

The objectives of the proposed study were sophisticated and at that stage we were concerned about the way in which we could explain why we wanted to do the study and why we were asking the villagers for active cooperation and help. The first contact with the villagers was during a party held in each village in the public hall. These parties were attended by Professor Samueloff and Dr Lehmann together with local officials. All the villagers in the 20–30 age group received individual invitations and very considerable efforts were made to ensure that all attended. At the party a film was shown, illustrating studies of human physiology, the general plan of the study described and questions answered. Frequently provocative points were raised such as, what benefit will the villagers gain from this study. On such occasions, unusual aspects of scientific investigations were successfully discussed with the subjects. Participation in a research study was pointed out as an opportunity to meet people from the 'outside' – Jerusalem and London – to see new equipment and techniques, to make new social acquaintances and to take part in activities which could be fun and as such different from the routine everyday life in a small village. Following the party, dates were fixed for the clinical examinations.

The second step, and in the event a most important one for the success of the study, was the establishment of an intermediate liaison between the local

228

population and the research team. It soon became evident that the best choice for this job would be a local man, or even better, the local nurse or male nurse. The existing professional contacts between the clinic and regional hospital helped in recruitment of the local man. In the Negev a good example of a most successful choice in this respect was Mr S. Salech, the male nurse in charge of two of the village clinics. Mr Salech, an Iraqui Jew, spoke good English and was widely respected not only in the villages where he ran the clinics but was invaluable in contacting individual subjects and explaining the whys and wherefores. The recruitment of a local nurse as a member of the research team greatly facilitated the clinical examinations in all villages.

The clinical examinations were held in the village clinic in the evenings. It was expected that 30 subjects could be examined each evening but in practice the numbers varied from under 10 to approximately 30. On several occasions it was necessary to visit subjects in their homes and bring them to the clinic. As a result, 76% of the total population in the age group 20–30 were examined clinically. There were fewer objections to venous blood samples than anticipated; on the other hand, there were more cases of genuine illness preventing attendance than predicted. These included a psychiatric group – patients suffering from various mental disorders whose existence was not established until the studies had been under way for some months. It was not a question of concealment; it was assumed that we would not be interested in those who were so abnormal they were unable to work. Some of the other demographic problems will be mentioned below.

It was necessary to put a lot of energy into planning and promoting at the onset; it soon became clear that to maintain interest, continuous reinforcement of the initial build-up was equally essential. Following the clinical examination there was a gap of several weeks before field work began, and some of the potential subjects thought the study was already over and their interest had to be reawakened. This was largely achieved by having evening sessions in the village clinic for recording the measurements of body size and dimensions, skin colour, dermatoglyphs, handclasping and a standardised photograph of each subject. These evenings soon developed into lively social events, not always confined to the actual subjects but including older and younger members of the family; so the team and the subjects became acquainted. Since the establishment of close and preferably cordial relationships between the members of the team and subjects was rightly regarded (rightly in the light of experience) as the key to success in this venture, the narrative style of this present account will be changed by considering the problems involved and the lessons learned, looking back on the study as a whole. In what follows, it will be realised that problems were not immediately identified, nor were useful solutions always found at once. It would be tedious to go over the course of the study, week by week, charting failures and successes, changes of plan and the varying morale of the team members. Since field

O. G. Edholm and S. Samueloff

work of the types involved in the HA Section of the IBP is expensive in man-
power, time and money, and since the necessary skills are not generally
taught in universities, recording experience and conclusions reached about
such field work could be regarded as at least as necessary as the description of
methods or laboratory procedures. It is with such a belief that the present
comments have been written, and before the conclusions of this account are
reached it is necessary to state clearly a very obvious point – namely that the
problems peculiar to one field study may not exist in other studies, and the
solutions which worked well in Israel might be quite useless in India. Until a
body of knowledge and experience is built up, it would be dangerous to
generalise. No doubt, at this stage, the social anthropologist with field experience would give up any further reading of such naive comments, muttering
'surely the human biologists know something of the considerable experience
and detailed publications about field work which has always been considered
of crucial importance in the training of anthropologists, especially in the UK'.
Here one has to admit that the physiologist interested in examining man in a
variety of environments is, in general, extremely ignorant about cultural or
social anthropological studies, and in defence to argue that some of the ways
in which anthropological field studies are carried out do not appear to be
applicable to human biological work. By far the most important practical
point is the time factor: human biological field work has so far generally been
measured in weeks, sometimes in months and very rarely for periods as long
as a year, whereas a year's residence in a village (or its equivalent) would be
regarded as the minimum period for an anthropologist. Before attempting to
adjudicate on duration of study, and alternative strategies, the role of social
anthropology in human biological studies should be examined, however
superficially. At the planning stage of the Israel study, the potential value of
including anthropological work was discussed. At that time, it was realised
that this could be useful but it did not appear to have high priority, so in
view of the difficulties (and the cost) of finding and engaging an appropriate
social anthropologist, it was quickly decided it was not necessary, and only
anthropometric measurements of the subjects were included. At the time of
planning the second Israel study (see below), in the light of experience of the
first study the question was again discussed and the same decision reached.
After completion of the Israel studies and their detailed evaluation, some of
the advantages of including social anthropology or, rather, the effects of the
failure to include anthropology were evaluated. The details of the organisation of the village could have been learnt from the anthropologist: such
details as were of relevance were observed and appreciated by the members of
the field team, who then questioned the village secretary and/or the village
nurse for advice and interpretation. Nevertheless, it took time to get such
information and to learn how to interpret it. One point which was of relevance concerned the way in which farm work was carried out. The villages

230

were described as '*moshav*', i.e. village co-operatives, which meant, *inter alia*, central buying of seed and central selling of some products. It also meant co-operation in the harvesting of crops, but the details of such co-operation were complex. All the land in theory belonged to the State and was leased to the individual farmer; although land was not limited, water was rationed, and in the Negev, with very low rainfall, land without water was virtually useless. There was only a limited area which could be effectively irrigated, hence in practice a shortage of land soon developed, so when sons grew up there was no land for them or very little. Hence, those without or with little land had to have other work if they wished to remain in the village, and many worked as farm labourers, often on a day-to-day basis. If we had understood the working of the local labour market from the beginning, it would have been possible to plan more effectively the observation and assessment of individual habitual activity.

Economic aspects of the population were not of primary interest to the team, and were only relevant in so far as family income might affect food intake, or the kind of work the individual was prepared to do. However, the team did not know the income of any family, nor the differences in income between subjects or villages, or if there were income differences between Kurdish and Yemenite villages. Here, no doubt, some marginally useful information could have been provided by a social anthropologist.

It is difficult, even with hindsight, to know how far an anthropologist would have been able to advise about the conduct of the field work, so as to avoid conflicts or difficulties and increase co-operation as regards attendance in the laboratory for the standardised tests of responses to exercise and to a raised body temperature. The main problem, as stated on p. 228, and one which is certainly common to all field studies, is the need to obtain full and complete co-operation so that all the subjects who should be are studied, and the studies are complete. Success at the 100% level is probably non-existent, but actual studies show variations between 50 and 80%, few being above 85%. In the Israel studies, in both parts about 75–80% of all potential subjects in the age group 20–30 attended the clinical examination. There were only a few of these who did not have their body dimensions etc. recorded (anthropometry) and their genetic blood markers also, but there was a considerably lower percentage in whom habitual activity and energy expenditure was recorded and lower still for food intake. The lowest proportions were those who attended the laboratories. These numbers are set out in Table 8.1. The fall-off reflects (*e*) the difficulties, real or perceived, of any particular procedure as affecting the subject; and (*b*) the limitation in number of subjects with whom the team could deal. The social anthropologist might have been able to improve subject co-operation by explaining the objects of the exercise and answering questions, correcting misunderstandings with his background knowledge; the anthropologist could also have conveyed and explained the experimental findings

Table 8.1. *Number of subjects* (*Negev study*)

	Kurdish Jews		Yemenite Jews	
	♂	♀	♂	♀
Clinical examination	67	52	37	39
Anthropometry	56	41	34	38
Energy expenditure	48	23	32	21
Work capacity	19	12	20	12
Heat test	20	13	19	15

better than the members of the team could do. So it would appear that a principal role for the anthropologist would be to act as a channel of communication. In the present study this role was largely undertaken by the village nurse. In the Negev, Salem Salach was outstandingly effective in this role: in the Judean hills, help was obtained from two of the village nurses and was useful but not of the same order as in the Negev. In two other villages co-operation with the village nurses was poor. However, even with better communication and understanding not many more subjects could have been studied given the size of the team and the time available. With our present information it is not possible to formulate precise values about team size; it is obvious that too big a team causes severe disturbance within the community studies, and it is certain that even one stranger can produce changes, although probably for a short time only and possibly insignificantly. But at what stage the disturbance is sufficient to distort data, and how long a team can remain or should remain is far from certain. In the Israel study there were also language difficulties; the English members of the team all learnt some Hebrew, and one or two were quite fluent. A number of villagers knew some English or French, or German. In practice, there were no great linguistic problems and the team members in many cases were accepted as friends. Some of the villagers were less friendly and in some cases resented the whole investigation, but these were few and did not seriously affect the study as a whole. If a similar study was planned in the future, a determined effort to get a suitable social anthropologist working in the village for some time before field studies began would be well worth while.

Frequency and duration of study

In the initial plan, four periods of study to coincide with the four seasons were envisaged. In the event, only summer and winter studies were carried out, largely for reasons of cost. However, observation had shown that there were no obvious seasonal variations of activity as regards farming. In the semi-arid desert region of the Negev irrigation has to be used, as rainfall is not only small, with 8 cm/year, but also occurs on a few days with quite heavy rain with a run-off into wadis and streams. The bulk of the farming could be described as market gardening, with emphasis on vegetables, fruit and flowers. There was no season of harvest or of sowing; throughout the year some ground was being prepared; on the other land, crops were harvested and some plants would be at various stages of growth, requiring different treatment such as hoeing or spraying with fertilisers or insecticides. There was evidence that energy expenditure by farmers probably did not differ much throughout the year, and subsequent detailed studies in summer and winter confirmed this evidence. In other regions with different farming practices, marked seasonal variations of activity could be expected, so no general principle can be laid down except that observations which can be relatively casual should be made at two- to three-month intervals throughout the year to determine when and to what degree variations occur. Here, too, the taking of histories from village secretaries and from departments of agriculture would help to complete the picture. In some studies, some members of the investigating team have remained in or near the village throughout the year, and this would seem the ideal solution if cost is not critical. In the Israel study, a senior member of the technical staff of the Arid Zone Research Institute at Beersheva made weekly visits to all five villages in order to check the recordings made in some houses of temperature and humidity. In this way, a useful contact was retained throughout the year with a minimum disturbance.

It would be tedious to list all the individuals and authorities who were visited and consulted prior to and during the study, especially as local conditions must influence the need and the extent of such contacts. The only generalisation is a trite one: make sure you visit every possible official you can. If consulted they are generally helpful, if not they can be extremely obstructive. Special mention of all those who assisted or advised in any way has been made in the papers published about the study.

Methods and results

Since detailed accounts have been published in the *Philosophical Transactions of the Royal Society* (Edholm & Samueloff, 1973) of methods, results and discussion of the results, only a relatively brief resumé will be given in this report together with comments on the methods and the problems involved in the use of the methods. The object of this section will be not so much to repeat the findings and the technical discussion, as to criticise the way in which the investigation was carried out and what lessons might be learned for use in future field work. The discussion will be first on the village field studies and then the laboratory work.

The clinical studies could be regarded as satisfactory; they were carried out by medical personnel who were, in most cases, well known and respected by the villagers, and attendance was good since a knowledge of one's state of health was prized by the great majority. There was far less objection to the taking of blood samples than expected, and in those cases when subsequently a second sample was requested to confirm doubtful findings most subjects agreed readily. Provision of transportation to bring subjects to the clinic proved to be rewarding: however, a number attended with reluctance. Although much effort was devoted to persuading individually every one to attend, and personal invitations were delivered at all households, probably much energy could have been saved here by the advice and guidance of an anthropologist.

In one respect the clinical examination could be criticised; not enough was done to provide villagers with the results of the examination, which would have increased the personal interest of the participants in the study. This is a question which needs to be discussed in detail with the clinicians responsible for carrying out the examination, to determine the best way this can be done without interfering with the patient–local doctor relationship, and so the discussion and consequent actions should be agreed *before* the clinical examination takes place. A primary object of the clinical examination was to find and exclude individuals whose state of health was such that they should not be subjects for the laboratory studies. It was agreed that the decision would rest with the senior physician, Dr Lehmann. Inevitably, this involved a degree of subjective judgement, and when the whole study was repeated in a different area of Israel with another senior physician in charge, the judgements leading to inclusion or exclusion were not necessarily precisely the same. It was not expected that this would be a serious problem as the population concerned was all in the age range 20–30, living in rural conditions, for the most part engaged in farming and with good nutritional standards. It was one of the unexpected results of this study that in spite of such favourable conditions there was a relatively large number of rejects, approximately 16% of all subjects. In addition, it was agreed that no pregnant women would be included in the laboratory tests, although there was no evidence to suggest that these tests would be in any way harmful. This decision caused difficulties as in the 20–30 age group a number of women were pregnant at the time of the summer study, and when the winter study began a majority of those who were studied in the summer were pregnant. This meant that seasonal comparisons were difficult to make, and the total number of women studied was relatively small. In future, in order to make comparisons between groups, seasonal changes may have to be assessed by drawing on much larger populations or inferences drawn by comparisons between one group studied in the summer and a partially different group in the winter.

There was no significant difference between the Kurdish and Yemenite Jews as regards the incidence of disease. In both groups the men appeared to be more healthy than the women. Both communities had good teeth but the incidence of decayed, missing or filled teeth (DMF Index) was just significantly greater amongst the Kurdish Jews.

No one suffered from any form of respiratory disease, in striking contrast to the high incidence of tuberculosis amongst the Yemenite Jews when they entered Israel in 1949–50. Other points of interest were the very low incidence of visual defects (in the Negev only one subject wore spectacles) and the relatively common enlargement of the spleen associated on occasion with an enlarged liver. The cause or causes of the enlargement was not clear.

As mentioned above, 16% of those examined were rejected for the laboratory tests. The majority of these were not seriously ill, as they were all working, as housewives or as farmers, etc. In addition, there were some of the inhabitants who were not examined as these did not present themselves. Subsequently, great efforts were made to trace all the missing subjects, and it is probable that information (sometimes limited) was obtained about everyone from the secretaries' records, and the clinical records in the village clinics. The largest number of missing male subjects were in the Army; these were assumed to be 'healthy'..There were also those incapacitated by disease who were either in hospital or at home, unable to work. These included subjects suffering from a variety of psychiatric disorders, and several cases of congenital dislocation of the hip. It was estimated that not more than 5–10% of the total group were unwilling to co-operate. In order to gain their interest, very considerable further efforts would have been required to overcome indifference or, in a few cases, hostility. In general, the response was regarded as satisfactory. Some of the ways by which this was achieved have been indicated; they included a small gift of wine for each subject who attended the clinical examination as a sign of personal attention. Subjects who were asked to take part in the laboratory tests were given 40 Israeli pounds as compensation for their absence from their work for a day. At that time this was equivalent to £5 sterling and was approximately the same as the average daily income.

The results of the laboratory study in the Negev have been published in detail (Edholm & Samueloff, 1973) and will only be summarised here. The main conclusion was that the Yemenite and Kurdish Jews were genetically markedly dissimilar but had virtually identical physiological responses and, since environmental conditions for the two ethnic groups were shown to be the same, the physiological identity was attributed to environmental factors. In the determination of physiological characteristics it would appear that environmental factors were prepotent over genetic factors. Since this conclusion is of fundamental interest it was decided to repeat the study to determine whether this conclusion would be supported or refuted.

O. G. Edholm and S. Samueloff

The second stage of the Israel study

Once more, the subjects were Kurdish and Yemenite Jews living in the Judean hills not far from Jerusalem. The design of the study followed as closely as possible the lines of the first observations in the Negev. The differences were mainly the omission of some procedures, due in most cases to a smaller research team. In the planning stage it was assumed that with the experience gained in the Negev study the team work in the Judean hills would prove easier. Although in the event there was more confidence about carrying through the project successfully, nevertheless there were a number of problems arising from the new conditions. These in turn were due in part to the differences in farming between the Judean hills and the Negev and in part to social changes in Israel. Following the Six-day War in 1967, Jerusalem had grown in size and population. Hence villages which in the first surveys, before 1967, had been relatively isolated and largely dependent on agriculture were now (1971–72) virtually suburbs of Jerusalem, with good public transport. One consequence was that many villagers now had a variety of jobs in Jerusalem, in shops, offices and factories, and commuted to and from the villages. In a careful survey it proved impossible to identify agricultural villages similar to those in the Negev. Four villages were eventually selected, two inhabited by Kurdish Jews and two by Yemenite Jews. Two of the villages were in the hills, 13 and 15 km west of Jerusalem, at an altitude of approximately 650 m; two were at a lower altitude, approximately 300 m where the Judean hills give way to the coastal plain. These two were about 25–30 km west of Jerusalem.

The choice of the four villages was the result of a compromise: the lowland villages were primarily agricultural but their climate was in many respects similar to the Negev, although with rather more rainfall. The highland villages had a climate closely resembling that of Jerusalem, with a cool, even cold, winter, but with only a minority engaged in farming. The problems encountered here resulted from the difficulties of studying subjects commuting daily to Jerusalem. This required more time devoted to each subject by the observers and placed limitations on the measurements which could be made. In the Negev the SAMI heart rate integrator had been used as one method to assess daily energy expenditure. The results were difficult to interpret, so daytime recording of heart rate was omitted in the second study which was termed, for convenience, the Judean Hills project.

The farming was affected by the geography of the region. In the Judean hills the country is broken by steep valleys, the sides are terraced and there are small fields in the flat valley bottom. The main produce is fruit, plums, peaches, apples, and some crops of vegetables, including tomatoes and cucumber. In the lowland villages, grapes are important in the summer and flowers in the winter. There are much larger fields and the crops raised are

236

in many cases the same as in the Negev. However, in all four villages many subjects worked in factories or offices in the nearby towns. There had also been other social changes since the first study, of which the most obvious was the widespread ownership and use of television sets. Instead of the busy social life in the evenings, which had been such a feature of the Negev villages, there were few people to be seen in the evenings but in nearly every house the television screen was on and the family was watching. This change in habits simplified the recording of habitual activity – nearly all the subjects spent every evening at home sitting and watching television.

The villagers in general appeared more prosperous than in the Negev, with larger houses which were better furnished. A feature common to all four villages was the raising of poultry, mainly for eggs. The hens were kept in batteries, up to 2000–3000 in number. The work involved in feeding, watering, cleaning, as well as collecting the eggs etc., was usually shared by husband and wife. The income from the sale of eggs was considerable, and when large numbers of hens died during an outbreak of Newcastle disease (prior to the winter study) many households were in difficulties.

The field work was carried out in a manner as comparable with the Negev study as possible. Some imposed differences have already been mentioned. In addition, it was not possible to have all the medical examinations carried out by the same physician; in any case, it is not easy to compare one clinical study with another when different physicians are involved. The same number of subjects was examined as in the Negev but there was a better balance between Kurds and Yemenites. As already mentioned, the monitoring of daily activities was considerably more difficult than in the Negev, and fewer subjects were studied. Respiratory measurements (FEV and FVC) were made only on a limited number. The estimation of food intake proved to be on the whole unsatisfactory and the quantitative aspects were considered to be unreliable.

The detailed results of the studies in the Judean Hills have not yet been published. In general terms they confirm the findings of the Negev study. There were some differences: the clinical findings indicated about the same level of morbidity as in the Negev but the disease pattern was different, with fewer cases of hepatosplenic enlargement and more cases of cardiac and respiratory disease.

The patterns of habitual activity and of energy expenditure were rather more varied than in the Negev but the average levels were more similar than expected. Although there were climatic differences between the upland and lowland villages these were relatively small. The genetic differences between the Yemenite and Kurdish Jews was as large as had been found in the Negev. However, when the gene frequencies were calculated it was evident that there were some differences between the Kurdish Jews in the Negev and the Judean Hills, and so there were between the Yemenite Jews in the two localities.

237

These differences appear to result from differences in place of origin, i.e. Kurdistan or Yemen. In the former there is a west–east cline, and in the latter a north–south cline. There were also some anthropometric differences; although the contrast between the Yemenite and Kurdish Jews observed in the Negev was also seen in the Judean Hills it was less marked.

In the laboratory studies the methods used for the estimation of oxygen consumption during work on the bicycle ergometer were the same as in the Negev study, and the same team was involved in both studies. However, some changes were made in the assessment of responses to a raised body temperature. There were complaints about the test by a number of subjects in the Negev, and when the tests were repeated in the winter some subjects who had undergone the test in the summer were unwilling to repeat it in the winter. One way to make the test more acceptable was to make it shorter, and Dr R. Fox worked out a new programme in which hyperthermia was maintained for 30 minutes instead of one hour. Satisfactory results were obtained with the new regime but it was not possible to compare the Negev and Judean Hills figures precisely.

However, in the Judean Hills study, the sweat rates of Yemenite and Kurdish Jews were on the average very similar as in the Negev studies and as was also observed in the Negev the women had approximately half the sweat rate of the men. The summer sweat rates were more than twice as high as the winter sweat rates. The measurements of oxygen consumption showed that the Yemenite men had slightly higher maximum oxygen consumptions than the Kurdish men, a difference which was just significant.

Conclusion

This study of Yemenite and Kurdish Jewish villagers living in Israel has shown that environmental factors are prepotent in determining physiological responses. Although genetic differences between the examined population groups were very considerable, these differences did not appear to affect the physiological characteristics that were examined. However, it is realised that such broad generalisations cannot be accepted without some qualifications, some of which have already been mentioned elsewhere (Edholm *et al.*, 1973). The individual subjects showed considerable variation, and only some of the variation could be explained by differences in environmental exposure.

It may be pertinent to ask whether studies of the kind carried out in Israel should be continued or encouraged; are they likely to yield useful information? Are they likely to provide specific answers to the questions raised? It is common practice in recent biological research to search for solutions by using a simple experimental model in order to obtain specific answers to specific questions. In human biology the relationship between the various parameters is of a much more complicated nature and the approach to complex phenomena could be by interpretation of research from different disciplines. In the Israel study, genetic, physiological and environmental variables have been inter-related. It can be said that this study was a modest trial of a multi-disciplinary approach for investigation of problems of human biology. The results are of interest to a variety of related disciplines, and these might be considered to justify such a long, laborious, time-consuming investigation. However, the central question of the relative role of genetic and environmental factors in the development and maintenance of physio-logical characteristics may not be answerable by further work of this kind. The difficulties include the need for large numbers of subjects to find out if there are significant differences between two groups. Since the physiological tests available today are complex, the number of subjects who can be examined by the physiological team is limited. What is required is the development of physiological tests which are simple and short and acceptable by the subject. Amongst the HA studies some have shown that physiological differences do exist between different ethnic groups, but so far there have also been large environmental differences as well. It would be worth while in a country like Israel to do extensive screening tests to find if, under circumstances where environment may be similar, physiological differences exist. What has been learnt from the Israel study is that it is possible to carry out investigations of acceptable scientific standards in the field on relatively large numbers of subjects. It would be sad if such experience was not further exploited.

References

Edholm, O. G. (1966). Southwest Asia, with special reference to Israel. In *The Biology of Human Adaptability*, ed. P. T. Baker & J. S. Weiner, pp. 357–94. Clarendon Press.

Edholm, O. G. & Samueloff, S. (1973). Biological studies of Yemenite and Kurdish Jews in Israel and other groups in Southwest Asia. I. Introduction, back-ground and methods. *Philosophical Transactions of the Royal Society*, B, **266**, 85–95.

Edholm, O. G., Samueloff, S., Mourant, A. E., Fox, R. H., Lourie, J., Lehmann, H., Lehman, E. E., Barly, S., Beaven, G. & Even-Paz, Z. (1973). Biological studies of Yemenite and Kurdish Jews in Israel and other groups in Southwest Asia. XIII. Conclusions – summary. *Philosophical Transactions of the Royal Society*, B, **266**, 221–4.

Goldschmidt, E. (ed.) (1963). *The Genetics of Migrant and Isolate Populations*. New York: Williams & Wilkins.

Weiner, J. S. & Lourie, J. (ed.) (1969). *Human Biology. A Guide to Field Methods*. Oxford & Edinburgh: Blackwell.

9. A comparative survey of African people living in the northern semi-arid zone: a search for a baseline

J. HUIZINGA

From 1964 onwards, human biological studies of several African groups have been made by subsequent teams associated with the Institute of Human Biology at Utrecht. The populations concerned live in the climatic zone between the tropical deserts and semi-deserts to the North and the dry savanna belt to the south, i.e. in the northern semi-arid zone, in the Republic of Mali (Dogon), in Upper Volta (Kurumba) and in North Cameroon (Fali and Fulani).

These studies were planned, first, to provide some insight into possible mechanisms of biological adaptation to the relatively extreme environment to which these groups supposedly have been exposed for 'a long time'; and secondly, to offer a description of several aspects of these populations at some point in time which (in later studies) these populations may be taken to have passed during the unavoidable process of subsequent urbanization and westernization (a 'baseline').

In the three regions concerned, the human biological observations were made in January and February, i.e. the middle of the dry season. This period not only offers excellent conditions for field work but as agricultural activities are minimal most local people are available for examinations and interviews.

The Dogon studies, including the collection of data on the extinct Tellem, took place in 1964, 1965, 1966 and 1971; the observations on the Kurumba were made in 1966 and 1967 and the Fali and Fulani were visited in 1968 and 1970.

In order fully to appreciate possible genetic or non-genetic, adaptational or non-adaptational effects of certain environmental or ecological factors on a population, some estimate of the duration of exposure to these influences is necessary. In African practice, if one wants to obtain information about the duration of stay of a population in a certain region, one often has to rely on indirect evidence presented by oral tradition. As far as the time depth of this kind of information is concerned, one is warned by Fage (1965) who feels that beyond 500 years the 'reliability' of oral tradition is generally low.

In the *Dogon* case the analysis of the relevant content of the great amount of information collected during decades by the Griaule–Dieterlen group indicates that between the tenth and thirteenth century Dogon moved from Dyigou (near Tombouctou) to a region called Mandé, or, at least some families, to Hombori, more to the east (Griaule & Dieterlen, 1965, pp. 16–17). At that time Mandé (located southwest of modern Bamako) constituted the central region of the later powerful Mali empire. In the fourteenth to fifteenth century, the Dogon are said to have continued their migration and to have moved from Mandé to their main present-day location in the south central part of Mali. This region is bisected by the well-known high sandstone cliff (the *'falaise'*), an escarpment along which a number of Dogon villages are grouped.

An additional and interesting indication of the time of arrival of the Dogon in their present habitat (the beginning of the fifteenth century) was given by the presence of nine so called Sigi-masks in a cave at Ibi, a village at the foot of the high cliff. Such a mask is used only once at the occasion of the Sigi-celebration, an event that takes place every 60 years (Griaule, 1938, pp. 28 and 245). The possible duration of stay of the Dogon in their present location has been elucidated by our elaborate archaeological investigations of several of the natural caves which honeycomb the sandstone cliff near Sanga (Bed-aux, 1972).

Detailed comparative studies of the abundantly available skeletal remains made it clear that before the fifteenth century this region was inhabited by a now extinct population (Tellem) which clearly differed morphologically from the Dogon (Huizinga, Birnie-Tellier & Glanville, 1967; Huizinga, 1968a, Knip, 1971). The study of remains of material culture strongly supports the view that before the fifteenth century the Dogon had not yet arrived in the region. It may well be stressed that this case is one of the rare instances in which information based on oral tradition could be confronted with, and substantiated by, archaeological and physical anthropological evidence.

There is no reason to assume that the migrations described (which do not have the character of real nomadism) made the Dogon experience appreciable environmental changes.

This also applies to the *Kurumba*, Upper Volta, whose migration history, however, is less clear. The view that they should be considered to be the descendants of the Tellem who originally inhabited the high cliff and subsequently would have been expelled by the arriving Dogon (Schweeger-Hefel & Staude, 1972, pp. 14–16) is not supported by our archaeological and physical anthropological findings (Huizinga, 1968a). Ethnographical evidence as given by Izard (1970, pp. 118–19) suggests that the Kurumba might have immigrated in the beginning of the fifteenth century from a region near Niamey and Say, i.e. to the east of their present location. At that time this area most probably was occupied by Dogon and Nyonyosi, the autochthonous population.

The oral tradition of the *Fali* people suggests a relatively recent immigration into their present-day mountainous environment. Like their predecessors in Northern Cameroon, the so-called 'Sao', they were expelled from different localities, mainly in the north, and found a temporary refuge in the mountains. This forced migration of the Fali dates back as late as the end of the eighteenth and beginning of the nineteenth century (Gauthier & Jansen, 1973, pp. 15–16).

Solid archaeological evidence concerning Fali migrations is lacking. Physical anthropological study of skeletal remains excavated at a site (as yet undated but undoubtedly late) near N'Goutchoumi did not throw light on this problem (Huizinga, unpublished). As in the cases of the Dogon and Kurumba, one can be fairly confident that, in the Fali case also, the migratory movements have not been associated with significant environmental changes.

Climate

In describing 'environment' in terms of simple parameters, climatic variables are often used.

In the case of Africa many climatic classifications have been advocated in which areas of similar climate are grouped, generally using the parameters of temperature and precipitation. It emerges from such classifications that vegetation and climate are closely related, without being identical, however.

The groups studied by us are found in the 'northern semi-arid zone', one of the basic zones according to Griffiths' (1972) climatic classification. They inhabit the 'thorn savanna belt' in Troll's (1965) classification, i.e. a biotope largely coinciding with the 'végétation épineuse' as described by Gourou (1970). The latter's vegetation map has been used by Crognier (1973) in his study on human ecological relationships in Africa, to which I will refer later.

The semi-arid zone is defined as a region having at least 50 mm of rain in each of three to six months. Of course there are climatic differences between any two places within this zone.

As Upper Volta (the country of the Kurumba) falls almost entirely within the zone concerned, the climatic table for Ouagadougou, its capital, may be taken to demonstrate some characteristics of the climate (Griffiths, 1972, p. 208, table x). There, the mean annual precipitation amounts to 897 mm, 85% of which occurs during the months June through September. Mean annual maximum temperature is 35 °C, the mean minimum temperature 22 °C. In certain months (April–May), extreme maxima of 50 °C may be recorded. The mean relative humidity per month is loosely associated with the rainfall and reaches its maximum mean (12.00 h) in August (67%). Minimum means are found in January–February (19%).

The geographical distribution taken by the different morphological forms

243

of the polytypic species man suggests that the ecological rules related to climatic variables followed by warm-blooded polytypic animal species may also be applicable to man.

The oldest of these rules was originally formulated (Gloger, 1833) to account for variations in the colour of birds (feathers) and hairy animals (fur). However, these differences (much less environmentally stable than skin colour in man) are not directly comparable to the differences in skin colour. The adaptive character of the human skin which eventually ought to form the basis of Gloger's rule may well be found in differential responses to ultra-violet radiation.

Bergmann's rule (1847) describes the association between 'size' and environmental temperature in homoiotherms: the size of a subspecies generally increases with decreasing mean temperature of its habitat. Two things are clear: not only that this rule appears to have many exceptions but also that most authors do not hesitate to explain the cause of this relationship in terms of simple physics as applied to thermoregulation.

Recently, MacNab (1971) argued that most homoiotherms (usually carnivores or granivores but *not* the browsers and grazers) that are large at northern latitudes and small at southern latitudes (an implication of Bergmann's 'rule') are thus for reasons other than simply the physics of heat exchange. This study indicates that, in carnivores, these latitudinal variations in body size result from latitudinal variations in the size of the prey, the latter mainly being determined by the frequency distribution of size of available food particles and the presence of other species that utilize the same food resource. Eventually, in trying to apply this alternative explanation for Bergmann's rule in the case of man, one should have to find other reasons for 'variations in size of the prey'. One is tempted here to think of type and amount of food supply or differences in habits.

There are valid reasons to suggest that indeed Bergmann's rule is applicable to man. Roberts' (1953, 1973) study on the geographical distribution of mean body weight in indigenous human populations clearly showed a statistically highly significant inverse relationship to mean annual environmental temperature. However, as made clear by Roberts (1953, 1973) such statistical data are not adequate to show whether the relationship is direct (immediate response to temperature) or indirect (acting through intermediaries of food supply, habits, genetics, etc.). In such cases, experimental work may help to solve the question. In this respect some of Schreider's studies are of importance (for review see Schreider, 1971).

As a third rule describing ecological relationships that may also be of concern to man, Allen's rule (1877) may be mentioned: in warm-blooded animals, protruding body parts (tails, ears, extremities, etc.) are relatively longer in the warmer regions of the range of the species than in the cooler districts. The usefulness of Allen's rule to describe the human situation has been evaluated only to a certain extent. On the basis of simple thermoregulatory principles, 'the long skinny legs and long gracile necks' and, not to be forgotten, 'the long arms with particular emphasis on the length of the forearm, and large hands with long fingers' of some black-skinned African peoples have been considered to provide the clearest case of adaptation to a given environmental situation (Coon, 1954).

Our study of African peoples, in particular as far as morphological aspects are concerned, has been influenced considerably by these and similar ecological considerations.

In this respect, skin colour is often mentioned as a possible example of an adaptive morphological feature. Data on our three main groups (EEL reflectance spectrophotometer) have been published into detail (Huizinga, 1965*a*, 1968*b*; Rigters-Aris, 1973). We have not yet been able to relate skin colour in a meaningful way to any other observation made in the groups concerned. Future work (including work on African albinos) will be directed so as to explore some possible relationships within the realm of thermoregulatory aspects of climatic adaptation.

Nutrition

An important agent by which environmental influences act on the individual and on human groups is nutrition. Not only the qualitative aspects but even the quantitative aspects of nutrition are largely unknown in the case of most African populations.

In 1961, Miracle stated with provocative confidence that considering the available evidence there was 'little or no basis for the assertion of widespread hunger in tropical Africa'. Even the regularity of occurrence of a 'pre-harvest hunger' and the degree of hunger appeared to him to be questionable. Since then, more studies on nutrient values of urban and rural diets have been made. In general the mean percentage of requirements provided by the daily food intake indicates that well-balanced and adequate diets virtually do not exist among Africans. However, calorific adequacy during the greater part of the year is relatively often found. If, on the other hand, consumption of animal protein were taken to be 'the most sensitive indicator of the quality of nutrition in a community because of its usually close association with the economic base and cultural background of the residents' (Buck, Anderson, Sasaki & Kawata, 1970, p. 104) the low quality of the nutrition in many African communities would become obvious.

Pales & Tassin de Saint Pereuse's impressive account (1954) on nutrition in French West Africa does not deal with the groups discussed here. However, their detailed study contains a wealth of useful general information.

Probably the most comprehensive study of nutrition in the region in which the Kurumba live dates back to December–February 1950–51, i.e. the information was collected in the early months of the dry season in which food is relatively abundantly available (Serre, 1950/51). At that time a frequently occurring picture was described (*ibid.*, p. 27): an adequate daily calorific intake (2920 cal) was found to be associated with a normal total protein intake (about 90 g; animal proteins practically absent, however) and high carbohydrate (521 g) and low lipid intake (about 46 g).

The deficient calcium intake (516 mg) does not seem to be particularly harmful. Schweeger-Hefel & Staude (1972) presented some information on non-quantitative aspects of the Kurumba nutrition as studied in 1961 but the 1950–5 picture described by Serre did not seem to have been basically changed. There is no special reason why the nutrition would have been much different at the moment of our investigations among the Kurumba (January–February 1966 and 1967).

Reliable quantitative data on nutrition in the cases of the Dogon and the Fali are lacking. Personal observations and our medical–biological research do not support Gauthier's (1969) statement that the Fali suffer from hunger and malnutrition. We feel that a study of quantitative and qualitative aspects of the Fali nutrition may well result in a Kurumba-type pattern mentioned above.

It is difficult to evaluate the information on Dogon nutrition as presented by Dieterlen & Calame-Griaule (1961) in terms of calorific and/or compositional adequacy.

Data in this respect are clearly deficient, but the existing knowledge does allow one to conclude that, in the large areas of subsistence farming, a regularly occurring pre-harvest shortage of resources, often resulting from the bad technical quality of methods of storage, severely affects the populations concerned. Miracle's (1961) 'caloric adequacy', perhaps in some cases even found during the period of relative shortage, does not exclude the possibility that during these 'hungry months' the resources available 'do not permit people to satisfy their hunger in the way prescribed by their culture' (Ogbu, 1973).

'Caloric adequacy' may be considered too biophysical and too simple an approach to the problems of the spatial and temporal distribution of nutrition and hunger. Moreover, changes in body weight (in this context an indicator of the availability of food) clearly reflect the fluctuating seasonal quantitative and qualitative nutritional differences (Fox, 1953; Nurse, 1968, 1975), paradoxically reversibly associated with the energy output. Nurse (1975) also draws attention to his observation that when the seasonal hunger comes to an end the 'morose and slow-moving and sometimes irascible people undergo a frequently strikingly sudden reversion to cheerfulness and good humour'.

Interestingly, it appears in the case of labourers from the Ncheu district, Malawi, that the most pronounced gain in body weight is associated not with a satisfying bulky calorific intake but with the time of the year when first-class animal protein is consumed abundantly (Nurse, 1968). It may be added that in most cases these observations have been made on males, i.e. dietetically the favoured section of the African family and community.

A detailed study of the biomedical phenomena associated with the fluctuating food supply among African subsistence farmers, comprising both sexes and children as well as adults, is needed.

As stated earlier our observations were made during January–February, i.e. following and during the period of relative nutritional adequacy. One may assume that, in the given context, the average nutritional state may be considered optimal.

Demography

To a certain extent health aspects of a population are also reflected in certain demographic data.

Despite recent progress in the collection of demographic data in Africa, vital registration still has a very limited coverage. Perhaps the most complete presentation is made by the Office of Population Research of Princeton University (see Brass *et al.*, 1968).

Detailed data relevant to the groups studied by us exist in the case of the Fali (Podlewski, 1966). Their total number (1958–59) amounts to about 42000 individuals. The people studied by us belong to the two smallest of the four major Fali subdivisions, i.e. the Tinguelin (numbering about 4300 individuals) and Kangou (numbering about 5900 individuals) groups. The demographic analysis was based on a sample of 2529 individuals. It appears that, as compared to the findings in traditional 'pagan' tribes from North Cameroon, the fertility among the Fali is low and resembles that of recently Islamized groups. Between 14 and 49 years of age, the average Fali women gives birth to an average of 3.97 living children. Mortality figures indicate that the Fali are 'subject to the hard law of the "pagans" from the mountains' (Podlewski, 1966, p. 145): half of the younger generation does not reach the age of 20. Detailed analysis of the data on fertility, mortality and population structure led to the conclusion that the Fali are a numerically declining population.

Palau Marti (1957, p. 6) draws attention to the various widely differing estimates of the total number of Dogon. Instead of Griaule's 1938 figure of 148898, based on data from the official administration, Palau Marti's informations from official sources add up to at least 200000 individuals scattered over a wide area in regions with widely varying population densities. We know of no other demographic data on Dogon relevant to our subject.

In the case of the Kurumba demographic information is almost non-existant. Tauxier (1917) estimated the number of Kurumba in Upper Volta

J. Huizinga

Table 9.1. *Fertility by age class and cumulative mortality as reported by the mothers* (*Kurumba, 1967*)

Ages of women (years)	N^a	Children ever born				Deceased children					Surviving children				
		♂	♀	♂+♀	Average age	♂	♀	♂+♀	(%)	Average age	♂	♀	♂+♀	(%)	Average age
19–29	23	41	26	67	2.9	16	7	23	(34)	1.0	25	19	44	(66)	1.9
30–35	24	67	46	113	4.7	34	19	53	(47)	2.2	33	27	60	(53)	2.5
36–47	24	73	66	139	5.8	40	40	80	(58)	3.3	33	26	59	(42)	2.5
48–75	24	81	68	149	6.2	40	39	79	(53)	3.3	41	29	70	(47)	2.9
Total	95	262	206	468		130	105	235			132	101	233		
Sex ratio		(127:100)				(124:100)					(131:100)				

^{a} In each age group one of the mothers included here reported to be childless.

as being about 100 000, but an examination of the more recent ethno-demographical maps as drawn by Le Moal (see Brasseur & Le Moal, 1963) leads to a considerably lower estimate.

We collected some information in 1967 on fertility and mortality by questioning about 100 women from the village of Roanga who volunteered to be examined. In this group the fertility reported (Table 9.1) approaches the potential reproductive capacity. The Kurumba do not use contraceptive drugs (although such drugs are known to exist in the area) or practise induced abortion on any appreciable scale (Schweeger-Hefel & Staude, 1972, p. 273). Moreover, sexual intercourse is allowed to take place only four months after the birth of a child. However, lactation sterility may have some influence. Polygyny does not appear to influence fertility in any systematic way (Gomila, 1969; Crognier, 1973).

Data comparable to ours and collected in the same way are given by Crognier (1973, p. 35, table X), in the case of the Sara (Chad). The fertility figures of women of comparable age groups appear to be of the same order of magnitude. The parities by age of the mothers in selected regions of West Africa as listed by van de Walle (see Brass *et al.*, 1968, p. 54, table 2.12) also indicate that the average fertility level fluctuates between five and six children in the case of mothers of menopausal age.

The quality and quantity of data available on fertility in African groups is such that at present no striking regional and/or climatic regularities, if they exist, can be shown.

The relatively high fertility in the Kurumba women is associated with high mortality as indicated in Table 9.1 in which cumulative mortality (the proportion dead of children ever born by age class of the mother), as reported by the mothers, is given. Compared to the Sara data (Crognier, 1973), cumulative mortality figures in the Kurumba sample tend to be somewhat

248

higher. Comparison with data on cumulative mortality in selected regions of West Africa (van de Walle, in Brass *et al.*, 1968, p. 81, table 2.21) reveals that the Kurumba data fit into the general West African picture of high cumulative mortality.

The high sex ratio, if not a chance variation, of the children ever born to our Kurumba mothers remains unexplained. However, one should not exclude the possibility of over-reporting of boys, as mentioned by van de Walle (see Brass *et al.*, 1968, p. 63, table 2.14) in the case of several retrospective studies in Africa.

Morphology

In the following presentation of our main physical anthropological findings* in the groups under study, some aspects of morphology will be dealt with first.

In previous articles, anthropometric data of both sexes of groups studied by us have been given (Dogon: Huizinga & Birnie-Tellier, 1966; Huizinga & de Vetten, 1967; Kurumba: Huizinga, 1968*a*, *b*; Fali: Huizinga & Reijnders, 1974*a*). Special attention has been given to some sex differences. In Table 9.2 a few measurements from all tribes have been assembled, together with some proportions of the body. The proportions calculated roughly describe *shape aspects* of the human body.

In previous papers, attention has been drawn to the remarkable inter-sexual similarities of shape as indicated by these simple proportions in our samples. It appears that in terms of shape of the body the 'average' female may be described as a harmoniously reduced male ('miniature man'). This phenomenon is equally found (Table 9.2) in the Fulani, i.e. a previously nomadic people said to be less negroid than any other group studied by us. In their case, however, the sex difference in relative pelvic breadth is some-what greater than in the case of other groups.

It may be stressed that this homogeneity of shape across the sexes extends over a wide area in which relatively severe climatic and dietary circumstances are more or less the same.

The 'adaptational value' of 'shapes' as roughly described by relative measurements has been indicated previously (Huizinga & Birnie-Tellier, 1966). One could imagine that when in certain environments certain shapes are biologically advantageous both males and females should at least show the tendency (given the limiting morphological consequences of the bio-logical role of the female) to share the 'advantages' of these shape aspects by

* Data on immunoglobulin markers (Kurumba and Fali) and blood groups (Fali) will be published by Drs Erna van Loghem and L. E. Nijenhuis (Central Laboratory of the Netherlands Red Cross Blood Transfusion Service, P.O.B. 9190, Amsterdam, The Netherlands) in *African Pygmies: an Investigation*, ed. L. L. Cavalli-Sforza, (in press).

Table 9.2. *Means of body measurements and proportions in Dogon, Kurumba, Fali Tinguelin, Fali Kangou, Fulani and Sara adults*

	Dogon				Kurumba				Fali Ting			
	N	♂	N	♀	N	♂	N	♀	N	♂	N	♀
Stature	82	168.6	48	158.0	107	171.0	131	159.6	116	169.3	41	159.2
Arm length	83	78.8	48	73.4	36	77.7	46	73.0	116	76.4	41	70.8
Crista height	84	104.8	48	98.7	35	107.4	46	101.2	116	104.6	41	97.6
Sitting height	82	84.3	48	80.3	79	85.5	97	80.5	58	85.6	41	81.4
Biacromial breadth	83	36.6	48	33.4	104	36.5	131	33.8	116	37.1	41	33.8
Pelvic breadth	84	25.6	48	24.7	103	25.4	131	25.0	116	25.1	41	24.9
Rel. arm length	—	46.5	—	46.2	—	45.4	—	46.4	—	45.1	—	44.5
Rel. leg length	—	62.3	—	62.5	—	62.8	—	63.4	—	61.8	—	61.3
Cormic index	—	50.0	—	50.8	—	50.0	—	50.4	—	50.6	—	51.1
Rel. shoulder breadth	—	21.8	—	21.1	—	21.3	—	21.2	—	21.9	—	21.2
Rel. pelvic breadth	—	15.2	—	15.6	—	14.8	—	15.7	—	14.8	—	15.6
Weight	60	59.1	27	51.4	105	59.0	130	51.3	116	58.1	40	49.3
S:W ratio	60	284	27	291	105	288	130	296	116	288	40	303
E:W ratio	29	6.36	—	—	35	6.24	46	6.72	116	6.28	40	6.90

Table 9.2 *cont.*

	Fali Kangou				Fulani				Sara (Crognier, 1973)			
	N	♂	N	♀	N	♂	N	♀	N	♂	N	♀
Stature	49	168.8	21	157.6	49	169.1	26	156.6	302	173.9	405	164.1
Arm length	49	76.6	21	69.7	49	76.8	26	71.9	236	80.6	332	73.8
Crista height	49	103.0	21	96.0	49	103.5	26	96.1	—	—	—	—
Sitting height	36	85.7	21	81.7	—	—	—	—	303	88.6	407	84.4
Biacromial breadth	49	37.0	21	34.6	49	36.5	26	33.5	303	38.1	408	34.3
Pelvic breadth	49	25.0	21	24.9	49	25.2	26	25.3	303	26.4	408	25.8
Rel. arm length	—	45.4	—	44.2	—	45.4	—	45.9	—	46.3	—	45.0
Rel. leg length	—	61.0	—	60.9	—	61.2	—	61.4	—	—	—	—
Cormic index	—	50.8	—	51.8	—	—	—	—	—	50.9	—	51.4
Rel. shoulder breadth	—	21.9	—	22.0	—	21.6	—	21.4	—	21.9	—	20.9
Rel. pelvic breadth	—	14.8	—	15.8	—	14.9	—	16.2	—	15.2	—	15.7
Weight	20	58.2	20	50.2	49	56.9	26	49.9	303	65.5	407	57.8
S:W ratio	20	287	20	297	49	291	26	298	—	273	—	280
E:W ratio	20	6.24	20	6.64	49	6.40	26	6.84	—	—	—	—

Rel. = relative, S = body surface area, W = body weight, E = total length of the extremities.

possessing similar proportions. In this way the size of sex differences may be used as 'a searching tool' to find indications for the possible adaptive significance of certain features. One could assume that the male and the female components of the groups studied differ less in that part of their genetic constitution that is related to long-term adaptations to similar environmental stresses, i.e. in the present case a climate which is hot and dry during most of the year. This makes one think in the direction of adaptational value of efficient thermolysis, especially by evaporation. Indeed, it has been assumed that the effect of this type of climatic stress may be reflected most directly in the high, and only slightly varying, values of the surface:weight (S:W) ratio (Schreider, 1953) and the extremities:weight (E:W) ratio, respectively, in most groups (the relatively heavy Sara being the exception). These measures of linearity of form are often supposed to represent indices of cooling capacity, although at present this interpretation of the associations found in some studies (Schreider, 1971) is not yet based on solid and confirmed experimental work.

However, it may well be that the importance of such simple indices as the S:W ratio for climatic adaptation studies has been overestimated (Schreider, 1963, 1971); the fact remains that on a world scale body proportions (and especially relative leg length, relative arm length and relative bicristal breadth) are closely related in both sexes to mean annual temperature (Roberts, 1973). The lack of variability of these proportions between the groups presented here (Table 9.2) illustrates this point.

An analysis of available data reveals that Schreider's (1963) view, supported by Lambert (1968), that the geographical distribution of the S:W ratio in Africa indicates that as a rule the average ratios in arid zones are high, needs reconsideration. Crognier (1973) has listed the relevant data on 74 African populations (including those presented here) according to seven main vegetation zones which are felt to be closely related to climatic differences. As the data of only one desert (Sahara) tribe are presented by Crognier we added those of seven more Sahara groups as listed by Lambert (1968, table 94, p. 259). Now, if one calculates the mean and standard deviation of the S:W ratio of the groups living in each of the zones distinguished, no appreciable differences between the means are found (Table 9.3). The variability of this ratio (expressed in terms of the coefficient of variation, c.v.) in equatorial forest people ($N = 23$; S:W ratio: 291 ± 14.5 cm²/kg) is the highest (5.0), closely followed by that of desert populations which combine a 'normal' mean S:W ratio ($N = 7$; 291 ± 12.1 cm²/kg) with a relatively high c.v. (4.2). In the peoples of the thorn savanna belt ($N = 10$; S:W ratio 284 ± 8.7 cm²/kg) the coefficient of variation amounts to 3.1, i.e. one of the two lowest variabilities.

It may be concluded that within the African subsaharan context the S:W ratio does not clearly reflect an overall morphological aspect of thermo-

Table 9.3. *S:W ratio in different African vegetation zones (based on Crognier, 1973, and Lambert, 1968)*

Vegetation type	No. of ethnic groups	S:W ratio (cm²/kg) Mean	S:W ratio (cm²/kg) ±S.D.	S:W ratio (cm²/kg) C.V.
Equatorial forest	23	291	14.5	5.0
Savanna with deciduous forest	9	295	13.5	4.6
	(8ᵃ	291	8.0	2.7)
Open forest	11	286	11.2	3.9
Tropical montane steppe	16	289	10.7	3.7
Thorn vegetation	10	284	8.7	3.1
Semi-desert vegetation	4	293	8.3	2.8
Desert	8	296	16.7	5.6
	(7ᵇ	291	12.1	4.2)

S:W = Surface:body weight.
ᵃ One extreme value (Kung: 325 cm²/kg) excluded.
ᵇ One extreme value (Toubou: 326 cm²/kg) excluded.

regulatory adaptation to the various climatic differences as related to the vegetation zones distinguished here.

Data on another index supposed to be related to cooling capacity, the E:W ratio, in which the total length of the four extremities is related to body weight, are very limited (Schreider, 1963; Lambert, 1968; Huizinga, 1968*b*; Hiernaux, in Crognier, 1973). Moreover, as the length of the lower extremity is represented by different measurements (e.g. crista height or height of symphysion) the data available are not directly comparable. In the groups studied by us this ratio is relatively high and ranges between 6.24 and 6.36 in males and between 6.72 and 6.84 in females (crista height). In case it would be proved that this ratio really represents a useful parameter of thermoregulatory capacity, these figures would suggest a slight 'advantage' from the thermolytic point of view in females. However, this type of comparative study does not lead to more than working hypotheses: in this field experimental work is badly needed. Some experimental findings as reported by Schreider (1971) suggest that there exists an inverse relationship between each of the two indices of cooling capacity (S:W and E:W) and the rectal temperature in the resting state, thus contradicting our belief that the ratios are only very distantly related to the thermoregulatory control system.

Morphological data on technologically simple populations (still to undergo major acculturation) may furnish us with a baseline against which to evaluate the effects of culture change upon human phenotypes. It may be anticipated that such changes will include changes in nutritional status as reflected in, for example, body fat. The measurements of some *skinfold thicknesses* offer acceptable estimates of body fat, although the use of predictive equations based on 'white' values may invalidate some of the conclusions drawn from the inter-racial comparisons (Huizinga & Reijnders, 1974*b*).

In a forthcoming monograph (Huizinga *et al.*) a detailed analysis of the nutritional status of the groups studied will be presented, based among others on the recommendations as given by Jelliffe (1966).

Only data on subscapular (representing 'the trunk') and triceps (representing 'the extremities') skinfolds will be discussed here, especially with regard to age changes as demonstrated in a transversal study and as yet unpublished. The measurements were made according to IBP recommendations (Weiner & Lourie, 1969).

In highly developed countries, subscapular skinfold thickness increases definitely with age until about 65 years of age in both male and female subjects (Montoye, Epstein & Kjelsberg, 1965). The skinfold thickness at the triceps site shows little change with age in males (Pett & Ogilvie, 1956; Montoye *et al.*, 1965); in females, however, some authors report a clear increase in triceps skinfold (Pett & Ogilvie, 1956; Montoye *et al.*, 1965) whereas other authors do not (Young, Blondin, Tensuan & Fryer, 1963; Shephard *et al.*, 1969).

The data on male subjects support the implications of the hypothesis of Albrink & Meigs (1964) that fat of the extremities is increased in inherited obesity but not in acquired obesity of middle life. Further evidence to support this hypothesis is given by Damon, Damon, Harpending & Kannel (1969) who showed with the aid of discriminant function analysis that large triceps skinfold thickness was associated with a decreased likelihood that males aged 35–49 would develop coronary heart disease (except myocardial infarction), whereas large subscapular skinfold thickness was associated with an enhanced chance of doing so. The analysis of the skewness of the triceps and subscapular skinfold distributions in American Negro and white children led Johnston, Hamill & Lemeshow (1974) to draw similar albeit tentative conclusions regarding the genesis of triceps and subscapular fat: differences in limb fat between the two groups concerned result from hereditary differentials; differences in subscapular fat result from the operation of environmental factors, thus supporting the suggestions made earlier by Robson (1964, 1971).

Yet, thicker triceps skinfolds may be associated with higher social levels (Fry, Chang, Lee & Ng, 1965; Underwood *et al.*, 1967) suggesting the possibility of environmental-nutritional differences acting during childhood. The findings of Brook, Lloyd & Wolf (1972) on obese and non-obese subjects suggest that the period during which the rate of fat cell multiplication (resulting in an increased total number of adipose cells) is most affected by overnutrition extends to about the age of one year. Thereafter cell multiplication proceeds at a normal rate (until puberty, when adult values in fat cell number are reached), regardless of nutritional circumstances, changes in which are

254

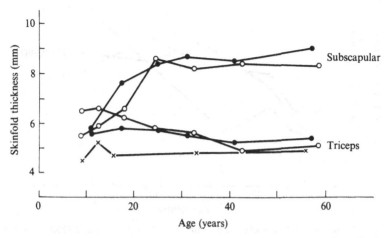

Fig. 9.1. Median skinfold thicknesses of Kurumba ($N = 140$, ○), Fali Tinguelin ($N = 169$, ●) and Dogon ($N = 112$, ×) males.

reflected by changes in the size of the adipose cells. At adult age the increase in body fat results from an increase in fat cell size rather than number (Sjöström, Smith, Krotkiewski & Björntorp, 1972).

Since the fattening process is clearly a matter of 'overeating' (Škerlj, Brožek & Hunt, 1953) the definite change in skinfold thickness with age after adolescence is characteristic of groups living under highly developed conditions. Skinfold thickness in adults living under primitive conditions remains constant throughout life (Glanville & Geerdink, 1970) or show only a slight increase (Johnston *et al.*, 1971; Barnicot *et al.*, 1972).

Data on skinfold thicknesses of African people from the hot–dry savanna are still rare. Although the small sample size and the uncertainties of age estimation are severe limitations, some evidence emerges of a constant subscapular skinfold thickness after adolescence, at least in our samples of Kurumba and Fali males (Fig. 9.1). In Kurumba and Fali males the triceps skinfold slightly decreases in thickness whereas in Dogon males a slight increase is shown. In Kurumba and Dogon females (no data on non-adult Fali females are available) a more irregular picture emerges (Fig. 9.2).

Only the triceps skinfold in Dogon females clearly increases with age after adolescence but it should be stressed that the underlying data show many irregularities.

Nevertheless it is obvious that the repeatedly found differential behaviour of the skinfolds concerned during ageing in males and females needs further study especially in the light of the possibility of a sexual difference in sensitivity to environmental (nutritional) influences, i.e. in terms of developmental canalization.

255

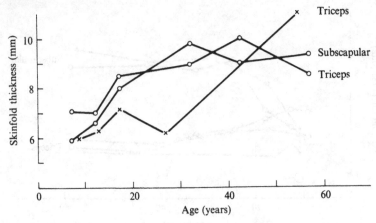

Fig. 9.2. Median skinfold thicknesses of Kurumba ($N = 125$, O) and Dogon ($N = 105$, ×) females.

Functional assymmetry

Within the context of this presentation of morphological aspects we may perhaps include some observations on functional asymmetry.

The best-known examples of functional asymmetry studied in man are handedness, hand clasping and arm folding. Inclusion of the last two traits in our programme is validated not only by some evidence of partial genetic control (Freire-Maia, A., Freire-Maia, N. & Quelce-Salgado, 1960; Leguebe, 1967), albeit that the inheritance of the mode of hand clasping is incompletely understood (Rhoads & Damon, 1973), but also by the fact that significant phenotype frequency differences exist between different ethnic groups, especially in the case of hand clasping (Singh & Malhotra, 1971). It has been suggested that hand clasping is a more useful trait than arm folding in population comparisons (Lourie, 1972; Rhoads & Damon, 1973).

The frequencies of the two alternative types of both traits have been reported in a number of populations throughout the world. Unfortunately, the data of African Negroes are still rather scanty. In addition to our data presented here (Huizinga, 1968b, and some unpublished material on the Fulani and Fali Kangou/Tinguelin studied in 1970) the only piece of information on both traits in an African group has been given by Freire-Maia, A. & de Almeida (1966). The main results may be summarized here. It appears that both in males and females of the three groups (Kurumba, Fali and Fulani) a significant age effect (< 20 years and > 20 years) on the frequencies of the *clasping patterns* is absent. Similarly, no trace of a sex influence on the frequencies of the respective phenotypes is detectable. These findings are in close agreement with the observations of Freire-Maia, A. & de Almeida

256

Table 9.4. *Hand clasping pattern in Kurumba, Fali and Fulani of both sexes and all ages (in brackets: frequency)*

Group	R	L	Total
Kurumba	129 (45)	160 (55)	289
Fali	163 (53)	145 (47)	308
Fulani	63 (52)	58 (48)	121

R, right-type; L, left-type.

(1966) on Angola Bantu Negroes and with the observations of the majority of previous studies (Lai & Walsh, 1965, on white Australians and New Guinean Aborigines).

In Table 9.4 the frequencies of both phenotypes in our three populations are given irrespective of age and sex. The frequencies of the R-type of hand clasping observed in our three African samples vary between 45 and 52% (the differences are not significant). According to the data published so far, the R-type frequencies seem to cluster around 50–55% in Caucasian populations, while a value of 62% has been found in the Angola Bantu. Freire-Maia *et al.* (1958) reported an R-frequency of 68.7% in Brazilian Negroes. As stated previously (Huizinga, 1968*b*) our findings (including some new data) of relatively low frequencies do not support the generalization of Freire-Maia, A. & de Almeida (1966) that 'Negroes present higher frequency of the R-trait than Whites'. Viewed in the light of the suggestion that hand clasping may be subject to selective forces acting differently in different ethnic groups and/or different environments, the finding of low R-frequencies in our groups from the thorn savanna belt and of high R-frequencies in Bantu tribes from Angola is not necessarily contradictory. Lourie (1972) in contrast with his findings in Kurdish Jews (R-frequency hand clasping 50%) observed a relatively high R-frequency (68%) in Yemenite Jews, which he feels to be consistent (based on the Angola data?) with the known 'African' contribution to the Yemenite gene pool.

A significant age effect on the frequencies of the *arm folding* pattern is absent in all three groups both for males and females. In the literature there is general agreement about the statistical non-significance of age influence on this trait. However, except in the Fulani, there is a tendency of R-frequencies to decrease with age.

Contrary to the findings in hand clasping a significant sex influence on the frequencies of arm folding types is found in all but one group (Kurumba). Thus, in Table 9.5 our data are presented according to sex. Although most other authors observed that the sex influence on this trait is not significant, the R-type frequency is commonly higher among males.

J. Huizinga

Table 9.5. *Arm folding pattern in Kurumba, Fali and Fulani males and females of all ages (in brackets: frequency)*

Group	♂ R	♂ L	♂ Total	♀ R	♀ L	♀ Total
Kurumba	89 (64)	51 (36)	140	96 (64)	53 (36)	149
Fali	129 (58)	93 (42)	222	38 (44)	48 (56)	86
Fulani	52 (64)	29 (36)	81	15 (38)	25 (62)	40

R, right-type; L, left-type.

Intergroup comparison shows no significant differences in R-frequency (58–64%) between the male groups whereas the Kurumba females show a significantly higher frequency (64%) in R-type compared to both other female groups (38–44%). From a survey of R-type frequencies in different ethnic groups it appears that in all but three groups (Russian immigrants to Brazil, Angola Bantu Negroes and Manipuris from India) these frequencies cluster around 40%. (Freire Maya *et al.*, 1960; Freire Maya, A. & Almeida, 1966; Singh & Malhotra, 1971.)

Our findings in the males support the observation on the 'exceptional' relative high R-type frequency in Angola Bantu-speaking Negroes; it may be remarked that two of our female groups showed the commonly observed low frequencies as do Brazilian Negroes (42%).

Man shares the capacity for bilateral variation in function with other non-human animals. But man is unique by the fact that a shift occurred in the normal distribution of differences in skill between the two symmetrical halves of the body towards the right. It is suggested that genetic influences operate on this right shift rather than on the basic distribution of differences between the sides (Annett, 1973). Usually, limb dominance has been defined as 'handedness'. A refinement of this definition is proposed by Collins (1961), who discussed total limb dominance in terms of the sum of the dominance patterns shown by the four joint levels involved (shoulder, elbow, wrist and thumb). Evidence is presented for association between handedness (subjective), armfolding (elbow level) and hand clasping (thumb level). Besides it is shown that, for each type of purposeful manipulation, one essential joint level is functionally dominant. It may be noted that in the data of most authors a significant correlation between the three traits is not detectable.

Quite often the genesis of these patterns is connected with a hypothesis about cerebral dominance (Chaurasia, 1974). It is suggested that left–right asymmetries in man's nervous system may have evolved along with the use of tools (and speech?) and thus became associated with the kinds of behaviour that are no longer restricted to his natural environment (Corballis & Beale, 1971).

258

Up till now, our knowledge about the genetic and/or environmental background of functional asymmetries is far from complete. Obviously, for African populations more data are needed but it may be observed that at this moment the examples of functional asymmetry discussed show that these are perhaps not the most promising traits to choose to differentiate between human populations.

Dermatoglyphics

Within the context of morphological aspects, dermatoglyphics may also be mentioned.

Both digital and palmar features have been studied in our West African groups (Huizinga, 1965*b*; Glanville & Huizinga, 1966*a, b*; Glanville, 1968; Rigters-Aris, 1975). More than as possibly adaptive features these multifactorially determined characters were used as anthropological markers to demonstrate genetic differences or resemblances between the populations concerned. The frequencies of the finger patterns do not clearly differentiate between about 15 West African groups on which comparable data could be collected nor could the existence of a gradient of increasing whorl frequencies from east to west in this part of the world, as postulated by Sunderland & Coope (1973), be demonstrated.

This lack of discrimination was also evident in the palmar features, although the data suggest the probable existence of a west–east gradient of increasing percentages of radial loops on the hypothenar area.

Physiological aspects

In our study of the physiological aspects of the populations concerned, some observations were made on the cardiovascular and the respiratory systems. As far as the cardiovascular system is concerned, casual blood pressure was determined in both sexes and different age groups in the cases of the Kurumba and the Fali. Moreover, heart rate changes during standardized exercise (a modified Master step test) were studied in the Fali.

Blood pressure

As outlined previously (Huizinga, 1972) the study of blood pressure offers one of many opportunities to study the biological effects on man of changes in his living conditions resulting from the processes of civilization.

Only in a few populations has it been possible to study longitudinally the reactions of the human organism to (often poorly defined) cultural change. A comparative study of different populations living under a variety of biologically relevant circumstances (sometimes vaguely called 'stages of civilization') may, however, indicate the kind and direction of biological change brought about by certain environmental changes.

J. Huizinga

In 1972, blood pressure data of populations in which age and sex had been reported were collected from the literature and the regressions of systolic and diastolic pressure (b(SP) and b(DP)) on age were calculated in both sexes. The main results of this study (Huizinga, 1972) may be summarized here.

It appeared that in 11 out of 12 subsaharan populations on which data are available, including the Kurumba and the Fali, the regression coefficients, both in the case of systolic and diastolic pressures, are larger in females than in males. The only exception noted was in case of the Fali: b(SP) in males (+0.31) was found to be larger than in females (−0.06). It appeared from our worldwide survey that in 68 out of the 76 non-European populations the female coeficients (b(SP)) exceeded those of the male groups; this phenomenon was present in all European groups surveyed.

A closer look at the 'deviant' non-European populations revealed that these groups cannot be grouped under one simple denominator and may perhaps be considered as chance variations on the theme: female blood pressure–age regression coefficients in any 'normal' population tend to be larger than those in males. However, in 27 out of the 76 non-European populations one or both regression coefficients b(SP) do not differ from zero (arbitrarily $\leqslant$ 0.20 in these generally small groups).

Analysis of these 27 populations revealed the interesting fact that, generally, either in both sexes systolic pressure does not appreciably rise with age (10 populations) or if it does it is only in the females (15 cases).

Perhaps more than ecological studies, physiological–endocrinological studies are needed to elucidate this varying behaviour of blood pressure during ageing. Longitudinal studies may throw some light on this problem, e.g. as published by Miall & Lovell (1967) in their study of two populations measured on three occasions. These authors performed a multiple regression analysis allowing simultaneous examination of the relative contributions of age and pressure in predicting change of pressure. Blood pressure changes were, on average, positive and appeared to be highly significantly related (a rectilinear regression fitting adequately) to mean pressures and only indirectly related to age. This finding suggests that perhaps a kind of 'self-perpetuating' mechanism may be implicated. The population data strongly suggest that this postulated mechanism is only triggered off at a certain pressure level. The low levels found in those populations in which a rise is absent apparently do not 'provoke' Miall's mechanism, either in males or in females. This may well represent the original situation in so-called primitive, non-Westernized, non-acculturated human populations.

At this stage of our knowledge it is impossible to say which environmental factor(s) was responsible for the increase of blood pressure in these populations. We have been unable to analyse the data on 'conditions of life' as presented in the literature since they form an almost unmanageable mixture of all types of information.

The important point, however, in our opinion, is that blood pressure must have risen in certain populations. The data suggest that the critical blood pressure levels SP and DP at which the postulated mechanism is triggered off may lie in the neighbourhood of 123 and 78 mm Hg, respectively.

As can be predicted from Miall's findings in the longitudinal study of European groups, at this relatively low pressure level males show only small increments during ageing or, as our data indicate, no increment at all.

Blood pressure in the female component of these populations, however, starts to rise: the 'self-perpetuating' mechanism works, and the mean pressure values in the populations concerned increase. It seems unlikely that in this stage of 'cultural change' (consisting of largely unknown factors) differential mortality, rather than a biologically determined sexual difference in reaction to this change, produces the phenomenon.

The data indicate that the next step in the evolution of population blood pressure was a general increase which also made younger males victim of the 'self-perpetuating' mechanism: the 'normal and Westernized' population has come into being.

In this view on the evolution of population blood pressure from the stage of constancy of low pressures during ageing to the Western situation in which ageing is associated with ever rising pressure levels, the factors initiating and/or perpetuating these increments cannot be listed as yet. It is clear that the systematic acquisition of relevant data is a matter of great urgency.

Heart rate

Another functional aspect of the cardiovascular system has been studied by means of observations on the heart rate response to standardized work.

Physical working capacity, defined in terms of the maximum oxygen consumption of the individual, can be considered as an important attribute of physical fitness and as such aroused considerable interest during recent years. Measurements have now been made of many subjects in different parts of the world. From these studies it has become clear that V_{O_2} max. is affected by various factors such as age, sex and habitual activity (Åstrand & Rodahl, 1970). It is, however, still controversial whether ethnic differences in physical working capacity are to be interpreted as biologically meaningful in the sense of having occurred through evolutionary natural selection (Davies *et al.*, 1972; Glick & Shvartz, 1974). However, the problem of how much of the variability in physical fitness observed between individuals is the result of genetic rather than environmental differences has been studied by cross-twin analysis where the mean intra-pair variance of physical fitness of MZ twins is compared to that of DZ twins (Klissouras, 1971, 1973). This evidence (still to be confirmed) indicates that heredity alone would account almost entirely for existing differences in physical fitness in a homogeneous population exposed to a common environment. In this context the study of physical fitness in investigations on climatic adaptation is of obvious importance.

During the 1970 visit to the Fali and Fulani the study of heart rate response to standardized exercise on a modified 'Master' step test was tentatively undertaken in order to obtain at least some experience of physiological work under difficult circumstances in the field. Recently, the results of this preliminary investigation, including general remarks on the scientific background of the method and on its usefulness in field studies, were reported (Huizinga & Reijnders, 1974a). Some of the main observations and impressions will be described here.

The exercise consisted in stepping on a modified Master step (height of each of the two steps = 23 cm) at progressively increasing but fixed rates during four periods of 3 min each. The rate was set by a metronome counting 54, 76, 100 and 120 beats/min. For this test, stepping at a rate of 54 means that the individual has made nine ascents per minute (see Weiner & Lourie, 1969, p. 226). Heart beats were counted at the ictus with a stethoscope during 20 sec in the resting state (f_0) and after each of the 3-min periods of exercise (f_{54} to f_{120}).

Although under difficult field situations the step test is the ultimate choice, its inclusion in the programme influenced the willingness to volunteer in other tests of our investigation in a negative way. This was especially clear in the case of the Fulani. Besides, they showed a rather high percentage of bad performers as compared to the Fali. For several reasons (Huizinga & Reijnders, 1974a) the original sample of 86 male individuals broke down to a highly non-randomly selected sample of 52 male subjects of different ages (range 16–47 years).

Analysis of the results demonstrated the linear relationship between heart rate and relative load well-known from laboratory studies. The regression coefficients may be used as some measure of fitness, i.e. a measure of the ability of the organism to resist deviation from the resting state during strenuous exertion as indicated by certain parameters of functional activity. In our sample, differences in fitness (regression coefficient) between the individuals were not related to differences in age or in body weight. However, we observed that relatively fat individuals ('fatness' being indicated by the sum of the four classic skinfolds, measured according to IBP recommendations) were less fit than the lean ones.

Maximum oxygen consumption was predicted according to the nomogram of Åstrand & Rhyming (1954), based on observations on Swedish subjects. In our design the work during descent was taken to be 31 % of the total load.

Davies *et al.* (1972) demonstrated that Africans (Yoruba) reach similar maximum cardiac frequencies as Caucasians and that their f_h versus V_{O_2} curves behave in an exactly similar way during exercise (bicycle ergometer).

From a survey of the literature on African populations where physical working capacity was measured during exercise on a step test, it appeared that our male subjects, divided into three age categories, showed an *exceptionally* high absolute and relative maximum aerobic capacity (3.26–3.70 l/min and 59.9–64.7 ml/kg/min). Against the background of the selection procedure this high degree of physical fitness may well be related to a relatively high level of habitual daily activity of our subjects. However, it should be realized that recording of the heart rate *after* the work load during 20 sec may contain a source of error as the rate tends to drop rather rapidly. Comparison of the results of this type of study with those obtained by more sophisticated means may eventually elucidate the methodological rather than genetic cause(s) of our deviating results in the Fali.

Lung function

Relatively simple methods of evaluating lung functions, that can be used in field studies, have been developed during the last two decades. Data on subsaharan Africans, however, are virtually non-existant.

Our measurements of lung function included forced vital capacity (FVC), which is defined as the volume of air which can be expelled during a forced breath from full inspiration to complete expiration, and forced expiratory volume in one second (FEV$_1$), which is the volume of air expelled in one second during a forced expiration starting from full inspiration. The forced expiratory volume ratio (FEV%) is the FEV$_1$ expressed as a percentage of the FVC. In the absence of obstructive processes the FVC may be considered to represent the volume of air displaced from the alveoli.

The measurements were made with the spirometer described by Collins, McDermott & McDermott (1964). FVC and FEV$_1$ were measured at the same expiration. Observed volumes have been corrected for temperature and all results are expressed in litres BTPS. FEV$_1$ is an indirect measure of the speed of expiratory flow and, as such, related to the peak flow values as, for example, obtained by Wright's peak flow meter.

Both measures are indicators of the presence of disturbances of lung function caused by obstructive or restrictive processes. In restrictive ventilatory defects, a relatively low FVC is associated with a normal FEV % value; an obstructive defect is characterized by a decreased FEV % in the presence of a normal FVC (Leiner *et al.*, 1963).

Our observations were made in the Kurumba in 1967 (Huizinga & Glanville, 1968) and in the Fali and Fulani in 1970 (as yet unpublished). In the case of the Kurumba, considerable efforts were made to obtain a sample representative of the entire population and to ensure an adequate number of individuals in the older age groups.

It appears from the limited demographic data available locally (the chief of the village keeps the records officially used for tax assessment and for

Table 9.6. Composition of Kurumba (K) and Fali/Fulani (FF) samples; mean stature, FEV_1, FVC and FEV% by age group

Age (years)	Number			Stature			FEV_1 (l BTPS)			FVC (l BTPS)			FEV%		
	K♂	F/F♂	K♀	K♂	F/F♂	K♀	K♂	F/F♂	K♀	K♂	F/F♂	K♀	K♂	F/F♂	K♀
7–8	5	1	9	139.8	134.4	134.6	1.5	1.5	1.2	1.7	1.8	1.5	84.1	83.3	80.0
9–10	16	16	3	136.1	137.4	133.7	1.4	1.5	1.1	1.7	1.8	1.6	81.3	85.2	71.9
11–12	11	11	9	141.7	146.7	144.1	1.5	1.9	1.5	1.9	2.2	1.5	81.1	86.3	82.6
13–14	11	8	6	146.3	155.3	150.3	1.6	2.2	1.6	2.0	2.4	2.0	82.7	87.5	82.0
15–16	7	10	3	153.9	157.4	157.0	2.0	2.4	2.0	2.4	2.9	2.5	86.5	85.1	79.1
17–18	3	13	9	159.4	166.6	156.8	2.1	2.9	2.0	2.8	3.4	2.5	75.9	88.9	78.2
19	3	3	2	165.8	163.1	154.2	2.7	3.2	1.9	3.2	3.6	2.4	83.5	89.7	79.7
Total	56	62	41	—											
20–29	31	47	26	171.0	168.8	159.8	3.1	3.4	2.3	3.8	3.9	2.9	81.8	85.7	78.9
30–39	22	32	34	170.4	169.6	159.2	2.7	3.3	2.0	3.6	3.9	2.7	73.3	83.2	73.7
40–49	16	29	18	170.0	169.2	159.5	2.5	3.0	1.8	3.5	3.6	2.6	72.1	82.1	71.9
50–59	9	15	13	167.6	169.1	160.3	2.2	2.7	1.9	3.1	3.3	2.7	70.6	80.3	70.9
60–69	3	4	6	171.3	174.7	159.7	1.8	2.7	1.7	2.8	3.8	2.3	62.7	70.2	73.1
70	—	2	—	—	161.9	—	—	2.4	—	—	3.2	—	—	76.5	—
Total	81	129	97	—											
Totals	137	191	138												

264

Table 9.7. *Regression equations of FVC and FEV$_1$ on stature (cm) and age (yr) in male Fali/Fulani aged 20 and over (n = 129)*

		±s.d.
1a FVC	$= +0.0576\,(\pm0.0073) \times$ stature-5.99	0.49
1b	$= -0.0173\,(\pm0.0044) \times$ age $+ 4.38$	0.56
1c	$= +0.0584\,(\pm0.0066) \times$ stature$-0.0170\,(\pm0.0034) \times$ age-5.52	0.44
2a FEV^1	$= +0.0393\,(\pm0.0076) \times$ stature-3.53	0.51
2b	$= -0.0235\,(\pm0.0038) \times$ age$+3.97$	0.49
2c	$= +0.0405\,(\pm0.0064) \times$ stature$-0.0243\,(\pm0.0033) \times$ age-2.85	0.43

military conscription) that our sample comprises between one-quarter and one-third of the local Kurumba population of both sexes in the age groups 'over 18 years of age' and '15 years of age'.

In the case of the Fali/Fulani the sample certainly is not representative of the local population. Our decision to include the step test in our 1970 programme (see p. 262) led to the inclusion of relatively many adult males and, for the time being, to the exclusion of females. As no significant differences in the variables concerned were found between the Fali and Fulani the data have been combined.

The ventilatory capacity has been analysed in terms of the three major parameters universally demonstrated to influence vital capacity and its derivatives: body size, age and sex.

Adult means (Table 9.6) not only reveal the relatively (as compared to Europeans) low values of both measures but also indicate that in both samples and, in the Kurumba, in both sexes a decrease in both FEV$_1$ and FVC occurs with increasing age. As may be judged from the decrease of FEV % during ageing this change is especially notable in FEV$_1$.

The Kurumba data suggest that the decrease in ventilatory capacity during adulthood may be more rapid in males than in females. These observations have been analysed in more detail by calculating linear regressions and multiple regressions of FVC and FEV$_1$ on stature (cm) and age (yr); the data on Kurumba adults and children have been presented elsewhere (Huizinga & Glanville, 1968). The highly significant positive correlation between both FVC and FEV$_1$ and stature, and the strong negative correlation between age and the respiratory parameters in Kurumba adults are also found in the Fali/Fulani adult group as is reflected in the regression equations given in Table 9.7.

In Kurumba and Fali/Fulani growing children both stature and age when examined separately are positively correlated with FVC and FEV$_1$. However, when the partial regressions are considered, it is apparent that stature is the more important determinant.

J. Huizinga

The finding of a relatively low 'lung volume' in both African groups is not too surprising: as long ago as 1869, Gould had already drawn attention to the low values of the vital capacity in Negro soldiers. The literature on Negro–White differences in pulmonary function has been reviewed by Damon (1966).

The low FVC in Negroes as compared to Europeans of similar stature, age and sex may indicate, indeed, that their lungs are smaller in absolute terms. However, also in cases where the normal residual volume is large FVC will be low. If a large residual volume is found more or less exclusively in populations adapted to hot–dry climates this would perhaps suggest a selective advantage: the lungs are less exposed to the dehydrating influences of hot–dry inspired air. We do not know of any direct observations on Africans which may support this view. Observations on Indians (Cotes & Malhotra, 1964) indicate that relative to the FVC the residual volume in Indians amounts to 31.5% (FVC: 3.49 l) whereas in Europeans only 26.2% (FVC: 4.06 l) were found. The only important Negro–White difference relevant to this problem, found by Damon (1966), was the smaller thorax expansion (difference between the circumferences after inspiration and expiration, respectively). Eventually this might be associated with a larger residual volume. Pertinent observations are still to be made, however, and will be included in our future programme.

These remarks on some physiological aspects conclude this necessarily brief survey of about 10 years work on the human biology of some African populations. It may be clear that the insight into adaptive mechanisms has not been dramatically deepened by our studies. One thing may perhaps be stressed, however: we came to 'know' some populations to such an extent that important changes in functional and/or morphological aspects caused by seemingly irreversible processes like urbanization and/or Westernization may be evaluated. The human biological and health aspects of such environmental changes are largely unknown. Yet, the prevention of the hazards resulting from deleterious environmental factors depends on such knowledge.

At least for some parameters our studies may perhaps provide a useful baseline. Further observations on more people of all ages and both sexes from populations on the verge of being caught by developments that lead at ever increasing rates to an unknown destiny are urgently needed.

The studies reported have been made possible by grants from the Netherlands Foundation for the Advancement of Tropical Research (WOTRO), The Hague (1965, 1967); the Wenner–Gren Foundation for Anthropological Research, New York (1966, 1967); the Boise Fund, Oxford (1966, 1967, 1971); the National Geographic Society, Washington, D.C. (1971).

We gratefully acknowledge the cooperative attitude of the Governmental Authorities of the Republics of Mali, Upper Volta and Cameroon.

Thanks are due not only to those who took part in the field work but also to those members of the scientific, administrative and technical staff of the Institute of Human Biology (Utrecht) who offered their skill in the processing of the data and in preparing the manuscripts, including the present one.

References

Albrink, M. J. & Meigs, J. W. (1964). Interrelationship between skinfold thickness, serum lipids and blood sugar in normal men. *American Journal of Clinical Nutrition*, **15**, 255–61.

Allen, J. A. (1877). The influence of physical conditions on the genesis of species. *Radical Review*, **1**, 108–40.

Annett, M. (1973). Handedness in families. *Annals of Human Genetics*, **37**, 93–107.

Åstrand, P. O. & Rodahl, K. (1970). *Textbook of Work Physiology*. New York: McGraw-Hill.

Åstrand, P. O. & Ryhming, I. (1954). A nomogram for calculation of aerobic capacity from pulse rate during submaximal work. *Journal of Applied Physiology*, **7**, 218–21.

Barnicot, N. A., Bennett, F. J., Woodburn, J. C., Pilkington, T. R. E. & Antonis, A. (1972). Blood pressure and serum cholesterol in the Hadza of Tanzania. *Human Biology*, **44**, 87–116.

Bedaux, R. M. A. (1972). Tellem, réconnaissance archéologique d'une culture de l'Ouest africain du moyen âge: recherches architectoniques. *Journal de la Société des Africanistes*, **42**, 103–85.

Bergmann, C. (1847). Uber die Verhältnisse der Wärmeökonomie der Thiere zu ihrer Grösse. *Göttinger Studien*, **3**, 595–708.

Brass, W., Coale, A. J., Demeny, P., Heisel, D. F., Lorimer, F., Romaniuk, A. & van de Walle, E. (1968). *The Demography of Tropical Africa*. New Jersey: Princeton University Press.

Brasseur, G. & Le Moal, G. (1963). *Cartes Ethno-démographiques de l'Afriquph Occidentale. Feuilles 3 et 4 Nord*. Dakar: IFAN.

Brook, C. G. D., Lloyd, J. K. & Wolf, O. H. (1972). Relation between age of onset of obesity and size and number of adipose cells. *British Medical Journal*, **2**, 25–7.

Buck, A. A., Anderson, R. I., Sasaki, T. T. & Kawata, K. (1970). *Health and Disease in Chad. Epidemiology, Culture and Environment in Five Villages*. Baltimore & London: The Johns Hopkins Press.

Chaurasia, B. D. (1974). Modes de croisement des mains et des bras dans le centre de l'Inde. *Anthropologie, Paris*, **78**, 425–30.

Collins, E. H. (1961). The concept of relative limb dominance. *Human Biology*, **33**, 293–318.

Collins, M. M., McDermott, M. & McDermott, T. J. (1964). Bellows spirometer and transistor timer for the measurement of forced expiratory volume and vital capacity. *Journal of Physiology, London*, **172**, 39–41.

Coon, C. S. (1954). *Climate and Race*. Smithsonian report for 1953, Publication no. 4156, pp. 277–96. Washington, D.C.: Smithsonian Institution.

Corballis, M. C. & Beale, I. L. (1971). On telling left from right. *Scientific American*, **224**, 96–104.

Cotes, J. E. & Malhotra, M. S. (1964). Differences in lung function between Indians and Europeans. *Proceedings of the Physiological Society. Journal of Physiology, London*, **177**, 17–18.

Crognier, E. (1973). Adaptation morphologique d'une population africaine au biotope tropical: les Sara du Tchad. *Bulletins et Mémoires de la Société d'Anthropologie de Paris*, série XIII, **10**, 3–151.

Damon, A. (1966). Negro–White differences in pulmonary function (vital capacity and expiratory flow rate). *Human Biology*, **38**, 380–93.

Damon, A., Damon, S. T., Harpending, H. C. & Kannel, W. B. (1969). Predicting coronary heart disease from body measurements of Framingham males. *Journal of Chronic Diseases*, **2**, 781–802.

Davies, C. T. M., Barnes, C., Fox, R. H., Ojikutu, R. O. & Samueloff, A. S. (1972). Ethnic differences in physical working capacity. *Journal of Applied Physiology*, **33**, 726–32.

Dieterlen, G. & Calame-Griaule, G. (1961). L'alimentation Dogon. *Cahiers d'Études Africaines*, **1**, 46–89.

Fage, J. D. (1965). Some thoughts on migration and urban settlement. In *Urbanization and Migration in West Africa*, ed. H. Kuper, pp. 39–49. Berkeley: University of California Press.

Fox, R. H. (1953). A study of the energy expenditure of Africans engaged in various rural activities. With special reference to some environmental and physiological factors which may influence the efficiency of their work. Ph.D. Thesis, University of London.

Freire-Maia, A. & de Almeida, J. (1966). Hand clasping and arm folding among African Negroes. *Human Biology*, **38**, 175–9.

Freire-Maia, A., Freire-Maia, N. & Quelce-Salgado, A. (1960). Genetic analysis in Russian immigrants. *American Journal of Physical Anthropology*, **18**, 235–40.

Freire-Maia, N., Quelce-Salgado, A. & Freire-Maia, A. (1958). Hand clasping in different ethnic groups. *Human Biology*, **30**, 281–91.

Fry, E. I., Chang, K. S. F., Lee, M. M. C. & Ng, C. K. (1965). The amount and distribution of subcutaneous tissue in Southern Chinese children from Hong Kong. *American Journal of Physical Anthropology*, **23**, 69–80.

Gauthier, J.-G. (1969). *Les Fali de Ngoutchoumi, Montagnards du Nord-Cameroun*. Oosterhout: Anthropological Publications.

Gauthier, J.-G. & Jansen, G. (1973). *Ancient Art of the Northern Camercons: Sao and Fali*. Oosterhout: Anthropological Publications.

Glanville, E. V. (1968). Digital and palmar dermatoglyphics of the Fali and Bamiléké of Cameroons. *Proceedings, Koninklijke Nederlandse Akademie van Wetenschappen*, C71, 529–36.

Glanville, E. V. & Geerdink, R. A. (1970). Skinfold thickness, body measurements and age changes in Trio and Wajana Indians of Surinam. *American Journal of Physical Anthropology*, **32**, 455–62.

Glanville, E. V. & Huizinga, J. (1966a). Palmar dermatoglyphics of the Dogon and Peul of Mali. *Proceedings, Koninklijke Nederlandse Akademie van Wetenschappen*, C69, 528–39.

Glanville, E. V. & Huizinga, J. (1966b). Digital dermatoglyphics of the Dogon, Peul and Kurumba of Mali and Upper Volta. *Proceedings, Koninklijke Nederlandse Akademie van Wetenschappen*, C69, 664–74.

Glick, Z. & Shvartz, E. (1974). Physical working capacity of young men of different ethnic groups in Israel. *Journal of Applied Physiology*, **37**, 22–6.

Gloger, C. L. (1833). *Das Abändern der Vögel durch Einfluss des Klimas*. Breslau: Schulz.

Gomila, J. (1969). Note sur la polygamie et la fécondité respective des hommes et des femmes chez les Bedik (Sénégal Oriental). *Bulletins et Mémoires de la Société d'Anthropologie de Paris*, série XII, **5**, 5–16.

Gould, B. A. (1869). *Investigation in the Military and Anthropological Statistics of American Soldiers*. Hurd & Houghton: New York.

Gourou, P. (1970). *L'Afrique*. Paris: Hachette.

Griaule, M. (1938). *Masques Dogons*. Paris: Institut d'Ethnologie. *Travaux et Mémoires de l'Institute d'Ethnologie*, 33 (Reprinted, 1963).

Griaule, M. & Dieterlen, G. (1965). *Le Renard Pâle*. Tome 1. *Le Myth Cosmogonique*. Fasc. 1. *La Création du Monde*. Paris: Institut d'Ethnologie. *Travaux et Mémoires de l'Institut d'Ethnologie*, 72.

Griffiths, J. F. (ed.) (1972). *Climates of Africa. World Survey of Climatology*, vol. 10. Amsterdam, London & New York: Elsevier.

Huizinga, J. (1965a). Reflectometry of the skin in Dogons. *Proceedings, Koninklijke Nederlandse Akademie van Wetenschappen*, C68, 289–96.

Huizinga, J. (1965b). Finger patterns and ridge counts of the Dogons. *Proceedings, Koninklijke Nederlandse Akademie van Wetenschappen*, C68, 398–411.

Huizinga, J. (1968a). New physical anthropological evidence bearing on the relationships between Dogon, Kurumba and the extinct West African Tellem populations. *Proceedings, Koninklijke Nederlandse Akademie van Wetenschappen*, C71, 16–30.

Huizinga, J. (1968b). Human biological observations on some African populations of the thorn savanna belt. I & II. *Proceedings, Koninklijke Nederlandse Akademie van Wetenschappen*, C71, 356–90.

Huizinga, J. (1972). Casual blood pressure in populations. In *Human Biology of Environmental Change*, ed. D. J. M. Vorster, pp. 164–9. London: IBP.

Huizinga, J. & Birnie-Tellier, N. F. (1966). Some anthropometric data on male and female Dogons. I & II. *Proceedings, Koninklijke Nederlandse Akademie van Wetenschappen*, C69, 675–95.

Huizinga, J., Birnie-Tellier, N. F. & Glanville, E. V. (1967). Description and carbon-14 dating of Tellem cave skulls from the Mali Republic: a comparison with other negroid groups. I & II. *Proceedings, Koninklijke Nederlandse Akademie van Wetenschappen*, C70, 338–67.

Huizinga, J. & de Vetten, A. L. (1967). Preliminary study of the foot of the Dogon. *Proceedings, Koninklijke Nederlandse Akademie van Wetenschappen*, C70, 97–109.

Huizinga, J. & Glanville, E. V. (1968). Vital capacity and timed vital capacity in the Kurumba from Upper Volta. *South African Journal of Science*, 64, 125–33.

Huizinga, J. & Reijnders, B. (1974a). Heart rate changes during exercise (step test) among the Fali of North Cameroon. *Proceedings, Koninklijke Nederlandse Akademie van Wetenschappen*, C77, 283–94.

Huizinga, J. & Reijnders, B. (1974b). Skinfold thickness and body fat in adult male and female Fali (North Cameroon). *Proceedings, Koninklijke Nederlandse Akademie van Wetenschappen*, C77, 496–503.

Izard, M. (1970). *Introduction à l'Histoire des Royaumes Mossi*, vols. 1 and 2. *Recherches Voltaïques*. Paris & Ouagadougou: CNRS-CVRS

Jelliffe, D. B. (1966). The assessment of the nutritional status of the community. With special reference to field surveys in developing regions of the world. *WHO Monograph Series*, no. 53.

Johnston, F. E., Gindhart, P. S., Jantz, R. L., Kensinger, K. M. & Walker, G. F. (1971). The anthropometric determination of body composition among the Peruvian Cashinahua. *American Journal of Physical Anthropology*, 34, 409–16.

Johnston, F. E., Hamill, P. V. V. & Lemeshow, S. (1974). Skinfold thicknesses in a national probability sample of U.S. males and females aged 6 through 17 years. *American Journal of Physical Anthropology*, 40, 321–4.

Klissouras, V. (1971). Heritability of adaptive variation. *Journal of Applied Physiology*, **31**, 338–46.

Klissouras, V. (1973). Genetic aspects of physical fitness. *Journal of Sports, Medicine and Physical Fitness*, **13**, 164–700.

Knip, A. S. (1971). The frequencies of non-metrical variants in Tellem and Nokara skulls from the Mali Republic. I & II. *Proceedings, Koninklijke Nederlandse Akademie van Wetenschappen*, C74, 422–43.

Lai, L. Y. C. & Walsh, R. J. (1965). The patterns of hand clasping in different ethnic groups. *Human Biology*, **37**, 312–19.

Lambert, G. (1968). *L'Adaptation Physiologique et Psychologique de l'Homme aux Conditions de Vie Désertiques*. Paris: Hermann.

Leguebe, A. (1967). Hand clasping: Etude anthropologique et génétique. *Bulletin de la Société royal belge d'Anthropologie et de Préhistoire*, **78**, 81–107.

Leiner, G. C., Abramowitz, S., Small, M. J., Stenby, V. B. & Lewis, W. A. (1963). Expiratory peak flow rate. Standard values for normal subjects. Use as a clinical test of ventilatory function. *American Review of Respiratory Diseases*, **88**, 644–51.

Lourie, J. A. (1972). Hand clasping and arm folding among Middle Eastern Jews in Israel. *Human Biology*, **44**, 329–34.

McNab, B. K. (1971). On the ecological significance of Bergmann's rule. *Ecology*, **52**, 845–54.

Miall, W. E. & Lovell, H. G. (1967). Relationship between change in blood pressure and age. *British Medical Journal*, **2**, 660–4.

Miracle, M. P. (1961). 'Seasonal hunger': A vague concept and an unexplored problem. *Bulletin de l'Institut Français d'Afrique Noire*, série B, **23**, 273–83.

Montoye, J. J., Epstein, F. H. & Kjelsberg, M. O. (1965). The measurement of body fatness. *American Journal of Clinical Nutrition*, **16**, 417–27.

Nurse, G. T. (1968). Seasonal fluctuations in the body weight of African villagers. I & II. *Central African Journal of Medicine*, **14**, 122–7, 147–50.

Nurse, G. T. (1975). Seasonal hunger among the Ngoni and Ntumba of Central Malawi, Africa. *Africa*, **45** 1–11.

Ogbu, J. U. (1973). Seasonal hunger in tropical Africa as a cultural phenomenon. *Africa*, **43**, 317–32.

Palau Marti, M. (1975). *Les Dogon*. Paris: Presse Université de France.

Pales, L. & Tassin de Saint Pereuse, M. (1954). *L'Alimentation en A. O. F. Milieux – Enquètes – Techniques – Rations*. Dakar: ORANA.

Pett, L. B. & Ogilvie, G. F. (1956). The Canadian weight–height survey. *Human Biology*, **28**, 177–88.

Podlewski, A.-M. (1966). La dynamique des principales populations du Nord-Cameroun (entre Bénoué et Lac Tchad). *Cahiers O.R.S.T.O.M.*, série Sciences Humaines, **3** (4).

Rhoads, J. G. & Damon, A. (1973). Some genetic traits in Solomon Island populations. *American Journal of Physical Anthropology*, **39**, 179–83.

Rigters-Aris, C. A. E. (1973). Réflectométrie cutanée des Fali (Cameroun). *Proceedings, Koninklijke Nederlandse Akademie van Wetenschappen*, C76, 500–11.

Rigters-Aris, C. A. E. (1975). Dermatoglyphics of three west African tribes (Fali-Cameroon, Kusasi-Ghana, Baoule-Ivory Coast). I. Digital patterns. II. Palmar dermatoglyphics. *Proceedings, Koninklijke Nederlandse Akademie van Wetenschappen*, C78, 47–57, 298–309.

Roberts, D. F. (1953). Body weight, race and climate. *American Journal of Physical Anthropology*, **11**, 4, 533–58.

Roberts, D. F. (1973). Climate and human variability. *Addison-Wesley Module in Anthropology*, **34**.

Robson, J. R. K. (1964). Skinfold thicknesses in apparently normal African adolescents. *Journal of Tropical Medicine and Hygiene*, **67**, 209–10.

Robson, J. R. K., Bazin, M. & Soderstrom, R. (1971). Ethnic differences in skinfold thickness. *American Journal of Clinical Nutrition*, **24**, 864–8.

Schreider, E. (1953). Régulation thermique et évolution humaine. Recherches statistiques et expérimentales (1). *Bulletins et Mémoires de la Société d'Anthropologie*, série X, **4**, 138–48.

Schreider, E. (1957). Ecological rules and body-heat regulation in man. *Nature, London*, **179**, 915–16.

Schreider, E. (1963). Physiological anthropology and climatic variations. In *Environmental Physiology and Psychology in Arid Conditions*, Proceedings of the Lucknow Symposium 1962, pp. 37–73. Paris: UNESCO.

Schreider, E. (1971). Variations morphologiques et différences climatiques. *Biométrie Humaine, Paris*, **6**, 46–69.

Schweeger-Hefel, A. & Staude, W. (1972). *Die Kurumba von Lurum*. Monographie eines Volkes aus Obervolta (Westafrika). Vienna: Verlag A. Schendl.

Serre, A. (1950/51). *Enquêtes Alimentaires en Haute Volta. Région de Ouahigouya*. Dakar: ORANA.

Shephard, R. J., Jones, G., Ishii, K., Kaneko, M. & Olbrecht, A. J. (1969). Factors affecting body density and thickness of subcutaneous fat. *American Journal of Clinical Nutrition*, **22**, 1175–89.

Singh, N. R. & Malhotra, K. C. (1971). Hand clasping and arm folding among the Manipuris (India). *Human Heredity*, **21**, 203–7.

Sjöström, L., Smith, U., Krotkiewski, M. & Björntorp, P. (1972). Cellularity in different regions of adipose tissue in young men and women. *Metabolism*, **21**, 1143–53.

Škerlj, B., Brožek, J. & Hunt, E. E. (1953). Subcutaneous fat and age changes in body build and body form in women. *American Journal of Physical Anthropology*, **11**, 577–600.

Sunderland, E. & Coope, E. (1973). The tribes of South and Central Ghana: a dermatoglyphic investigation. *Man*, n.s., **8**, 228–65.

Tauxier, L. (1917). *Le Noir du Yatenga*. Paris: Larose.

Troll, C. (1965). Seasonal climates of the earth. The seasonal course of natural phenomena in the different climatic zones of the earth. In *World Maps of Climatology*, by H. E. Landsberg, H. Lippmann, K. H. Paffen & C. Troll. New York: Springer Verlag.

Underwood, B. A., Hepner, R., Cross, E., Mirza, A. B., Hayat, K. & Kallue, A. (1967). Height, weight and skinfold thickness data collected during a survey of rural and urban populations of West Pakistan. *American Journal of Clinical Nutrition*, **20**, 694–701.

Weiner, J. S. & Lourie, J. A. (1969). *Human Biology: a Guide to Field Methods*. Oxford & Edinburgh: Blackwell.

Young, C. M., Blondin, J., Tensuan, R. & Fryer, J. H. (1963). Body composition studies of 'older' women, thirty to seventy years of age. *Annals of the New York Academy of Sciences*, **110**, 589–607.

10. Biological research on African Pygmies

L. L. CAVALLI-SFORZA

The Pygmies were so named by the Greeks, who may have first learned about them from the Egyptians. They are one of the 30 or so known groups in existence (or only recently disappeared) who still live mostly by hunting and gathering (Lee & DeVore, 1968). They may have formerly occupied the largest part of tropical Africa, and presumably the fraction of it covered by the tropical forest. This area must have suffered considerable restriction, as a consequence of agricultural development (in the last few thousand years) favoring transformation into savanna, and perhaps also of climatic changes that have taken place in the last 10000 years. Because of this, the number of Pygmies is likely to have shrunk considerably. Today there still exist, however, a few relatively large isolated Pygmy groups, all of them in the Congo basin, and many smaller groups, some of whom have changed so radically that their identification as Pygmies may be difficult. Groups usually identified as such are all characterized by small stature and life in the forest, mostly of hunting and gathering, but a few of the smaller groups have taken some other occupation. Some time ago, the somewhat arbitrary definition was suggested that the word Pygmy should be limited to populations in which the average male stature is below 150 cm. But the similarities that are found at a sociological level between groups of Pygmies which are located very far one from the other (and some of which are slightly below, others slightly above this arbitrary stature threshold) seem sufficiently important that sociological considerations should take priority over the strictly physical. Moreover, it is not impossible that environmental factors are changing in stature, making the suggested limit even more unsatisfactory.

The major groups of Pygmies still living are limited to areas where forests are still dense and essentially virgin over vast stretches. Of the large groups, the smallest in stature are the Mbuti Pygmies, a group living in the Ituri region (northeastern part of Zaïre) which may number about 50000. I refer to them as 'eastern'; and call 'western' the group that lives over a large forest region on the northwestern part of the Congo basin, sometimes called Babingas. Politically, this area is shared by the Cameroon (the southern part of this country, CAR (the Central African Republic, and exactly the southeastern part of it), and the Popular Republic of Congo (the northern part). It is not clear if there are now Pygmies in Gabon, where there were some until a relatively short time ago, and some may be left in the Republic of Equatorial Guinea. The group of 'western' Pygmies may be slightly less numerous than

Fig. 10.1. A Pygmy camp in Central African Republic.

that of the 'eastern' Pygmies, counting between 30000 and 40000 individuals. The tallest in stature are in the region of Lake Leopold (in the Republic of Zaïre). They are often called Twa but I will refer to them as 'central'. They are probably the most numerous group, numbering perhaps 100000. Murdock's book (Murdock, 1959) can be consulted for a few smaller groups he also describes.

At the periphery of the forest one can note almost every degree of acculturation of Pygmies to an agricultural life. In recent years, vast efforts have and are being made by some African countries to favor or force the shift of Pygmies from the hunting and gathering economy to a non-nomadic agricultural life. These efforts are only partially successful, but have certainly affected the life style of many Pygmies. Even previously, however, contact with the farmers had induced changes of some magnitude, which make it impossible to consider present Pygmies as an example of totally unaltered pre-agricultural life. It is difficult to say for how long this contact has been going on, but it must go back for several hundreds of years, probably longer in some places. The large forests where Pygmies live are crossed by rivers and by a few trails opened by farmers who have pioneered the occupation of the forests. There still are large tracts of forests where farmers are very few, if any, but there probably are no Pygmy groups of appreciable size that have never been in contact with African farmers and in fact that have not established fairly close social connections with them.

Among consequences of these early contacts with African farmers, the following should be noted:

(1) At the technological level: although Pygmies do not normally make iron tools or pottery, they use them currently and obtain them through barter from the farmers. It may be remembered that iron was in use very early in Africa.

(2) Practically without exception, a special social relationship is found between Pygmies and African farmers. This is a situation of hereditary servitude in which Pygmies are the servants, and the farmers the masters (Turnbull, 1965). Pygmies usually work for their masters for a fraction of the year, usually in the months that are important for plantation work, and spend a variable part of the year away in the forest, hunting. Even when they live near the farmers, however, Pygmies usually have camps of their own, which are sometimes very close to the villages of the farmers, but are more frequently found a few kilometers away.

(3) This 'symbiosis', as it is sometimes called, seems to be usually biassed in favor of the farmers who take from it more advantage than the Pygmies, but has not perhaps modified very deeply many of the social customs of the Pygmies. In any case, the social structure of Pygmies remains distinct from that of the farmers and it seems reasonable to assume that it has remained more or less unchanged in its most typical aspects.

L. L. Cavalli-Sforza

(4) The languages of Pygmies are still fairly poorly known but, in all cases studied, a considerable influence of farmers' languages is noted. In fact, the language spoken by Pygmies is often very similar to one of a farmers' tribe with which they may have been associated in the past. It does not usually correspond with that of the farmers with whom they are associated at present, probably because of relatively recent movements of the Pygmies and/or of the farmers. At least three and possibly more mutually unintelligible languages are spoken in each of the two areas, the 'eastern' and the 'western' (Migliazza, unpublished).

Our investigations took place between 1966 and 1971 and covered mostly 'western' Pygmies in the Central African Republic and Cameroon (Cavalli-Sforza, 1971). We also collected samples in the Ituri forest (Zaïre) where Motulsky and his collaborators (Motulsky, Vandepitte & Fraser, 1966) also carried out investigations before the Congolese Civil War. All Pygmies examined were given a physical examination; blood as well as urine, stool and saliva samples were obtained. A range of genetic markers was studied and parasitological analyses were carried out.

Demography

Some demographic information was collected, but this kind of analysis is extremely difficult in populations like Pygmies and most other primitive ones which do not keep a record of their ages. In a few cases we have been able to obtain more reliable information by ranking individuals of a camp by birth order, which is almost always possible, and then assigning them to age blocks by comparison of births with external events. It is worth noting that there are two aspects of demography in which the geneticist is especially interested. One of them is the size, location, and migratory exchange of 'subgroups', however defined. Less crucial but undoubtedly important are conventional demographic quantities like birth, death and marriage rates, age-specific mortality and fertility, etc. The absence of reliable age information makes the latter set of data extremely unsatisfactory. Even the first set, however, is wrought with complications.

In the anthropological literature there appear two 'magic' numbers for the demography of hunters and gatherers (Lee & DeVore, 1968): the number 25 for the size of the band and the number 500 for the size of the tribe. We have a good correspondence between the first number and that of the average size of Pygmy camps. Unless they have undergone some acculturation, Pygmies always live in camps made of a number of small semi-spherical huts, each corresponding to a small nuclear family or part of it. The site, size, and composition of a camp can vary considerably from place to place and with time. We have not seen camps with less than 10 or more than 100 individuals. However, occasionally the camps are very close to each other;

276

up to four such have been noted. Further, there is a concentration of camps around larger farmers' villages which somehow have attracted more Pygmies. Around the cluster of villages located at Bagandou, south of M'Baiki (Central African Republic), which has from 4000 to 6000 farmers, up to 25 Pygmy camps have been counted, all occupied at the same time for a total of almost 800 Pygmies. But these camps were spread over a fairly wide range, essentially a 10 or 11 km radius around the village. This was an exceptionally high concentration of Pygmies in any one place, and in the majority of other cases there were only one or a few camps in the vicinity of each farmers' village. It is also true that the majority of farmers' villages are much smaller in size and seem to fluctuate around an average which in one tribe is close to 200 (Thomas, 1963). In a sub-prefecture of the Central African Republic (Bambio) where an official census was available, the ratio of farmers to Pygmies was four to one (Cavalli-Sforza, 1968).

We found no correspondence for the other 'magic' number, 500. The linguistic groups in Pygmies are much larger, perhaps by a factor of 10 or more. The major source of this number is the Australian tribe, which is usually also a linguistic unit. The size of the linguistic unit among Pygmies is larger and closer to that of New Guineans, whose economy, however, is considerably different. It is possible that if there were originally smaller social groups than the linguistic unit, they have been disrupted by the association with the farmers.

The band is formed by a number of small nuclear families; children once independent, especially if married, usually belong to other bands. The average kinship in the band, apart from relationships in the nuclear families, tends to be small. The band has a turnover, with individuals or families temporarily or permanently leaving or joining the band; and is subject to splits and fusions. It is therefore a relatively labile unit. The size is determined mostly by the necessity of net hunting, the major activity for 'western' Pygmies and part of the 'eastern' ones. It is thus not the ideal unit for the population geneticist. In fact, in a non-sedentary population such as the Pygmies, it is very difficult to define units as permanent as one can for the village of settled farmers. Because of hunting rights and territories, and social contacts with the farmers, most Pygmies seem to spend most of their life in a limited territory; most are born usually not far from where they live, but some of them, especially those born in the hunting season may have a birthplace very far away from the place where they usually live. The permanence of hunting territories, which do not seem to change much during the lifetime of a Pygmy, and the tendency to remain around certain farmers' villages, create areas of relatively limited size in which the life of an individual is spent. The extension of such an area and its individual variation has not been studied in detail, and is worth accurate investigation. Some idea of it is given by the distribution of distances between birthplaces of husbands and wives (Cavalli-Sforza & Bodmer, 1971).

Fig. 10.2. A net hunter in the forest (Central African Republic).

Nuclear families are small, polygyny is relatively rare (rarer than among farmers). The interval between successive births is of the order of four years, a long one, perhaps correlated with the need for freedom of movement on being displaced. One would expect birth rates to be low, as a consequence, but direct estimates are not available. Age at marriage can be very low. The

278

average age of the population is very low (of the order of 22–23 years); old people are rare. There probably is a high death rate among adults, but the age distribution of the population is a poor index of birth and death rates as it depends on a balance between them. Pre-reproductive death rate is not necessarily higher than among farmers, although it may be a little higher. The long birth-interval suggests that there exists some kind of birth control, the nature of which is very uncertain and requires further cultural and biological study. The effects on birth rates of transition to agriculture (does the passage to sedentary life increase birth rates?) is also a subject well worth future investigation.

Genetic studies

Our study of genetic markers has shown that Pygmies of all groups have higher frequencies than African farmers of those markers that were considered as most typically African (especially alleles of the Rh, Fy, Gm systems). This suggests that Pygmies belong to a protoafrican type which has remained more isolated than other African groups. There are genetic differences, however, between Pygmy groups, 'western' and 'eastern' being those on which most information is available: the 'western' group seems to be somewhat intermediate between 'eastern' Pygmies and African farmers (Cavalli-Sforza, *et al.*, 1969). This fact, and the somewhat higher stature of the 'western' group may be interpreted by assuming that the 'western' group has had more admixture with farmers. This interpretation, however, has difficulties. All authors report a social bar to gene flow from farmers to Pygmies. Some farmers' groups accept marriage with Pygmy women, but these women and their progeny are then absorbed into the farmers' culture. The flow of genes into Pygmies can only occur because of extramarital relationship; in only one case did we find a Pygmy woman who had divorced a farmer and returned with her children to the forest. Further direct analysis of the hypothesis of admixture, both at the global level for 'western' Pygmies (Cavalli-Sforza *et al.*, 1969), and for single subgroups (Cavalli-Sforza, 1971), gives unsatisfactory agreement. Perhaps much of the admixture, if any, is old and partially masked by later change; or the difference between 'western' and 'eastern' Pygmies is part of an old cline. Differences between local groups defined by geographic areas comprising a number of bands tend to be small, and not highly correlated with geographic distances probably as a consequence of the migration pattern. Local genetic variation ('microdifferentiation') depends on population density and subsequent migration, decreasing when migration increases. The population density of Pygmies is lower than that of the farmers, but migration, especially long-distance migration, is higher and probably overcompensates for the lower population density.

A few markers indigenous – or almost so – to Pygmies have been described.

279

Some of them, e.g. two alleles for peptidase C (Santachiara-Benerecetti, & Negri, 1970), a phosphoglucomutase allele (Santachiara-Benerecetti, Modiano & Negri, 1969) and two delta hemoglobin markers (Flatbush and Babinga (De Jong & Bernini, 1968) reach frequencies that are clearly (or almost) polymorphic while they are low or absent elsewhere. This indicates a substantial isolation for a very long period, in agreement with the differences in gene frequencies for standard polymorphic markers.

Parasite load

An extensive study, which is mostly unpublished, was made of the parasites affecting Pygmies (Pampiglione & Ricciardi, 1972). Like most other Africans, they show an overload with parasites of all types. Not unexpectedly, Pygmies show a high eosinophilia (incidences of up to 30% and higher have been observed) but it was not clear if any particular parasite was associated with it. Almost every sample of Pygmy blood contains microfilarias, some of them pathogenic. The study of correlations of the presence of parasites and genetic markers was largely unsuccessful. The initially observed association of ABO and yaws (Cavalli-Sforza, 1971) was rejected as a result of later data. Yaws is endemic in all Pygmy groups examined. It was suggested to me that this might be because of peculiarities of the skin of Pygmies, favoring entrance of the parasite. This point may deserve further analysis. When compared with farmers of the same areas, 'western' Pygmies showed a lower frequency of malarial parasitization; but 'eastern' Pygmies, however, have a very high percentage of heterozygotes for hemoglobin S, one of the highest observed, in partial contradiction with the hypothesis that agriculture has increased malarial prevalence.

During these investigations, Pampiglione & Ricciardi (1972) discovered a new human parasite, *Strongylocentroides fullebornii*. First detected among Pygmies, it was soon found to be frequent in many other African groups. It was originally discovered in chimpanzees, and had until then escaped detection among humans, in spite of its rather high prevalence, probably because of rarity of eggs in stools and their resemblance with those of another common parasite.

The causes of the presumed high death rate among adult Pygmies, which may keep the number of aged people in a population at a low level, are obscure. But the load of parasites may be responsible for an increased wear of the human organism, making it more susceptible to intercurrent disease. Pygmies have enormous amounts of gammaglobulin in the blood, possibly connected with their parasite load, a situation now being studied by Siecardi. The frequency of anahaptoglobinemia (Cavalli-Sforza et al., 1969) is inordinately high (more than 50%) indicating hemolytic anemia probably caused by parasites.

Nutrition

Superficially, Pygmies seem to enjoy very good health. Apart from teeth and especially gums which are usually in poor condition, there are no external signs of suffering. Most values seem to be normal and usually signs of nutrition are good. An analysis of amino acids in plasma, however, revealed some abnormalities as compared to commonly accepted standards, the most characteristic of which was a doubling of the phenylalanine level. The general picture was somewhat reminiscent of kwashiorkor (Paolucci, Spadoni, Pennetti & Cavalli-Sforza, 1969). Several cases of this disease have been found among Pygmy infants. Observations of both the children and adults were made on Pygmies living in villages and therefore committed for a period of weeks or months to the farmers' diet (manioc and bananas), which is especially poor in proteins. The hypothesis that Pygmies have not developed genetic resistance to a high carbohydrate diet seemed a natural one. Humans have first been confronted with a high-carbohydrate and low-protein diet mostly since their transition to agriculture.

The high phenylalanine concentration in plasma was found also in some of the farmers, many of whom, however, had more frequently lower or normal values. In order to examine environmental factors, a group of Pygmies was subjected to a balanced controlled diet (Paolucci *et al.*, 1973): the level of phenylalanine normalized under these conditions, showing that the effect was nutritional and easily reversible. The possibility of a genetic liability to the adverse effects of a protein-poor diet nevertheless remains and is worth further investigation.

Stature

It would be frustrating not to study the problems of stature in a research on Pygmies. Thanks to the collaboration of D. Rimoin, and others (Rimoin *et al.*, 1968, 1969), it was possible to make estimates of growth hormone in Pygmy blood. The assays gave normal values; the possibility remained that small stature might result from either an altered hormone or from peripheral decreased responsiveness to it. Results of physiological tests were in agreement with the latter hypothesis. But the genetics of stature in Pygmies remain largely unsolved. For many reasons, this solution is difficult to find. Mixed marriages occur only in a few places and only in one direction, as already noted. Only a very few farmers' tribes 'indulge' in intermarriage with Pygmies, who are considered 'inferior' and therefore are disdained as marriage partners (but very much appreciated as serfs). In the area we surveyed we found only one tribe that intermarried, and only in a few villages: the Lissongos (or Issongos) of the M'Baiki region. This Bantu-speaking tribe is somewhat smaller than the average height of the farmers. Whether this is a

Fig. 10.3. Pygmies dancing in a forest village (Central African Republic).

consequence of past intermarriage or is a condition favoring it is difficult to say. Most hybrids which are easy to trace are children, and the growth of Pygmies is likely to be very different from that of farmers anyway. Thus an analysis of the inheritance of stature and of growth-hormone reaction is fraught with complications. Two adult hybrids have been traced and their

282

reaction to growth hormone was found to be normal. This indicates a recessive inheritance of the 'Pygmy' gene but is far from being sufficient for a fully satisfactory conclusion. It may be added, however, that there are some vague indications that average stature of Pygmies more than 50 years ago was smaller than at present. Have Pygmies increased in stature, as all other groups have? Sociological data do not seem to indicate admixture in recent times. But contacts with farmers are likely to have increased in the last decades. Work on plantations, and with it greater exposure to the sun, may have increased in recent times and may be responsible for an increased stature, if indeed the phenomenon is a real one.

Conclusion

Pygmies are not samples of Paleolithic hunters that have done us the favor of surviving until today totally unchanged, so that we can form an idea of the life of our remote ancestors. They are, however, perhaps the closest approximation to it that exists or at least the largest group that has survived today in relatively unchanged conditions. Naturally there must have been a great variety of modes of life among Paleolithic hunters, as a function of the great variety of different environments they inhabited. The tropical forest is only one such environment. However, similarities between Pygmies and the Bushmen, who must have inhabited a drier environment, are great. So far I have found no major difference when comparing our experience with that reported from research on Bushmen.

Genetic results have confirmed the African nature of Pygmies. They have pointed out the relatively low local heterogeneity of hunters/gatherers. Even so, differences between widely separated Pygmy groups, say 'western' and 'eastern', are relatively large. Nutrition studies seem to indicate that the Pygmy who lives with farmers is not particularly healthy. This, and even more the large parasite load, may account for the mortality, probably high also in adult age. Pygmies' stature is perhaps best interpreted by assuming a major gene controlling sensitivity to growth hormone, but there are also important polygenic and environmental contributions which are hard to dissect.

This work was dedicated to biological aspects. Throughout it I have felt the necessity of greatly expanding ethnographic and socio-cultural investigations of various kinds. Pygmies all over Africa are under the impact of more advanced economics trying to convert them to an agricultural mode of life. They reject the transition, in which they do not seem to see any advantages. Total destruction of the forest will eventually make the dilemma one of 'change or perish'. Very many Pygmies have already faced this dilemma. They have disappeared, either physically or in the anonymity of small villages which remain somewhat isolated socially only until their Pygmy origin is remembered. Those who remain as 'sociologically Pygmy' offer a unique chance of investigation, destined to last a few more years, or at best decades, in a way of life which will inevitably disappear. Even if somewhat contaminated by the agricultural transition, Pygmies are still among the closest representatives of an earlier, widely different, and certainly very charming and attractive mode of life.

References

Cavalli-Sforza, L. L. (1968). Recherches Génétiques sur les Pygmées Babingas de la République Centrafricaine. *Cahiers de La Maboke*, **6**, 19–25.
Cavalli-Sforza, L. L. (1971). Pygmies, an example of hunter-gatherers, and genetic consequences for man of domestication of plants and animals. In *Proceedings of the IVth International Congress of Human Genetics*, Paris. pp. 79–95.
Cavalli-Sforza, L. L. & Bodmer, W. (1971). *The Genetics of Human Populations*. San Francisco: Freeman.
Cavalli-Sforza, L. L., Zonta, L. A., Nuzzo, F., Bernini, L., De Jong, W. W., Meera Khan, P., Ray, A. K., Went, L. N., Siniscalco, M., Nijenhuis, L. E., Van Loghem, E. & Modiano, G. (1969). Studies on African Pygmies. I. A pilot investigation of Babinga Pygmies in the Central African Republic (with an analysis of genetic distances). *American Journal of Human Genetics*, **21**, 252–74.
De Jong, W. W. W. & Bernini, L. F. (1968). Haemoglobin Babinga. A New Delta Chain Variant. *Nature, London*, **219**, 1360.
Lee, R. B. & De Vore, I. (1968). *Man the Hunter*. Chicago: Aldine.
Motulsky, A. G., Vandepitte, J. & Fraser, G. R. (1966). Population genetic studies in the Congo. I. G6PD deficiency, hemoglobin S and malaria. *American Journal of Human Genetics*, **18**, 514–37.
Murdock, G. P. (1959). *Africa, Its Peoples and Their Culture History*. New York: McGraw-Hill.
Pampiglione, S. & Ricciardi, M. L. (1972). Experimental infection with a human strain of *Strongylocentroides fullebornii* in man. *Lancet*, **1**, 663.
Paolucci, A. M., Spadoni, M. A., Pennetti, V. & Cavalli-Sforza, L. L. (1969). Serum free amino acid pattern in a Babinga Pygmy adult population. *American Journal of Clinical Nutrition*, **22**, 1642–59.
Paolucci, A. M., Spadoni, M. A. and Pennetti, V. (1973). Modifications of serum free amino-acid patterns of Babinga adult Pygmies after short-term feeding of a balanced diet. *American Journal of Clinical Nutrition*, **26**, 429.
Rimoin, D. L., Merimee, T. J., Rabinowitz, D., Cavalli-Sforza, L. L. & McKusick, V. A. (1968). Genetic aspects of isolated growth hormone deficiency. *Proceedings of an International Symposium on Growth Hormone*, Milan, Italy, pp. 418–32.
Rimoin, D. L., Merimee, T. J., Rabinowitz, D., Cavalli-Sforza, L. L. and McKusick, V. A. (1969). Peripheral subresponsiveness to human growth hormone in the African Pygmies. *New England Journal of Medicine*, **281**, 1383–8.
Santachiara-Benerecetti, S. A. & Negri, M. (1970). Studies on African Pygmies. III. Peptidase C polymorphism in Babinga Pygmies: a frequent erythrocytic enzyme deficiency. *American Journal of Human Genetics*, **22**, 28.
Santachiara-Benerecetti, S. A., Modiano, G. & Negri, M. (1969). Studies on African Pygmies. II. Red cell phosphoglucomutase studies in Babinga Pygmies: a common PGM_2 variant allele. *American Journal of Human Genetics*, **21**, 315–21.
Thomas, Jacqueline M. C. (1963). *Les Ngbaka De La Lobaye*. Paris: Moulton.
Turnbull, C. M. (1965). *Wayward Servants. The Two Worlds of the African Pygmies*. New York: Natural History Press.

11. Human adaptability in Papua New Guinea

R. W. HORNABROOK

Inhabitants of the temperate and Mediterranean regions have long regarded life in the humid tropics with curiosity and apprehension. The high morbidity and mortality experienced by Europeans in the tropics during the epoch of colonial expansion in the eighteenth and nineteenth centuries gave the area an evil repute. Exotic and varied natural and mineral products were small reward for those expatriates destined to spend part of their lives in the region. European and American literature abounds in accounts of disease, lethargy, physical and mental strain – the tropical torpor and *ennui* which led, in many cases, to the physical or psychological destruction of the subjects. The fact that in most cases the indigenes had a less developed civilisation was taken as confirmation that man was not naturally a part of the tropics. Ethnographic distinctiveness suggested that there were also racial or genetic explanations for the failure of tropical man to thrive and develop complex civilisations. Central America and the Indian subcontinent being two notable exceptions to this generalisation.

What were the characteristics of the tropics which were the basis for their inclemency? Much could be attributable to disease. Physicians trained in London and Paris encountered disorders with which they were unfamiliar; neither their personal experience nor professional training gave them an understanding of the background in which strange diseases occurred, or the tools suited to their analysis. Tropical medicine came to encompass exotic diseases caused by protozoan or metazoan parasites, and often transmitted by arthropods. These constituted an insurmountable hurdle to human development.

There was another school of thought which accepted that climate itself was unfavourable to man achieving his full potential. The enervating humidity led to lassitude and weariness and was in ill-defined ways basically unhealthy, sabotaging existing drives and undermining the efficiency of the industrious. Climatic stresses often resulted in psychological consequences with addiction to alcohol, depression and neurosis.

Failure of advancement of the indigenes could be attributed to climate aggravated by defective diet and an inferiority of genetic (racial) origin. There is still an element of mystery and puzzlement concerning man's role in the tropics. Gourou (1964) has offered a sophisticated re-statement of the traditional view that the tropics were basically ill-suited to the development of high civilisation, but he focussed greatest attention on specific ecological characteristics which hindered productivity. These involved the high rainfall, poor soil, low nutrients, rapid soil exhaustion and poor reserves of biological energy. His work is associated with an increasing attention to the general ecology of man in the tropics and, in recent years especially, to cultural patterns which have adaptive significance.

In all these speculations, Europeans have seen the tropics primarily as a place for the transplantation of European culture and as a resource for development. During and after the Second World War, an increasing number of Europeans have been required to live and work actively in the tropics to a greater degree than the previous generation of colonial entrepreneurs. The need to perform physical work has led to a closer look at adaptation and acclimatisation.

The problem of human adaptability in the tropics has now assumed unusual significance with the recognition of the importance of over-population, inequitable distribution of wealth and lack of opportunity, and the tensions that these have created. Two issues emerge: first the problem of adaptation of temperate man to the tropics, and secondly the adaptive characteristics of the indigenes. Research into either may contribute to a bettering of human existence.

All authors agree that there is a remarkable paucity of information concerning the adaptation of aboriginal man to the tropical rain forest areas, recent work on South American Indians excepted (Lowenstein, 1973). The rapid changes of the present era have threatened the very existence of the aborigines and an element of urgency is conveyed to this research.

The joint Australian/United Kingdom human adaptability study, undertaken under the auspices of the IBP proposed to look more closely at various facets of human adaptation in the humid tropics. It was hoped to mount an integrated multi-disciplinary study of the environmental stresses, and the human biological devices employed to circumvent them. The project appropriately was destined to be undertaken in Papua New Guinea, the last major colonial territory in the world, and for long the last frontier of European exploration of the tropics. Perhaps significantly, the history of European contact with New Guinea epitomises the history of Europe in all the tropical world. The early days of German colonisation were characterised by sickness and failure. The Neu Guinea Kompagnie suffered a spectacular attrition of its European staff, mainly on the north coast of New Guinea. In the years 1886–98, of 224 employees, 41 died and 133 resigned or were dismissed (Souter, 1963). The hazards of settlement in New Guinea went some way to secure for the indigenes a degree of insulation or isolation from world trends. After the region was placed under Australian administration, an era of exploration and plantation colonialism ensued. Then came occupation by Japan and a war in which thousands of Japanese and European soldiers lived in the tropical forests. Since the early 1950s a new era of progressive decolonisation and sweeping technological changes have gained accelerated momentum, culminating in the birth of an independent Papua New Guinea on September 16 1975.

In Papua New Guinea there are a great variety of tropical ecosystems which include lowland and montane rain forest, swamp, riverine, maritime and savannah. Each ecosystem presumably poses particular stresses. At a series of planning meetings the joint Australian/United Kingdom committee decided to base the human adaptability project on a comparison between two populations: one typical of the warm tropical rain forest of the lowlands, the other typical of the cooler low montane rain forests of the central *cordillera*. A series of planning meetings were held with the Public Health Department of Papua New Guinea and later with the Papua New Guinea Institute of Medical Research (at that time known as the Institute of Human Biology). During the course of these meetings, it was decided to base the lowland study on a village on Kar Kar Island, some 50 km off the northwest coast of the New Guinea mainland in latitude 4° 30' S and longitude 145° 58 ' E. It was later decided to base the montane study at Lufa, 6° 20' S 145° 15' E, altitude 2000 m, a small patrol post lying some 60 km south and east of the town of Goroka in the central *cordillera*.

Location

The social and cultural conditions characteristic of the people in the two locations were quite different and their recent history profoundly dissimilar; there were also general environmental distinctions. Kar Kar is a volcanic island with a rich and fertile soil. Human activities have been concentrated on the lower slopes where most of the original dense lowland rain forest cover has been felled and replaced by large plantations of coconut, cacao and extensive areas of native garden. The centre of the island rises to a typical volcanic cone and is precipitous, rugged and uninhabited; the lower slopes, where the population dwell, fall gently towards the sea. Today these slopes are a chequer-board pattern of plantation, garden and secondary forest growth, with small sections of forest along streams and in inaccessible areas. Although the Kar Kar people live in proximity to the sea, marine resources are of little significance to them, giving rise to the curious paradox of an island population which cannot be regarded as being maritime. The staple taro is supplemented by a variety of forest nuts and green vegetables. Imported canned food and rice are regularly consumed. Kar Kar has a hot (30–23 °C), humid climate, with little seasonal variation, although the rainfall (420 cm/ year) tends to be higher during the period of the northwesterly trades (*Taleo*) from December to April. The south-easterly trade winds (*Rai*) bring less moisture. The Kar Kar people have long been subject to European and other alien influences. It seems certain that travellers frequently made contact with

287

R. W. Hornabrook

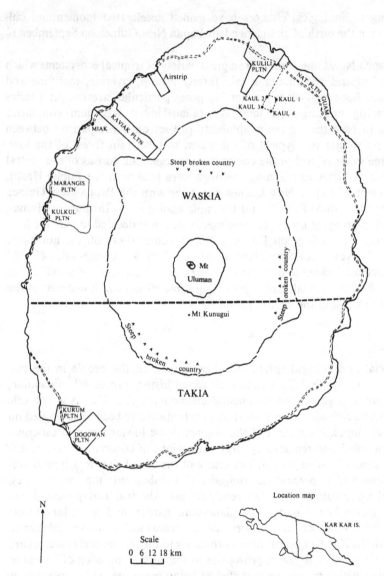

Fig. 11.1. Map of Kar Kar Island indicating the principal European and native-owned (Nat.) plantations (Plnt), Kaul village and the site of the IBP field work.

the island because of its prominent situation on the north coast of New Guinea. Dampier sighted it during his voyages, and Mikluho-Maclay (1973), who resided on the neighbouring Rai coast (during 1871–77), was visited by Kar Kar islanders. Early on, Kar Kar was placed under control of the Neu Guinea Kompagnie and then later under German colonial administraiion,

288

Fig. 11.2. Kaul village (from Hornabrook, 1970).

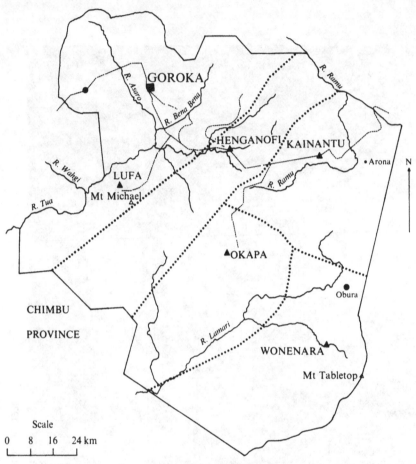

Fig. 11.3. Map of the Eastern Highlands of Papua New Guinea, showing principal roads
(– – –) and subprovincial boundaries (···).

and coconut plantations were established prior to the 1914–18 war. Christian
missions were pioneering by 1890 and they brought a measure of education
and health care which by the 1950s was in advance of that prevailing on most
of the neighbouring mainland. Thus, at the time of the human adaptability
studies the Kar Kar people had experienced some 80 years of European
contact and witnessed large-scale military activities during the 1940s. They
were familiar with the cash economy and, although largely subsistence
agriculturalists, had also worked as labourers on European-owned plantations. Some enterprises were already being run by indigenous entrepreneurs.
The people of Kar Kar are almost evenly divided into non-Austronesian-
speaking Waskia, living on the north and west, and the speakers of the

290

Fig. 11.4. Lufa – a typical village scene. (Photograph by H. O. M. King.)

291

Austronesian Takia language on the south of the island. They dwell in villages ranging in size from 100 to 1500 inhabitants. Houses are built on piles some 0.6 to 1.2 m in height with thatched sago-palm frond roofs and woven split-bamboo walls.

Lufa is entirely different. The Lufa terrain (altitude 2000 m) is hilly or mountainous, largely covered in wild cane and kunai (*Imperata*) grass which merges into low montane rain forest on the slopes of Mount Michael (3700 m) which rises immediately behind the villages. The climate is cooler; although not subject to great seasonal differences, diurnal variation is greater ranging from 26 to 7.2 °C. A distinct dry season between June and November is sometimes associated with seasonal fluctuations in food supplies. Annual rainfall is about 250 cm. Lufa people are almost entirely subsistence agriculturalists whose staple food is sweet potato. Only on irregular occasions is pork consumed. In the vicinity of the patrol post, a dialect of the Yagaria language is spoken, this being a non-Austronesian (Papuan) member of Wurm's East Central language family (Wurm, 1970).

Lufa experienced its first contact with Europeans when explorers penetrated the region in the late 1930s; only in the 1950s was the area placed under control of the central government. Both health and educational services have made their advent more recently. The Lufa adult population was reared in a 'pre-contact' situation. Although many adult men have worked as indentured labourers on lowland plantations, land alienation and expatriate settlers have not been seen in the area. Traditional practices are still of prime significance. Most housing is of the traditional 'beehive' type with walls and roof of kunai grass and the floor of dried earth. The villages range in size but usually comprise 20–40 dwellings housing 100–200 people. Very recently coffee has been introduced as a cash crop and a limited amount of food is now occasionally purchased in trade stores.

Methods

The human adaptability proposals involved the study of the population of a single lowland village and a comparable investigation of a similar-sized community in the central highlands. The United Kingdom and Australian organising committees outlined the general scope of the field investigations. These were to be undertaken in close compliance with the procedures laid down in the IBP Handbook of methods (Weiner & Lourie, 1969). A subsequent meeting of the participating scientists was held in Madang; the fields represented included demography, statistics, respiratory physiology climatic adaptation, physical anthropology, genetics, health and parasitology. It was agreed that each group of workers would conduct as complete a study as was feasible within their field and be responsible for its publication in an appropriate journal. The raw data would be transferred subsequently to computer tapes which would prepare the way for cross-disciplinary analyses. Detailed consideration was not given either to the objectives or the methods of this integrated analysis. In practice, as the project developed, each group of workers made a great deal of *ad hoc* modification to their research designs.

The demographic consultant, D. J. van de Kaa, was of the opinion that data concerning the population of the whole of Kar Kar might be of substantially greater interest than the demography of a single, albeit large, village. This opinion received support from the epidemiologists who saw little of value in a health survey of a single village community. The decision was therefore taken to census the whole population of Kar Kar and to follow this with an epidemiological and health survey of a sample of the whole island population. There were further advantages in collecting material for genetic markers from various communities on the island. An early divergence from the original research plan had already developed.

The demographic survey of Kar Kar Island was completed over the first quarter of 1969, and resulted in the compilation of records of every dwelling on the island, of the residents, and of data concerning their age, sex and marital status. The way was now clear to proceed with the health survey. This was initially undertaken in Kaul village as the centre of the IBP work on the island. In fact, Kaul was not a single village but a conglomerate of four distinct village communities which had merged, both for the convenience of government administration and by a process of natural growth. The population of Kaul was about 1500. A disused schoolroom was modified through the provision of hessian dividing screens and roof insulation. It served as an examination room and a field laboratory. Two houses of similar construction to those occupied by the villagers were erected in the village for the quarters of the medical and technical staff. A small generator was installed to supply electric power for refrigeration and lighting, and the running of centrifuges and other laboratory equipment. Galvanised iron roofing and tanks for water catchment and collection had to be installed. There is no doubt that the success of the whole project was, to a large degree, dependent upon the provision of these facilities within the village and of the residence of the research group there. The villagers were then able to identify with the project and become thoroughly familiar with the personnel and their procedures.

The resident population of Kaul were medically examined. Two specimens of venous blood were collected by vacutainer, one into anticoagulant. Containers for urine collections were distributed and saliva specimens obtained from the adults. In practice, about 50 individuals could be reviewed each day. At the end of the day, blood was examined for haematological indices, malaria slides stained and serum aliquots and haemolysates prepared. Blood specimens for the study of genetic markers were despatched on ice by air to Port Moresby, London and New Zealand. In this way the material for the genetic studies and the health survey was gathered at the same time. Every effort was made to avoid the disruption of village life which is often associated with such studies. Some anthropometric measurements were made on the same occasion, but as these procedures were very time-consuming there was frequently a necessity for subjects to report again for this examination. The data so-gathered was recorded on a proforma which was designed so that some quantitative estimations might be noted and transcription to computer cards facilitated. R. Harvey was responsible for the collection of anthropometric parameters, including dermatoglyphs, measurements of skin pigmentation and photography of the subjects included in the medical survey. A series of five additional villages on Kar Kar, which were considered to be representative of both linguistic groups and of locations adjoining the sea as well as inland, were selected for similar surveys. These Kar Kar surveys were concluded by the middle of 1969. The survey team was then transferred to Goroka, and similar procedures were followed in the investigation of the Lufa population.

At Lufa, village populations were considerably smaller than Kaul. To achieve sufficient numbers all the villages within easy walk of the patrol post were included, providing a population of about 2000 people. The Lufa Sub-District has a population of about 24000 distributed over many kilometres of rough country; those villages selected for the study were slightly atypical in that they were subject to a greater degree of influence from the patrol post. However, any attempt at extending the investigation to the whole district would have been quite impractical. As the Lufa people were drawn from a number of villages, it was not possible to accommodate the team in a village, and this was a decided disadvantage. Some existing government-owned buildings at the patrol post were modified for housing the research group and the laboratories. A detailed demographic investigation was not pursued. Only ages and some fertility data were recorded. In all other aspects the same procedures as those adopted on Kar Kar were followed.

On completion of the collection of medical, genetic and anthropometric data at Lufa, a description of the lowland and montane populations who were to be the subjects of the physiological and nutritional investigations was provided. The physiological research included both time-consuming observations of the villagers in their day-to-day activities and, at the same time, experimental laboratory procedures which were dependent upon the functioning of complex electronic equipment. The operation of the latter in a field situation, where facilities for technical servicing were non-existent, posed considerable problems. In order to provide the requisite laboratory space it was necessary to erect a suitable laboratory building and to install air-conditioning. It was neither practical nor financially feasible to install this equipment in the village. Eventually, a laboratory was erected at the government patrol post at Miak, some 30 km by road from Kaul village. A diesel generator was hired from the government, permitting the continuous use of adequate electric power. The subjects required for the experiments were then transported from Kaul village to Miak by a taxi service which the research team organised and provided. Again, because of the siting of the laboratory at Miak, it was necessary to erect there a further two bush-material houses for the scientists, technicians and their families. The actual observations were concluded in a relatively short time of about three months, but they did necessitate a very large infrastructure. The provision of the facilities was made more difficult on Kar Kar by the location of the project on an island, with the consequent need to transport equipment by sea from Madang. Twelve months of adminis-trative wrangling and a heavy expenditure were required before the Miak laboratories and accommodation were complete. It was concluded that such laboratory facilities could not be provided at Lufa. As a matter of expediency, business premises were rented in the town of Goroka and then modified to serve as a laboratory, and the subjects for study were transported there from Lufa by road each day – a return journey of about 160 km.

The work of the nutritionists did not necessitate the continuous use of electric power or air-conditioning. Norgan and Ferro-Luzzi were able to utilise the existing dwellings and laboratory in Kaul village, and later the medical facilities at Lufa. Whilst their procedures did not demand the same level of logistic support, the measurement of the energy balance required a protracted period of field work which made considerable demands on the researchers. Local village assistants were recruited to assist in such procedures as measurements of physical activity and the weighing of food. Nutritional observations were continued over some 10 months at both locations.

The physiological and nutritional investigations concluded the field work of the HA study in New Guinea. A significant factor in enhancing the value of the programme has resulted from a variety of follow-up studies which have been undertaken by the Papua New Guinea Institute of Medical Research. This organisation has maintained a continuing interest in the programme, and has had opportunities for amplification and extension of original data collec-tions, allowing for a degree of continuity of observation and an in-depth investigation of many facets which were inadequately or incompletely rep-resented in the initial programme. It has, to some degree, transformed a cross-sectional study into a long-range prospective investigation.

Results

The Bibliography of IBP publications following this account includes a list of all publications which have so far been completed, both in regard to the initial Papua New Guinea HA study and to follow-up investigations. Some work is still in press, and in other instances analyses and data collection are still in progress. It is evident that a substantial contribution to knowledge of man in New Guinea has emerged from the overall programme. I shall now briefly relate the principal findings. For the purpose of this brief review, the data may be grouped into three categories. First, contributions which describe the two populations according to their genetic and physical characters, their health and demographic structure. Secondly, analysis of certain aspects of adaptation, e.g. the physiological response to environmental stresses or the consequences of defects in adaptation. Finally, there have been a series of integrated cross-disciplinary analyses of relationships between genetic, anthropometric and health parameters.

The Kar Kar and Lufa populations

In Papua New Guinea it has long been accepted that there is a good deal of variation in the physical proportions of people from different regions, the Highlanders particularly conforming to a certain physical type. Harvey's work on Kar Kar and at Lufa has provided a firm quantitative measure of the distinctions between the two peoples. Furthermore, he has shown that this does not depend on nutritional status and presumably is related primarily to genetic differences. In this regard it is interesting that the Kar Kar population showed a greater heterogeneity, perhaps reflecting greater opportunities for genetic movement in the past. The adult Highlanders possessed a greater body mass, a more robust skeleton, the biacromial and bicondylar diameters being significantly larger, the trunk shorter, the legs longer and the musculature more highly developed. The cranial measurements showed a greater bizygomatic diameter and the morphological face height and nasal index were larger, the distinctions being more pronounced in males than in females and being foreshadowed during the years of childhood growth. The distinctive genetic composition of the two populations was reflected in the distribution of those polymorphic systems measurable in the blood. The following systems were investigated – ABO, rhesus, MNS, P, Lewis, secretor, Duffy, Kidd, acid phosphatase, phosphoglucomutase, 6-phosphogluconate dehydrogenase, glucose 6-phosphate dehydrogenase, adenosine deaminase, malate dehydrogenase, haptoglobin, transferrin, haemoglobin-J (Tongariki) and ulex. Haemoglobin-J(Tongariki) was not present in the Highlands but occurred on Kar Kar in low incidence. There is independent evidence (Gajdusek et al., 1967) that this is a reflection of recent migratory movement from Polynesia. Gerbich blood group was present on the island, as in other sites in north New Guinea, but was entirely absent from the Highland samples. In rhesus system the allele D^u did not occur at Lufa but was present in low frequency on Kar Kar. The prevalence and nature of the genetic polymorphisms in the populations studied in the first surveys of Kar Kar and Lufa have now been complemented by surveys of an extensive area of the central *cordillera* and the north coast of the New Guinea mainland. As yet unpublished results reveal clear genetic distinctions between the Highlanders and the lowland peoples. They also disclose a great heterogeneity, with each village or clan tending to have certain distinguishing characters, presumably a reflection of genetic drift and the founder effect. These phenomena may be expected to change with social transformations which are now affecting all communities.

Demography

The demographic structure of the population may be regarded as a quantitative measure of man's ability to reproduce and survive. As such it reflects physical and genetic fitness and measures the outcome of cumulative response to varied stresses. It would seem to reflect particularly the interplay of social and cultural forces on biological adaptation to the environment. The collection of demographic data at Lufa was restricted geographically and limited in scope but it was sufficient to reveal that the population structure was evidently quite distinctive.

The Kar Kar population has shown a large increase in the years since 1940, rising from 4000 to 17000 in 1970. Of this population, 48% are aged under 50 and 18.6% are less than five years old. The increase appears to result from a reduction in mortality rather than any intrinsic change in fertility. In fact on Kar Kar there was a higher level of spinsterhood and widowhood. Kar Kar women tend to marry later and 38% of women aged 20–24 years have never been pregnant. Of those in the 25–29 year old age group, 11% had not conceived. The comparable figures from Lufa are 21% and 4%. Kar Kar women overall bore more live children than Lufa women, having on an average two more children in a lifetime, and the children had a greater chance of survival. The menopause occurred between 40 and 49 years of age at Lufa, but there was little evidence that it had occurred before 45 years on Kar Kar. Active reproduction was occurring among Kar Kar women during their fourth decade.

There is every indication that the Lufa population is more stable and probably remains very similar to that prior to European contact. Interestingly, polygamy appears to have a depressing effect on fertility at Lufa. As in all other New Guinea populations that have been studied, there is an excess of males over females in all age groups, and this was most marked in the late child-bearing years when the male/female sex ratio was 1.05. It is significant that the massive population increase on Kar Kar has occurred in the presence of endemic malaria. The pattern discernible on the island must surely reflect the profound changes in social and cultural practices which have now been operating for many years.

Nutrition

Although the New Guinea Nutritional Expedition of 1947 (Hipsley & Clements) drew attention to the limited protein intake and restricted diet of

Papua New Guinea villagers, the dietary practices of the indigenes had been commented on by explorers much earlier. Reports of European administrators, prospectors and explorers in the early 1900s drew attention to the remarkable physical fitness of a people who subsisted on the unattractive, monotonous and apparently defective diet of yam, sweet potato, taro or sago. Discussion of adaptation of New Guineans to a protein-deficient diet has led to suggestions that a negative nitrogen balance may be corrected by nitrogen fixation in the gastrointestinal tract or a recycling of urea nitrogen (Oomen *et al.*, 1968; Hipsley, personal communication).

The investigations of Durnin, Ferro-Luzzi and Norgan on the energy expenditure and nutrient intake in Papua New Guinea diets which constituted an important section of the human adaptability project, have provided a firm basis for future nutrition research in the country. Although earlier investigations (Venkatachalam, 1962; Bailey & Whiteman, 1963; Bailey, 1967) had been concentrated among the Highlanders who appear to have suffered from the greatest dietary stringencies, the new data has drawn attention to a surprising and unexpected situation among the apparently prosperous lowlanders. The energy content of the diet of the Kaul men was 1940 kcal per day, of women 1420 kcal per day. The calorific intake at Lufa (men 2523 kcal per day and women 2105 kcal per day) was more satisfactory and more closely approached generally accepted standards relative to the energy demands of the subjects. On Kar Kar there was a considerable difference between the energy intake and the energy expenditure. This discrepancy was not apparent among the Highlanders. In the case of Kaul, the energy intake was on an average 403 kcal per day less than the expenditure, and for the women 407 kcal per day less than the expenditure. The protein and fat components of the diet in both populations were low. Protein, as a percentage of total energy, was 6.7% and 6.0% for Kaul men and women, while for Lufa men and women the figures were 6.5% and 7.2%. There was also a low content of fat. The corresponding figures being 9.8, 9.1% for Lufa men and women and 17.2% for both Kar Kar men and women. These findings are somewhat higher than those of earlier surveys, perhaps suggesting that a measurable change has already occurred in the dietary intake. The observations are certainly going to be subject to an appreciable change in the future. Indeed, since 1970 there has been a greatly increased intake of bought foods in both localities. In 1970 at Lufa the younger men had approximately 10% less energy than older men from vegetable staple crops, a reflection of this age groups' greater participation in the cash economy.

Several surprising facts which emerged from the nutrition studies are significant. Neither on Kar Kar nor at Lufa was there evidence of an increase in energy intake at times of increased physiological demand. For example, energy intake did not increase in pregnancy nor in most lactating women.

The apparent negative energy balance on Kar Kar had led Norgan *et al.*

(1974) to examine their figures critically. They have been unable to offer any plausible explanation to account for the discrepancy. The same procedures were adopted at Lufa as on Kar Kar, and the Lufa energy balance conformed quite well to generally accepted standards. They appear to have dismissed the possibility that a Kaul subject ate less food for the duration of the study period, or that there were technical errors or misjudgements in the field observations.

In regard to the protein intake, the findings suggested that whilst the gross dietary protein was low as was the protein to energy ratio, the amino-acid score as recommended by FAO/WHO were acceptable, and although there was a low absolute quantity of protein in the diet, both groups probably had adequate intakes. Only Lufa women showed a slight reduction below the recommended level. It is evident that there are important questions which await investigation in the field of adult nutrition. The results of the investigation of the child nutrition, which has vital significance in selection and survival, have not yet been reported.

The nutritional status of the population must have great significance to health, reaching far beyond the usual clinical parameters of specific protein, vitamin and calorific deficiency. It is unfortunate that the human adaptability studies did not specifically enquire into the immunological state of the community as this may be expected to be influenced by dietary factors. Growth also serves as a parameter of nutritional adaptation, and the anthropometric measurements confirmed the existence of a phenomenon which appears to play an important role in the adjustment of the New Guinean to his diet. Harvey (1974) has confirmed the observations of Malcolm (1970) on the slow rate of growth of Melanesians. This was evident both on Kar Kar and at Lufa, but especially so in the Highland group. The retardation in growth, with respect to generally accepted growth rates, is associated with a slowness in physiological maturity evidenced by a menarche at 15.56 ± 1.1 years at Kar Kar, and of 16.5 years at Lufa. Nutrition may also be responsible for the negative correlation of physical proportions with age. In both sexes this was again particularly conspicuous at Lufa where there was a striking decrease in body mass as age increased.

Medical work

The medical work of the International Biological Programme fell into two phases. In the first place a cross-sectional survey of the study populations at Kar Kar and Lufa was completed. This provided a background of medical information on subjects selected for physiological and nutritional study. We have discussed elsewhere the serious limitations which beset such cross-sectional investigations. They provide scant information on acute or sub-acute illnesses, particularly the majority of acute infectious diseases and, as these form the most important threat to health, this is a significant shortcoming.

299

The surveys did confirm the absence of the common degenerative diseases or urban and technologically sophisticated societies. Blood pressures remained within normal limits and tended to decrease with age; there was no diabetes or glycosuria; gout was absent and the consequences of arteriosclerosis not detected. The population studied was too small to give a clear indication of the prevalence of less common degenerative diseases. Positive findings were a high incidence of hypochromic anaemia which affected almost everyone on Kar Kar and a significant number of the people at Lufa. Both populations exhibited a significant incidence of chronic non-specific respiratory disease which was apparent in the middle-aged, and progressively increased in prevalence with advanced years. As might be expected, there were a number of distinctions between the health status of the Kar Kar people and those at Lufa. Most of these would appear to relate to the warm, moist climate of Kar Kar, promoting proliferation of certain pathogens and arthropod vectors. A large percentage of the Kar Kar people had malaria parasites in their blood and this was correlated with splenic enlargement. Malaria was very rare at Lufa, being detected only in individuals who had recently travelled to the lower altitudes of the Tua Valley or to coastal plantations. Both the clinical and serological testing revealed that treponematosis probably yaws, was very common on Kar Kar but much less prevalent in the Highlands. On Kar Kar there was a far greater evidence of fungal skin disease. On the other hand, a usually symmetrical enlargement of the parotid glands was prevalent at Lufa. The explanation for this hypertrophy remains obscure. It has been linked in the past with protein deficiency in the diet, but it is such a common finding in the Highlands that there may be some other explanation. In certain areas on Kar Kar there was a high prevalence of goitre and thyroid disease was not observed at Lufa. This is in accord with experience in Papua New Guinea where there are many foci of endemic goitre associated with iodine-deficient soils.

A second phase of the medical studies has involved enquiries into certain specific findings of the cross-sectional surveys. These studies have given a clearer view of the overall epidemiological characteristics of lowland and montane New Guinea populations.

It is a relatively simple matter to identify the presence of anaemia, but more complex testing is required to determine its basis and nature. Haemoglobin levels in men, women and children in the Highlands were 4.7, 4.05 and 3.4 g/100 ml higher than on the coast. Notwithstanding the compensation for altitude, some 30% of the Lufa population were anaemic; 99% of Kar Kar subjects had a significant lowering in haemoglobin levels. Crane has shown that there are a number of factors responsible for this anaemia. Malaria contributes: 36% of children, 16% of women and 19% of men on Kar Kar had positive films for *Plasmodium* parasites. At Lufa the overall figure was only 1.1%. Crane & Kelly (1972) have shown that malaria control will effectively improve haemoglobin levels, but this step alone does not restore the blood to normality. Intestinal helminthiasis on Kar Kar was not regarded as making a significant contribution to the anaemia. Beta-thalassaemia, present in about 3.5% of the Kar Kar population, may have made a minor contribution. Crane has also shown that folate levels are sometimes sensitive, particularly to short-term increased physiological demands. The predominant factor responsible for the anaemia appears to be iron deficiency, and this may be the result of defects in iron absorption, perhaps through its binding to food materials in the small intestine.

The physiological significance of anaemia and its relevance to fitness of the population is an entirely different matter. It is true that coastal and lowland Papua New Guinea populations have haemoglobin levels substantially below the accepted WHO standards. But is this deleterious either in regard to selection, survival or reproductive capacity? It is conceivable that a relative anaemia in the humid tropics might convey some benefit.

The absence of all overt signs of arteriosclerosis and the trend for blood pressures to fall with age had previously been investigated in some detail in Highland people by Sinnett (1975), Whyte (1958) and Whyte & Yee (1958). The same phenomenon was observed in the IBP study. Sinnett & Whyte (1973) have undertaken studies into this aspect of cardiovascular disease on Kar Kar. The investigations are still incomplete, but it seems that the situation there is similar to the Highlands, the influence of diet and physical activity playing a prominent role in protecting the people from these degenerative diseases. A trend of increasing prevalence is already apparent in those Papua New Guinean populations most exposed to Western influences.

The most prevalent chronic pathological processes in both Highland and lowland adult populations was the high incidence of chronic non-specific respiratory disease. In the course of follow-up investigations, Anderson (1974) has defined the natural history, clinical features and physiological factors of this disease. The causes remain obscure. It is probable that the seeds for this progressive disease are sown in early childhood, perhaps through the damaging effects of acute infectious illness, and only meticulous follow-up studies over a long period are likely to provide the key to prevention. Certainly, smoking native tobacco appears to aggravate the established disorder, but suggestions that the smoke-filled atmosphere of the traditional Highland houses is an important cause is open to question in the light of the fact that the disorder has the same prevalence in the coastal people, in spite of the fact that they live in a warm climate in which there is no necessity for smoky fires.

Micro-ecological influences may cause surprising variations in health in superficially homogeneous populations and general epidemiological trends are also influenced to a marked degree by socio-cultural experience and economic development. Alimentary parasites, both helminthes and protozoa, and blood parasitic diseases such as malaria and filariasis varied in prevalence on Kar Kar. There was a high prevalence of malaria, filaria and hookworm among the southern Takia people, but *Ascaris* was more prevalent among the Waskia on the north and west of the island. Yaws was more prevalent on the Waskia side of Kar Kar. This area is further from the Lutheran Mission Hospital at Gaubin and the diminished prevalence near the hospital may be a reflection of the availability of penicillin treatment in the neighbourhood.

The fact remains that both populations appear to be remarkably healthy. The greatest selective disease forces operating probably make their impact in early childhood. There is sound evidence that in this group malaria makes its greatest demands and so too do the acute infectious illnesses. It is a significant omission that child health was not studied in other than a very superficial way in the HA project.

Although tuberculosis was present on Kar Kar it did not appear to be a very serious problem and it was absent at Lufa. There was nothing to suggest that either population was suffering from the debilitating effects of recently introduced diseases.

Physiological studies – climatic stresses

The physiological studies involved in the HA project concentrated attention on investigating both the factors which permitted physical exertion in the humid tropics and the nature and manner of response to heat stresses.

Fox and his colleagues undertook field observations on the heat stress which was experienced by people in the course of life and work in the two locations. These observations were complemented by laboratory experiments.

The field workers measured the heat production of volunteers, both Papua New Guinean and European, during the day and calculated heat production, at the same time recording the air temperature, humidity, wind speed and the mean radiant temperature in sunshine and shade. Temperatures were continuously monitored in the subjects' houses. This allowed the external heat stress to be measured and with the subject's heat production, related to the efficiency of physiological cooling mechanisms. It was surprising that thermal stress at Lufa was very comparable to that experienced in the lowlands, the increased radiant temperature compensating for the cooler ambient temperature at the higher altitudes. The subjects in both areas appear to respond comparably in terms of sweat production, an average of about two litres of sweat being produced in an eight-hour day.

In the laboratory the response of the temperature-regulating system to a measurable heat stress was recorded by the use of a specially devised standardised test in an air-conditioned bed. The deep body and skin temperatures, blood flow in the hands, heart rate and sweat production were measured whilst the environmental temperature was controlled for specific periods of time. There was little variation between the lowland and Highland subjects, sweat production being very comparable in the two groups. The Papua New Guinean produces less sweat than the acclimatised European living on Kar Kar; elsewhere a similar relationship has been observed when Africans and Indians have been compared with acclimatised Europeans. Presumably the maintenance of a comfortable body temperature in the tropics with less demand on sweating has adaptive advantages. These studies did not identify the factors which are responsible for this ability, but Fox *et al.* (1974) suggested that Melanesians may lose a greater amount of heat from the vascular tree of the trunk, as they were unable to demonstrate any significant variation in circulation through the extremities during the heat test. Unacclimatised Europeans had similar sweat production to the Papua New Guineans, female Papua New Guineans having a somewhat higher sweat production than their unacclimatised European counterparts. Daily urine production was low in Papua New Guineans who usually finished with a negative fluid balance, the bulk of the body water being derived from food rather than the ingestion of water.

The various parameters of lung function, such as ventilatory capacity, forced expiratory volume and forced vital capacity, total lung capacity and the transfer factor for carbon monoxide were measured, as well as the ventilatory and cardiac frequency during submaximal exercise. The average size of the total lung of New Guineans was less than that of Europeans, although in the Highlands the figures corresponded more closely. Increased lung ventilatory capacity was attributed to increased daily activity rather than to the relatively low altitude of 2000 m.

The investigations were associated with measurements of the respiratory response to carbon dioxide. Papua New Guineans showed a greater tolerance to increased carbon dioxide levels than Europeans and this may be of adaptive significance, permitting a greater and more efficient adaptation to work in circumstances where carbon dioxide levels are increased.

The investigation of cardiovascular and respiratory response to exercise revealed that the Highlanders again had a greater aerobic capacity, in many ways resembling the performance of trained athletes. The Highlanders had a greater stroke volume and a slower heart rate at oxygen uptakes of $1 \, 1/min$ than the coastal people. Patrick & Cotes (1974) felt that the increased cardiac contraction which the Highlanders revealed on response to effort contributed to the relatively larger aerobic capacity.

In general it would appear therefore that differences in respiratory and cardiovascular function in Papua New Guinean Highlanders may be attributed to their increased habitual activity rather than to the altitude or to genetic factors. Investigations are in progress to determine the role of the latter. Both Highlanders and lowlanders were similarly distinct from Europeans in their smaller lung size, which appears to be a common attribute of non-European communities.

Multi-disciplinary studies

Cross-disciplinary integrations in which the influence of genetic factors on health, the relationship of fertility and reproductive capacity with physical anthropometry, and other multi-variant analyses, are incomplete. It has always been speculated that genetic polymorphism may affect fitness and survival. As a basis for human natural selection the question remains substantially unanswered, largely because of the difficulty in obtaining sufficient quantitative data. Limitations of sample size have placed such investigations in a precarious position. In the course of the IBP work in Papua New Guinea polymorphic systems were assessed in over 2000 medically examined subjects on Kar Kar, and a similar number at Lufa. An association between the ABO blood group and goitre has been reported, and additional investigations suggest that other genetic systems may be associated with the presence of several disease conditions. These, not unexpectedly, apply particularly to *Plasmodium* infections. The presence of malaria parasites and the existence of splenomegaly had a number of associations with genetic markers. Thus, acid phosphatase, glucose-6-phosphate dehydrogenase and haemoglobin-J(Tongariki) are all associated with a greater or lesser incidence of parasitaemia. The results confirm the existing well-known fact that glucose-6-phosphate dehydrogenase deficiency affords protection against malaria. In the haptoglobin system there was strong evidence that *Hp1* homozygotes were much more likely to show signs of hepatomegaly and splenomegaly than *Hp1-2* and *Hp2* homozygotes considered together. In the case of haemoglobin-J the abnormal phenotype was much more likely to be parasitised.

Further investigations in which integration and correlation between the data obtained during the course of the HA project are in progress.

Secular changes

We have already alluded to the fact that nutritional investigations showed some findings which were rather different to those observed some 10–12 years previously by other workers. The most likely explanation for the distinctions is that changes in dietary practices are already emerging. The influence of change on health and adaptation is of great significance to Papua New Guineans, and the project provided an opportunity for enquiries to be initiated before changes were advanced. The socio-economic status of individuals on Kar Kar varied according to the degree to which Western culture had been adopted. High-status individuals have adopted European practices to a greater degree than low-status individuals. During the medical assessment, individuals at both Lufa and Kar Kar were graded, using a variety of criteria, into high, average and low socio-economic status. On Kar Kar, men of high status were heavier, had no borderline malnutrition, and consumed more alcohol than low-status men. People of low status had more respiratory disease and on Kar Kar were more likely to be anaemic. At the same time, high-status women were heavier, had no borderline malnutrition and less skin disease than low-status women. Thus some distinctions were apparent although there were no measurable differences in serum biochemistry or in the blood pressure. It would seem that much may be gained from continued monitoring of these populations as these changes are likely to become more marked in the future.

The human adaptability study in perspective

The objectives of the Papua New Guinea HA project were ambitious. They have been only partially attained. We have discussed elsewhere some of the limitations and shortcomings of the project. For a pioneering study, the scope was overambitious and some of the failings may be attributed to this. At the heart of the difficulty lies the omission of a precise definition of aims, particularly those which involved the integration of various disciplines. Although this was to be a collaborative multi-disciplinary study, breaking new ground through cross-disciplinary analysis, in practice the enthusiasm of individual research groups tended to pull them in diverging directions. The result has been a conglomerate of discrete projects in a loose matrix of genetic and epidemiological fact, lacking a close and intimate web of integration. Too great a reliance was placed on the capacity of the computer to achieve a synthesis of the different disciplines, and insufficient attention to prior formulation of the manner in which such integration might be achieved. Each discipline has its limitations when engaged in cross-sectional studies of human biology. These effect some disciplines to a greater degree than others. We have mentioned some of the limitations of cross-sectional medical surveys. One might refer also to the rather restricted scope of physiological studies which deal with selected samples of healthy young men. Some areas of great significance to adaptation in the humid tropics were entirely omitted from the project. These include child health and socio-cultural anthropology. All the studies would have been greatly enhanced had the arrangements for follow-up been carefully planned at the outset.

The failure to incorporate advice in the planning and execution of the research from behavioural and socio-cultural scientists was a persistent handicap. It is evident that man, to a substantial degree, uses cerebral processes in his adaptive responses to environmental and other stresses. It might be said that socio-cultural practice serves as the interface between the biological

305

characteristics of the population and the environment. Any study of adaptation or human ecology which fails to consider this all-pervasive and uniquely human phenomenon will be incomplete and in some respects superficial.

Competent ethnographic and social anthropological advice would have enormously enhanced and facilitated work in such diverse fields as fertility, population genetics, nutrition and activity investigations.

On an entirely different plane, projects such as the human adaptability study in Papua New Guinea involve political and ethical issues which are of considerable importance. These may assume proportions and significance which can overshadow any scientific rewards that might be gleaned. On the one hand there is the ethical and humanitarian background to the investigations, created by the study of relatively unsophisticated villagers by an educated elite, either of national or foreign origin. The objectives of the IBP were explained to a group of villagers who could have little comprehension of the scientific importance of the project. Although no procedures which could conceivably have hazarded health were involved, there is no doubt that the villagers did experience a degree of inconvenience which they might not have accepted were they fully cognizant of all facets of the project. At Kar Kar, and to a lesser degree at Lufa, there exists today a degree of disillusionment with the research, which might have been avoided if greater care had been exercised in the initial preliminary discussions. At Kaul village, a large number of expatriate visitors met the villagers and explained the nature of the research project. Inevitably there was a tendency to enhance the benefits which might be derived from it. An acquaintance with the history of Melanesian societies, particularly these in the Madang region, should have alerted the IBP committees that no circumstance would be more likely to induce a 'cargo cult philosophy'. Indeed, the expectations which were aroused were unlimited; in Kaul village it was confidently expected that a general hospital would arise on the foundations of the IBP. It would have been more appropriate to exaggerate the inconvenience and limited benefits that might be incurred and to obtain co-operation on this basis rather than to create a situation where anticlimax and disappointment were the prevailing sentiments.

At a national level the 'expedition' concept needs to be critically examined, particularly when its objectives are concerned with the examination of a culturally distinct human society. The IBP failed to identify with Papua New Guinean aspirations. This was tangibly demonstrated by the title of the project – the Australia/United Kingdom Human Adaptability Study of Papua New Guinea. Insufficient attention was given to identifying the project with Papua New Guinea as a joint venture rather than as a convenient source of raw material for the intellectual endeavours of foreigners. The unfortunate consequence of this has been to enhance the suspicion which visiting research workers attract, and to be prejudicial to the useful role which expatriate researchers might fulfil in the future development of Papua New Guinean science.

In retrospect these shortcomings are obvious enough. They were not so evident when the IBP project was planned. Provided we benefit from our mistakes their recognition is an important product of the study.

In the final instance, the IBP will be judged in Papua New Guinea by its implications to the health and welfare of the people of the country. At the moment there is a tendency to consider this in terms of immediate relevance and gain, but it is more reasonable to regard the IBP in a much longer perspective. A great deal of data concerning the normal biological characteristics of Melanesians, both in the Papua New Guinea Highlands and in the lowlands, have been accumulated. Normal standards ranging from those established by the respiratory physiologists and the nutritionists to the serological and biochemical variables in the blood will constitute a firm basis to future clinical and medical work in the country. The Papua New Guinea human adaptability studies also provide some insight into the biological attributes of aboriginal man there. Accelerating socio-cultural changes have been conspicuously obvious following the conclusion of the IBP, emphasising that the opportunity for undertaking the study was a once-only event; the same data could never be collected again. The population and the environment have already changed and the changes will be more profound in the coming years In this regard, the data will be of lasting scientific value and may form a basis for future work which we can, at the moment, not clearly envisage.

Those aspects of the studies which are of immediate relevance to Papua New Guinea are primarily derived from the medical and nutritional work. Whilst the epidemiological observations confirm that the Melanesian was remarkably well adapted to his environment, they stress the importance of malaria as a debilitating infection which must prevent large sections of the population from realising their full potential. Certain important problems have been defined which await clarification and should be the basis of long-range research, particularly the chronic non-specific respiratory infections and the iron deficiency anaemia. Although introduced infections do not appear to play a major role in reducing the adaptation of Papua New Guineans to their environment, there are hints that a change in social and cultural practices may induce undesirable biological consequences. The absence of arteriosclerosis and other degenerative diseases may not be sustained in the face of significant changes in diet and in habitual activity. It is possible that the data which have now been accumulated may provide a basis from which continuing studies can assist in identifying those undesirable factors in the environment which may be amenable to control, and any steps taken to control or prevent the appearance of these diseases in Melanesia would be of inestimable value to the community in the future.

The Papua New Guinean study also suggests that an explanation for the

307

R. W. Hornabrook

failure of European colonisation was the presence of disease, particularly the result of the introduction of non-immune subjects to a region where malaria was endemic. The physiological research indeed suggests that there is little to distinguish the capacity of European temperate man to climatic stresses from the native-born Melanesians. Here again, however, it is probable that psychological and socio-cultural factors may have played an essential role in mitigating against the success of European communities in the humid tropics.

References

Bailey, K. V. (1967). Composition of New Guinea Highland foods. *Tropical Geographical Medicine*, 20, 141–6.

Bailey, K. V. & Whiteman, J. (1963). Dietary studies in the Chimbu. *Tropical Geographical Medicine*, 15, 377–88.

Crane, G. G. & Kelly, A. (1972). The effect of malaria control on haematological parameters in the Kaiapit Subdistrict. *Papua New Guinea Medical Journal*, 15(1), 38–73.

Gajdusek, D. C., Guiart, J., Kirk, R. L., Carrell, R. W., Irvine, D., Kynoch, P. A. M. & Lehman, H. (1967). Haemoglobin-J (Tongariki) (α 115 alanine-aspartic acid); the first new haemoglobin variant found in a Pacific (Melanesian) population. *Journal of Medical Genetics*, 4(1), 1–6.

Gourou, P. (1964). *The Tropical World – Its Social and Economic Conditions and its Future Status*, 3rd ed. London: Longmans.

Hipsley, E. & Clements, F. (1947). *Report of the New Guinea Nutrition Survey Expedition*. Canberra: Department of External Territories. Sydney: Government Printer.

Hornabrook, R. W. (1970). International Biological Programme Investigation on Kar Kar Island. *South Pacific Bulletin*, 20, 15–17.

Lowenstein, F. W. (1973). Some considerations of biological adaptation by aboriginal man to the tropical rain forest. In *Tropical Forest Ecosystems in Africa and South America. A Comparative Review*, ed. B. J. Meggers, E. S. Ayensu & W. D. Duckworth, pp. 293–310. Washington: Smithsonian Institute.

Malcolm, L. A. (1970). *Growth and Development in New Guinea. A Study of the Bundi People of the Madang District*. Institute of Human Biology, Monograph Series No. 1. Madang.

Miklouho-Maclay, N. (1973). *New Guinea Diaries 1871–1883*, transl. C. L. Sentinella. Madang, Papua New Guinea: Kristen Press.

*Norgen, N. G., Ferro-Luzzi, A. & Durnin, J. V. G. A. (1974). The energy and nutrient intake and the energy expenditure of 204 New Guinean adults. *Philosophical Transactions of the Royal Society*, B, 268, 309–48.

Oomen, H. A. P. C., Hipsley, E. H., Corden, M., Strengers, Th. & Kruijswijk, H. (1968). The human intestine as a potential contributor of utilizable protein to the human diet. International Symposium on New Protein Sources in the Human Diet, Amsterdam. Papua New Guinea, Department of Public Health. File No. 54–14–4.

*Patrick, J. M. & Cotes, J. E. (1974). Anthropometric and other factors affecting respiratory responses to carbon dioxide in New Guineans. *Philosophical Transactions of the Royal Society*, B, 268, 363–73.

Sinnett, P. F. (1975). *The People of Murapin. The Ecology of a New Guinea Highland Community*. Goroka: Papua New Guinea Institute of Medical Research, Monograph No. 4.

308

Sinnett, P. F. & Whyte, H. M. (1973). Epidemiological studies in a Highland population of New Guinea: environment, culture and health status. *Human Ecology* **1**(3), 245–77.

Souter, G. (1963). *New Guinea. The Last Unknown.* Sydney: Angus & Robertson.

Venkatachalam, P. S. (1962). *A Study of the Diet, Nutrition and Health of the People of the Chimbu Area (New Guinea Highlands).* Port Moresby, Papua New Guinea: Department of Public Health Monograph No. 4.

Weiner, J. S. & Lourie, J. A. (ed.) (1969). *Human Biology – A Guide to Field Methods.* Oxford & Edinburgh: Blackwell.

Whyte, H. M. (1958). Body fat and blood pressure in natives of New Guinea: reflections on essential hypertension. *Australasian Annals of Medicine,* **7**, 36–46.

Whyte, H. M. & Yee, I. L. (1958). Serum cholesterol levels of Australians and natives of New Guinea from birth to adulthood. *Australasian Annals of Medicine,* **7**, 336–9.

Wurm, S. A. (1970). Indigenous languages. In *An Atlas of Papua and New Guinea,* ed. R. G. Ward & D. A. M. Lea, pp. 16–19. Glasgow & Harlow: University of Papua New Guinea and Collins Longman.

Bibliography of IBP publications *(see also starred References)*

Anderson, H. R. (1971). Chronic lung disease at Goroka Base Hospital. *Papua New Guinea Medical Journal,* **14**, 139.

Anderson, H. R. (1974). The epidemiology and allergic features of asthma in the New Guinea Highlands. *Clinical Allergy,* **4**, 171–83.

Anderson, H. R. (1974). Smoking habits and their relationship to chronic lung disease in a tropical environment in Papua New Guinea. *Bulletin de Physiopathologie Respiratoire,* **10**, 619–33.

Anderson, H. R. (1974). Allergic aspects of asthma in the highlands of Papua New Guinea. *Clinical Science and Molecular Medicine,* **47**, 11–17.

Anderson, H. R. (1974). The prevalence and nature of chronic lung disease and asthma in Highland Papua New Guinea. M.D. thesis, University of Melbourne.

Anderson, H. R. (1976). A clinical and lung function study of chronic lung disease and asthma in coastal Papua New Guinea. *Australian and New Zealand Journal of Medicine,* **5**, 329–36.

Anderson, H. R., Anderson, J. A. & Cotes, J. E. (1974). Lung function values in healthy children and adults from highland and coastal areas of Papua New Guinea: prediction nomograms for forced expiratory volume and forced vital capacity. *Papua New Guinea Medical Journal,* **17**, 165–7.

Anderson, H. R. & Cunnington, A. M. (1974). House dust mites in the Highlands of Papua New Guinea. *Papua New Guinea Medical Journal,* **17**, 304–308.

Beavan, G. H., Hornabrook, R. W., Fox, R. H. & Huehns, E. R. (1972). The occurrence of heterozygotes of α-chain haemoglobin variant Hb-J(Tongariki) in New Guinea. *Nature, London,* **235**, 46–7.

Beaven, G. H., Fox, R. H. & Hornabrook, R. W. (1974). The occurrence of haemoglobin-J(Tongariki) and of thalassaemia on Kar Kar Island and the Papua New Guinea mainland. *Philosophical Transactions of the Royal Society,* B, **268**, 269–77.

Booth, P. B. (1974). Genetic distances between certain New Guinea populations studied under the International Biological Programme. *Philosophical Transactions of the Royal Society,* B, **268**, 257–67.

Booth, P. B., McLoughlin, K., Hornabrook, R. W. & MacGregor, A. (1972). The Gerbich blood group system in New Guinea. III. The Madang District, the Highlands, the New Guinea Islands and the South Papuan Coast. *Human Biology in Oceania*, 1, 267–72.

Budd, G. M., Fox, R. H., Hendrie, A. L. & Hicks, K. E. (1974). A field survey of thermal stress in New Guinea villagers. *Philosophical Transactions of the Royal Society*, B, 268, 393–400.

Cotes, J. E. Lung studies in New Guinea. *Thorax*, 26, 490 (Abstract).

Cotes, J. E. (1972). Why do New Guinea Highlanders have large lungs? *Thorax*, 27, 510.

Cotes, J. E., Adam, J. R., Anderson, H. R., Kay, V. R., Patrick, J. M. & Saunders, M. J. (1972). Lung function and exercise performance of young adult New Guineans. *Human Biology in Oceania*, 1, 316–17.

Cotes, J. E., Anderson, H. R. & Patrick, J. M. (1974). Lung function and the response to exercise in New Guineans: role of genetic and environmental factors. *Philosophical Transactions of the Royal Society*, B, 268, 349–61.

Cotes, J. E., Davies, C. T. M., Patrick, J. M., Reed, J. W. & Saunders, M. J. (1972). Cardiorespiratory response to submaximal exercise – comparison of young adults in New Guinea and the United Kingdom. *Ergonomics*, 15, 48.

Cotes, J. E., Saunders, M. J., Adam, J. E. R., Anderson, H. R. & Hall, A. M. (1973). Lung function in coastal and Highland New Guineans – comparison with Europeans. *Thorax*, 28, 320–30.

Crane, G. G., Hornabrook, R. W. & Kelly, A. (1972). Anaemia on the coast and highlands of New Guinea. *Human Biology in Oceania*, 1, 234–41.

Crane, G. G., Jones, P., Delaney, A., Kelly, A., MacGregor, A. & Leche, J. (1974). The pathogenesis of anemia in coastal New Guineans. *American Journal of Clinical Nutrition*, 27, 1079–87.

Fox, R. H., Budd, G. M., Woodward, Patricia M., Hackett, A. J. & Hendrie, A. L. (1974). A study of temperature regulation in New Guinea people. *Philosophical Transactions of the Royal Society*, B, 268, 375–91.

Fox, R. H., Hackett, A. J., Woodward, Patricia M., Budd, G. M. & Hendrie, A. L. (1972). A study of temperature regulation in New Guinea people. *Human Biology in Oceania*, 1, 310–13.

Garner, M. F., Hornabrook, R. W. & Backhouse, J. L. (1972). Prevalence of yaws on Kar Kar Island, New Guinea. *British Journal of Venereal Disease*, 48, 350–5.

Garner, M. F., Hornabrook, R. W. & Backhouse, J. L. (1972). Yaws in an island and in a coastal population in New Guinea. *Papua New Guinea Medical Journal*, 15, 136–8.

Harrison, G. A., Hiorns, R. W. & Boyce, A. J. (1974). Movement, relatedness and the genetic structure of the population of Kar Kar Island. *Philosophical Transactions of the Royal Society*, B, 268, 241–9.

Harrison, G. A., Boyce, A. J., Hornabrook, R. W. S. & Craig, W. J. (1976). Associations between genetic markers and biochemical variation in two New Guinea populations. *Annals of Human Biology*, 3, 557–68.

Harrison, G. A., Boyce, A. J., Platt, C. M. & Serjeantson, S. (1975). A note on body composition changes during lactation in a New Guinea population. *Annals of Human Biology*, 2, 395–8.

Harrison, G. A., Boyce, A. J., Hornabrook, R. W. & Craig, W. J. (1976). Associations between polymorphic variety and disease susceptibility in two New Guinea populations. *Annals of Human Biology*, 3, 253–67.

Human adaptability in Papua New Guinea

Harrison, G. A., Boyce, A. J., Hornabrook, R. W. & Serjeantson, S. (1976). Evidence for an association between ABO blood group and goitre. *Human-genetik*, **32**, 335-7.

Harvey, R. G. (1974). An anthropometric survey of growth and physique of the populations of Kar Kar Island and Lufa subdistrict, New Guinea. *Philosophical Transactions of the Royal Society*, B, **268**, 279-92.

Hornabrook, R. W. (1970). The Institute of Human Biology of Papua New Guinea. *Science, Washington*, **167**, 146-7.

Hornabrook, R. W. (1971). IBP Human Adaptability studies in New Guinea. *Australian Science Teacher's Journal*, **17**, 25-27.

Hornabrook, R. W. (1974). The demography of the population of Kar Kar Island. *Philosophical Transactions of the Royal Society*, B, **268**, 229-39.

Hornabrook, R. W., Crane, G. G. & Stanhope, J. M. (1974). Kar Kar and Lufa: an epidemiological and health background to the human adaptability studies of the International Biological Programme. *Philosophical Transactions of the Royal Society*, B, **268**, 293-308.

Hornabrook, R. W., Fox, R. H. & Beaven, G. H. (1972). The occurrence of haemoglobin-J(Tongariki) and the β-thalassaemia trait on Kar Kar Island and the mainland of Papua and New Guinea. *Papua New Guinea Medical Journal*, **15**, 189-93.

Hornabrook, R. W., Kelly, A. & McMillan, B. (1975). Parasitic infection of man on Kar Kar Island, New Guinea. *American Journal of Tropical Medicine and Hygiene*, **24**, 590-5.

Hornabrook, R. W., Serjeantson, S. & Stanhope, J. M. (1976). The relationship between socio-economic status and health in two Papua New Guinea populations. *Human Ecology* (in press).

Morris, P. J., Bashir, Helen, MacGregor, A., Batchelor, J. R., Case, J., Kirk, R. L., Ting, A., Hornabrook, R. W., Boyle, A., Dumble, L., Law, W., Lightfoot, A., Johnstone, J., Guinan, J. & Brotherton, J. (1973). Genetic studies of HL-A in New Guinea. In *Histocompatibility Testing*, ed. J. Dausset & J. Colombani, pp. 267-74. Copenhagen: Munksgaard.

Mourant, A. E. (1974). The hereditary blood factors of the peoples of New Guinea and the surrounding regions. *Philosophical Transactions of the Royal Society*, B, **268**, 251-5.

Norgan, N. G., Ferro-Luzzi, A. & Durnin, J. V. G. A. (1972). An investigation of a nutritional enigma. Studies on coastal and Highland populations in New Guinea. *Human Biology in Oceania*, **1**, 318-19.

Patrick, J. M. & Cotes, J. E. (1973). Cardiac determinants of aerobic capacity in New Guineans. In 25th *ICPS Satellite Symposium of Physical Fitness*, Prague 1973, pp. 309-14.

Schamschula, R. G., Barmes, D., Keyes, P. H. & Gulbinat, W. (1974). Prevalence and inter-relationships of root surface caries in Lufa, Papua New Guinea. *Community Dentistry and Oral Epidemiology*, **2**, 295-304.

Stanhope, J. M. (1971). Interrelations of culture, health and the environment in rural lowlands of New Guinea: an IBP human adaptability study. *Australian Science Teacher's Journal*, **17**, 29-32.

Stanhope, J. M. & Hornabrook, R. W. (1974). Fertility patterns of two New Guinea populations: Kar Kar and Lufa. *Journal of Biosocial Science*, **6**, 439-52.

Turner, K. J. & Anderson, H. R. (1973). The association of serum IgE levels with disease: contrast between populations in New Guinea and Australia. *Proceedings of ANZAAS*, Perth, 1973 (Abstract).

311

R. W. Hornabrook

Turner, K. J., Baldo, B. A. & Anderson, H. R. (1975). Asthma in the Highlands of New Guinea. Total IgE levels and incidence of IgE antibodies to house dust mite and *Ascaris lumbricoides*. *International Archives of Allergy and Applied Immunology*, **48**, 784.

Walsh, R. J. (1974). Geographical, historical and social background of the peoples studied in the I. B. P. *Philosophical Transactions of the Royal Society*, B, **268**, 223–228.

12. Adaptation to urbanization in South Africa

D. J. M. VORSTER

The Republic of South Africa had a number of different ethnic groups, as well as people of the same ethnic group, who were at different levels in the transition from a peasant rural life to an urban industrialized one. The transition could be associated with changes in climate, in nutrition, in levels of physical and mental activities and in the culture of the people. Thus new environmental stresses would introduce unfamiliar psychological demands, and sociological adaptations would be required which would affect the individual, the family and the group. Alterations in nutrition and in mental and physical activities could be expected to affect the rates of mental and physical growth in children, the mental abilities and the physique of adults, and consequently their capacities for mental and physical effort. The change could also alter the patterns of disease and the health of the communities. The adaptations made by different peoples to these changes in climate, nutrition, mental and physical activities, health and culture could, again, be influenced markedly by differences in the genetic constitutions of the peoples involved.

It was against a background of incomplete scientific understanding and in the knowledge that successful adaptation to environmental change by the various indigenous population groups of South Africa is of the utmost importance to the well-being of all those groups, that the project under discussion was conceived by the working group of scientists representing the fields of anatomy and genetics, physiology, nutrition, medicine, psychology, sociology and statistics.

Main objectives of the study

The principal aim of the project was to carry out inter- and intra-ethnic comparative studies of two distinct Bantu population groups which have members at both extremes of a traditional/rural to industrial/urban scale, with a view to identifying the characteristics which distinguish the subgroups from one another in the terms of all the participating disciplines.

The main individual objectives set by the participating disciplines were summarized as follows:

Nutrition

Comparative studies were to be made of the nutritional status of various groups both within and across ethnic classifications, and of nutritional change and its effect on disease patterns, physical working capacity, etc.

313

D. J. M. Vorster

Physiology

Studies were to be made of physical working capacity within and across the selected populations by measurement of the maximum oxygen intake of individuals on the basis of different rates of work performance.

Anatomy and genetics

By establishing the genotypic and phenotypic affinities and differences between selected urban and rural Bantu communities, it was hoped that a contribution could be made towards ascertaining to what extent genetic and non-genetic factors were involved in human adaptation to urbanization.

Psychology and sociology

Previous research in Africa had produced evidence of specific qualities in the structure and organization of the mental abilities of Bantu, which differed to a greater or lesser degree from other groups. It was also known that these differences or specific qualities were not fixed and unchangeable. An attempt was to be made to establish when, and under what circumstances, changes in the organization of mental abilities accompanying urbanization took place. Furthermore, it was known that the work motivation of the tribal Bantu differed widely from that of the European, yet that of the urbanized Bantu corresponded more or less with that of the European. A study of this adaptation process was to be undertaken.

Medicine

Although there was a lot of information about the patterns of disease amongst urban Bantu, very little was known about these patterns amongst rural Bantu communities. Morbidity and mortality patterns amongst the selected rural populations were to be studied and comparisons made with the extensive information already available for urbanized Bantu in Johannesburg.

A secondary objective was to make interdisciplinary comparisons of the distinguishing characteristics of the subgroups so as to establish interactions across disciplines.

Experimental design and methodology

The experimental design not only took into account the availability, geographical accessibility, local facilities, characteristics and cultural taboos of the experimental populations but endeavoured to meet the requirements of all the participating disciplines, each with its own methodology and limiting parameters. The only discipline excluded from this design was medicine since the survey of patterns of morbidity and mortality could naturally not be confined to the experimental groups.

314

The design made provision for four experimental groups of 200–300 subjects each drawn from rural or urban members of two distinct Bantu ethnic groups. Since all the measurements and techniques could not be readily adapted to females and/or children, adult males between ages 20–50 years were chosen. Subjects had to be physically sufficiently fit to participate in the examinations, which were spread over two days, preferably with no more than eight years schooling, living in the particular area continuously for at least three months, preferably with some occupational history, and as a group reasonably well spread out along a scale of individual urbanization. Methodology was required to conform, wherever possible, to that recommended for the various disciplines in the IBP guide books.

Demographic surveys

Restrictions on choice of ethnic groups

A decision had then to be made as to which of the 11 distinct South African Bantu ethnic groups should be selected for study. In arriving at this decision a number of natural restrictions had to be taken into account including the following.

(1) The rural area selected for study should preferably be fairly close to the headquarters of the research workers to reduce time spent on the field work and travelling and transport expenses.

(2) The terrain where the selected populations lived should be physically accessible, e.g. by means of a Landrover, to ensure a reasonable sampling of the people.

(3) The field work should preferably be carried out at a local hospital. This would facilitate the meeting of a number of requirements such as laboratory preparation of blood samples, etc.

(4) The population chosen should have a sufficient number of men, in the age group 20–50 years, who were normally present in the rural area and who would be available for study.

(5) The aim of the project was to study individual populations at different but comparable levels of the transition from rural to urban industrialized life. Such a comparative study would not be possible if the selected population were to consist mostly of rural or of urban people. Hence it would be useful to know the distribution of an index of 'degree of urbanization' and, if a population could be selected where this index had a fairly uniform distribution, this would be to the advantage of the proposed studies.

When these, and other more or less relevant considerations were taken into account, it became evident that two areas would be about equally suitable. These areas were Vendaland and Sekhukhuneland, the homelands of the

Venda and Bapedi people, respectively. Attention was therefore confined to these two ethnic groups.

General information about these two populations was readily obtained from government officials, superintendents of mission and other hospitals in the areas, and agricultural information officers. Quantitative data concerning population structure, age distribution, migratory labour movement etc. were, however, not available. To obtain this essential data, demographic surveys in both areas were initiated.

Sampling difficulties

In the reconaissance phase of the demographic surveys it became evident that it would be almost impossible, within the limits of an IBP project, to obtain a representative sample of either the Bavenda or Bapedi people. The areas were so vast and some regions so inaccessible that there could be no question of obtaining a random and representative sample from either rural population. Accordingly, it was decided to study only part of the total population and to make conclusions for such part only, accepting the fact that these conclusions could be relevant for the population as a whole.

The part of the population upon which it was decided to focus attention was obtained by using the natural system of stratification which exists in all the South African Bantu ethnic groups. Each ethnic group has a number of subgroups, each with its own natural leader as chief. In their original rural state the people pay allegiance to their own chief, who is considered a superior human being, and regarded with awe and admiration by all his subjects. The chiefs, among themselves, select a paramount chief who is responsible for the welfare of his tribe as a whole. The individual chiefs are geographically separated from each other and this enabled the research team to select, for the survey, the subjects of one chief from each group, i.e., from Vendaland and Sekhukhuneland, respectively. The areas of both these chiefs complied with the conditions set out above. The chiefs themselves were well-informed and well-disposed towards the study; an attitude which was indispensible if such a study was to succeed.

First rural Venda and Pedi demographic surveys

The two demographic surveys were conducted more or less as originally planned, concentrating mainly on family composition, age distribution, the migration pattern of labourers and completion of acculturation questionnaires. Bantu university students, members of the two ethnic groups, were trained and used as field workers. Many sampling difficulties occurred, however, despite meticulous planning, but no efforts were spared to make the sampling as representative as practical circumstances permitted. In short, the main conclusions were as follows.

(1) From an ethnological point of view there should be no difficulty in drawing a sample of either ethnic group since there were very few, if any, households from a different ethnic group to be found in the respective areas.

(2) With regard to availability of the population required for examination, it was found that one male person between the ages of 20 and 50 years could be expected to be present in about every second household in both areas studied.

(3) With regard to the migratory character of the target population, the surveys indicated that there were roughly two categories of male persons in the desired age range, viz. those 'usually present' and those 'usually absent' from home. There appeared to be no difference in this respect between the two areas studied.

(4) In respect of degree of acculturation no significant difference was found between the two areas studied although the general impression was formed that the Venda group may be slightly more rural.

Following on the investigations carried out during 1967, field work in the study of rural Venda commenced in Vendaland in May 1968.

Urban Venda demographic survey

It was established that the greatest concentration of urban Vendas were resident in the township of Tshiawelo southwest of Johannesburg. Since (*a*) the population consisted of 1595 Venda households, (*b*) it was known that each of the house owners had resided for a period of at least 10 years under urban conditions to qualify for residence, and (*c*) the names and house numbers of the heads of households appeared on the City of Johannesburg Urban Bantu Council Voters' List, it became evident that a demographic survey of the type conducted in the rural areas would be unnecessary. All that was required was to take a random sample on the basis of the Urban Council Voters' List large enough to allow for wastage on account of age, education, physical unfitness or non-response.

Field work on the urban Venda commenced on this basis in September 1968.

Second demographic survey of Bapedi in Sekhukhuneland

Preparations for the rural Bapedi study commenced early in 1969 when the co-ordinator for the HA project paid several visits to Sekhukhuneland. His discussions with government and tribal officials threw considerable doubt on the suitability of Chief Matlala's area, i.e. the area in which the first demographic survey was conducted. This led to further demographic studies of the area including actual counts of available subjects. It soon became evident that

317

D. J. M. Vorster

the conclusions of the 1967 demographic survey did not hold good and that the field work with this ethnic group would have to be postponed while other areas in Sekhukhuneland were being investigated to identify a suitable rural Pedi population. The main difficulty in the 1967 area turned out to be the unavailability of men in the 20–50 age group, although there was an abundance of men under 20 and over 50 years of age.

The subsequent search for a suitable population highlighted again the problems of tribal rivalries and prejudices which complicate the planning and execution of research amongst underdeveloped communities. Thus, the mere fact that a hospital, which was to serve as base for the field work, was geographically situated on the land of a neighbouring chief, could prejudice the participation of another chief's subjects.

Eventually, at a meeting of the Leola Tribal Authority on which all Pedi chiefs were represented, it was agreed that the study would be carried out on portion of the so-called Geluks Location in Sekhukhuneland and that several neighbouring chiefs, including the paramount chieftainess, would collaborate.

Field work in the study of rural Pedi commenced in July 1970.

Urban Pedi demographic survey

One shortcoming in the selection of the Venda urban sample was the fact that subjects could not be shown as belonging to the same Venda subgroup as the rural subjects. This was, however, not considered a serious weakness since the Venda were generally known to retain their ethnic identity to a very high degree and since they were, in any event, not a very large nation.

With the Pedi being a much larger and diversified group it was decided to ensure that the urban Pedi subjects would derive from the same subgroup as the rural sample, i.e. those Pedi occupying Sekhukhuneland 'proper'.

This, together with the fact that urban Bantu authorities included the Pedi within the larger ethnic grouping known as North Sotho, greatly complicated the identification of an urban population comparable with the rural population. Investigations in a number of Transvaal urban areas showed, first, that there was no township mainly occupied by Sekhukhuneland Pedi as was the case of Tshiawelo for the Venda, and secondly that municipal records of residents did not permit adequately accurate identification of tribal origin. Further investigations with municipal authorities in Johannesburg, Alberton and Germiston confirmed this but suggested that the Germiston townships probably had a greater density of this group. When, subsequently, a number of Sekhukhuneland Pedi with long service in urban areas were traced through their employers and interviewed, they indicated that it was common practice amongst their compatriots not to take up permanent urban residence even after qualifying for it, since they wanted to retain their ties with their homeland.

318

Adaptation to urbanization in South Africa

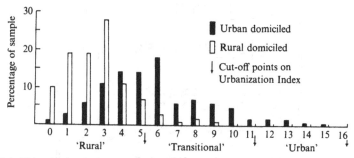

Fig. 12.1. Urbanization index scores (Venda).

Eventually it was decided that the urban Pedi sample would be identified from employers' records and then traced to the Germiston municipal township records for further screening in terms of minimum requirements: Pedi males, originally from 'Sekhukhuneland proper', between ages 16 and 65 years, physically fit for examination, employed and resident in the Germiston area for at least a period of six months prior to the study.

Composition of sample populations

The sampling models set up in the experimental design could not be fully met in that no completely rural or completely urbanized samples could be identified (Fig. 12.1). All population groups were in transition although at different levels. Random sampling models also proved impracticable, but the samples were considered representative of the de-facto populations falling within the specified demographic limits. The sample population did, however, satisfy most requirements. Samples numbered between 200 and 300 – the target size – and were drawn from two distinct population groups known as Venda or Pedi, occupying different geographic and climatic regions of the Transvaal. Because of the high degree of mobility amongst adult Bantu males the rigorous original requirements of continuous residence could not altogether be satisfied, but all subjects met the minimum requirements. Ideal age distributions could not be satisfied but the samples retained the essential characteristic of being in the labour market or employable. Educational requirements were satisfied by all groups (Table 12.1).

Results

The results of the study were presented in some 40 reports and publications listed separately in the Bibliography and for the purpose of this discussion only the general conclusions which could be reasonably well substantiated will be mentioned.

Table 12.1. *Age distributions of Venda and Pedi groups*

Age Group	Rural Venda N	Rural Venda %	Urban Venda N	Urban Venda %	Total Venda N	Total Venda %	Rural Pedi N	Rural Pedi %	Urban Pedi N	Urban Pedi %	Total Pedi N	Total Pedi %	Total N	Total %
Under 20	9	3.91	7	2.82	16	3.35	43	19.55	5	2.08	48	10.41	64	6.82
20–24	84	36.52	17	6.85	101	21.13	44	20.00	35	14.52	79	17.14	180	19.17
25–29	24	10.43	18	7.26	42	8.79	25	11.36	50	20.75	75	16.27	117	12.46
30–34	23	10.00	16	6.45	39	8.16	21	9.55	44	18.26	65	14.10	104	11.08
35–39	21	9.13	42	16.94	63	13.18	19	8.64	37	15.35	56	12.15	119	12.67
40–44	18	7.83	46	18.55	64	13.39	16	7.27	22	9.13	38	8.24	102	10.86
45–49	34	14.78	57	22.98	91	19.04	19	8.64	22	9.13	41	8.89	132	14.06
50–54	11	4.78	32	12.90	43	8.99	7	3.18	8	3.32	15	3.25	58	6.18
55–59	6	2.61	13	5.24	19	3.97	12	5.45	9	3.73	21	4.56	40	4.26
60+	—	—	—	—	—	—	14	6.36	9	3.73	23	4.99	23	2.45
Total	230		248		478		220		241		461		939	
Mean Age	32.80	—	41.07	—	37.09	—	31.38	—	34.42	—	32.99	—	35.13	—
± standard deviation	11.25	—	8.88	—	11.41	—	11.23	—	9.72	—	11.10	—	11.37	—

Table 12.2. *Comparison of the rural and urban populations*

System	(a) Venda			(b) Pedi		
	χ^2	N	P	χ^2	N	P
ABO	1.040	3	> 0.75	12.173	9	> 0.20
MNS	4.405	5	> 0.30	3.994	5	> 0.20
Rhesus	12.156	6	> 0.05	10.908	8	> 0 25
Kell	0.729	1	> 0.75	—	—	—
Duffy	1.770	1	> 0.10	0.257	1	> 0.50
Kidd	4.164	1	> 0.025	7.197	1	< 0.01
Secretor	0.057	1	> 0.80	—	—	—
Glucose-6-phosphate: dehydrogenase	1.325	1	> 0.20	0.070	2	> 0.95
6-phosphogluconate: dehydrogenase	—	—	—	4.312	2	> 0.10
Acid phosphatase	—	—	—	10.982	5	> 0.05
Phosphoglucomutase: PGM_1	—	—	—	5.752	3	> 0.10
PGM_2	—	—	—	0.007	1	> 0.90
Adenylate kinase	—	—	—	0.237	3	> 0.80
Haptoglobin	7.182	4	> 0.10	4.494	4	> 0.30
Transferrin	5.125	2	> 0.05	0.006	1	> 0.90
Haemoglobins: Sickle cell	0.069	1	> 0.75	—	—	—
HbB_2	0.397	1	> 0.50	—	—	—

Genetic and morphological variables

In terms of genetic constitution as determined by the examination of red-cell and serum protein polymorphisms (Table 12.2), all subgroups would appear to have been drawn from a genetically homogeneous population which is also genetically very similar to other Bantu speaking tribes studied previously. Comparisons of metrical and non-metrical morphological data also suggest an absence of genetic selectivity such as may occur in selective migration. The urban/rural differences found in morphological measures were largely confined to body weight, breadth, depth, circumference and skinfold measures, i.e. to bulk and girth. Since most subjects were, on urbanization, already fully adult, it was concluded that the differences were related to the greater calorific intake and physical activity imposed on the urban male by industrial work (Table 12.3).

Nutrition status

Shortly after the Venda studies were completed, reorganization of the participating Institute* resulted in a drastic reduction of the team participating in the IBP study. The medical, physiological and dietary aspects of the nutri-

* National Nutrition Research Institute, now National Food Research Institute and National Research Institute for Nutritional Diseases, S.A. Medical Research Council.

Table 12.3. *Comparative physical and physiological characteristics*

	Rural		Urban	
Characteristic	Venda $N = 241$	Pedi $N = 202$	Venda $N = 240$	Pedi $N = 223$
Average age (yr)	33	32	40	34
Average weight (kg)	56.7	56.2	64.1	60.6
Average maximum oxygen intake:				
(l min^{-1})	2.26	2.104	2.59	2.53
(l kg^{-1} min^{-1})	39.9	37.6	40.5	41.9

Table 12.4. *Mean values^a for the biochemical parameters: Venda and Pedi*

	Venda		Pedi	
	Rural: $N = 253$	Urban $N = 247$	Rural: $N = 204$	Urban: $N = 239$
	Mean ($\pm$ s.d.)	Mean ($\pm$ s.d.)	Mean ($\pm$ s.d.)	Mean ($\pm$ s.d.)
Cholesterol (mg/100 ml serum)	220 (45)	*232 (43)	169 (39)	*152 (31)
Triglycerides (mg/100 ml serum)	122 (53)	*135 (87)	107 (53)	102 (46)
Total protein (g/100 ml serum)	7.22 (0.60)	*7.67 (0.58)	7.94 (0.50)	*7.56 (0.45)
Albumin (g/100 ml serum)	3.81 (0.48)	*4.21 (0.39)	4.36 (0.42)	4.30 (0.47)
Total globulin (g/100 ml serum)	3.44 (0.49)	3.45 (0.42)	3.58 (0.42)	*3.27 (0.49)
α_1 globulin (g/100 ml serum)	0.25 (0.04)	*0.21 (0.04)	0.27 (0.04)	*0.23 (0.07)
α_2 globulin (g/100 ml serum)	0.65 (0.10)	0.67 (0.58)	0.71 (0.11)	*0.63 (0.12)
β-globulin (g/100 ml serum)	0.93 (0.15)	*1.06 (0.18)	0.87 (0.13)	*0.95 (0.19)
γ globulin (g/100 ml serum)	1.58 (0.32)	1.55 (0.31)	1.72 (0.33)	*1.42 (0.30)
Albumin as % of total protein	52.6 (5.3)	*55.0 (3.7)	55.0 (4.3)	*57.0 (5.2)
γ globulin as % of total protein	—	—	21.6 (3.6)	*18.9 (3.8)
Albumin/globulin ratio	1.13 (0.22)	*1.23 (0.19)	1.24 (0.21)	*1.37 (0.28)
N'-Me (mg/g Urinary) creatinine excreted)	3.94 (1.89)	*4.41 (2.76)	4.64 (1.76)	4.30 (1.69)
Urinary 2-pyridone (mg/g creatinine excreted)	3.65 (1.93)	*5.17 (4.39)	3.27 (5.32)	*4.79 (4.17)
2-pyridone/N-Me ratio	1.05 (0.61)	*1.28 (0.92)	0.67 (0.61)	*1.20 (1.02)
Urinary riboflavin (μg/g creatine excreted)	291 (170)	*487 (309)	335 (255)	317 (331)
Serum vitamin C (mg/100 ml serum)	0.47 (0.36)	*0.33 (0.26)	0.26 (0.18)	*0.47 (0.30)
Blood vitamin C (mg/100 ml serum)	1.18 (0.42)	*1.11 (0.29)	0.66 (0.20)	*0.73 (0.29)
Vitamin A (μg/100 ml serum)	52 (25)	*75 (39)	53.7 (22.1)	*33.4 (16.8)
Carotene (g/100 ml serum)	136 (76)	*75 (51)	56.2 (29.8)	*82.2 (42.5)

^{a} Significant differences between the means for the rural and urban groups at the 0.05 level, are indicated by an asterisk. N'-Me = N'-methyl-nicotinamide; 2-pyridone = N'-methyl-2-pyridone-5-carboxylamide.

Table 12.5. *Prevalence of nicotinic acid*

2-Pyridone (mg/g creatinine)	Deficient < 2.0	Low 2.0–3.9	High ≥ 4.0
Venda:			
Rural (%)	15.0	50.2	34.8
Urban (%)	15.3	31.6	53.0
Pedi:			
Rural (%)	44.1	30.9	25.0
Urban (%)	25.5	24.3	50.2

Table 12.6. *Prevalence of serum vitamin C*

Serum vitamin C concentration (mg/100 ml)	Deficient < 0.10	Low 0.10–0.19	Acceptable 0.20–0.39	High ≥ 0.40
Venda:				
Rural (%)	9.5	18.6	23.7	48.2
Urban (%)	13.4	27.5	27.5	31.6
Pedi:				
Rural (%)	5.9	35.5	44.8	13.8
Urban (%)	3.8	9.8	34.6	51.7

tional studies were discontinued, thus leaving only the biochemical analysis of nutritional status in the Pedi studies and thus for the final comparisons. Another somewhat limiting factor in the interpretation of the nutritional results is the fact that it was not possible to study all subgroups during the same season of the year. However, from the available data it would appear that the nutrition status of the urban groups is somewhat better than that of the rural groups although it is acceptable for all groups (Table 12.4). The main deficiency present was in nicotinic acid (Table 12.5) and to a lesser extent vitamin C (Table 12.6). Inter-ethnic comparisons slightly favoured the Venda groups, but not significantly so.

Physical work capacity

In both ethnic groups a significantly higher percentage of urban subjects were found to be capable of hard physical work than their rural counterparts. The percentages of urban Venda and Pedi capable of the different categories of work (light, moderate and hard) are very similar, but amongst the rural subjects a higher percentage of Pedi are in the light work category with a correspondingly smaller percentage in the hard work category (See Nutrition status above) (Table 12.7).

Table 12.7. *Distribution of working capacities*

	Rural		Urban	
	Venda %	Pedi %	Venda %	Pedi %
Light work category (<30 ml/(min/kg))	7.1	16.8	6.5	9.0
Moderate work category (30–45 ml/(min/kg))	66.0	65.4	57.1	57.0
Hard work category (>45 ml/(min/kg))	26.9	17.8	36.4	34.0

Social and psychological characteristics

On social/psychological grounds there would appear to be no absolute rural or urbanized subgroups of either of the two ethnic groups studied. All samples were attitudinally to a greater or lesser extent transitional. However, it was shown that, for both Venda and Pedi, education and to a lesser extent urbanization contribute towards improved performance in mental abilities as well as to a greater differentiation in the organization of such abilities (Fig. 12.2). Inter-ethnic comparisons of fully matched subsamples revealed no significant difference in the performance of the two urban samples, but the rural Pedi performed significantly better than the rural Venda. When matched literate and illiterate subsamples of both ethnic groups were compared, the Pedi performance was again better than that of the Venda on some tests. The significance of these latter findings is not yet clear, but it is suggested that the differences may be attributable to the differential work histories and experience of the two main groups rather than to tribal affiliation as such. In this respect it is interesting to note that urbanization, and to a lesser extent literacy, would seem to reduce the initial differences that were observed between the unacculturated Pedi and Venda. That is to say, acculturation would appear to work towards homogeneity of intellectual functioning. With regard to work motivation, unavoidable demographic differences present in the various samples precluded any firm conclusions with regard to the respective roles of ethnic grouping and acculturation in work motivation.

Statistical analysis

Apart from indicating that a number of variables used in the study had very little informative value on account of their statistical characteristics, the multivariate analysis of 73 variables suggested one important conclusion. This was that the multi-disciplinary study did not come up to expectations from the statistical point of view in the sense that no extremely interesting correlations seemed to exist between the variables of the different disciplines. This

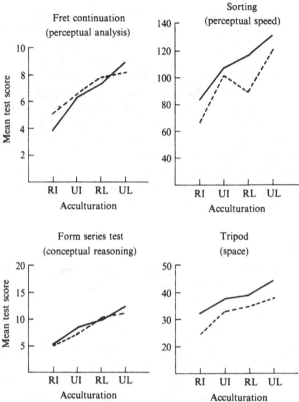

Fig. 12.2. Comparison of Venda (---) and Pedi (——) mean performance in four selected tests to demonstrate the effect of literacy and urbanization. *RI* = rural illiterates, *UI* = urban illiterates, *RL* = rural literates, *UL* = urban literates.

is perhaps not really surprising since current discipline-bound methodologies are specifically designed and constructed to exclude, as far as possible, variables extraneous to the particular discipline.

At an observational level, however, some interesting inter-disciplinary trends could be established. Thus it would seem that the better nutrition status (and calorific intake) of the urban groups, together with their greater exposure to physical exercise in industrial work, largely account for the higher working capacity of these groups as well as for the major morphological differences. To what extent these factors may be tied up with the better intellectual functioning of urban groups was not established.

Disease patterns

In this study, the disease patterns of female rural Pedi were compared with existing information in respect of urbanized Bantu and it revealed interesting differences and similarities. The similarities are probably mainly attributable to the fact that even urban-domiciled Bantu are still largely transitional. The rarity, in both communities, of conditions such as ischaemic heart disease and cancer of the lung or colon probably indicates insufficient exposure to many of the more noxious aspects of Westernization. Among the differences, some, such as the greater prevalence of goitre and certain infections, reflect exposure to hazards peculiar to rural areas. The majority of differences, however, concern a number of diseases which are commoner in city dwellers and which appear to be related to urbanization and Westernization. Most prominent were obesity and hypertension and to a lesser extent diabetes.

This study emphasized the importance of exploiting the present opportunities to study the early stages in the evolution of new diseases in developing populations exposed to the impact of urbanization and Westernization.

Overall evaluation

The more than 40 reports and publications which resulted from the project make many valuable contributions to the literature of several disciplines much of which has practical applications in health, education and labour utilization services in Africa. It is also true that the work was done at a fraction of the cost of the same studies, had they been undertaken separately, and generated great goodwill among the scientists involved. The principal disappointment is that the project eventually turned out to be a series of high-quality parallel studies using the same subjects at the same time instead of one closely integrated study of the 'whole person' and 'whole sample' and thus the resultant of the interactions between genetic, physiological, nutritional, medical, psychological and sociological components of the sample selected. We can therefore add little to our knowledge of the totality of man's interaction with, and adaptation to, the environmental changes inherent in the transition from a rural traditional way of life to an urban, industrialized society.

In trying to analyse the reasons for the very limited success achieved in this respect, one concludes that it lies mainly in inadequate experimental design notwithstanding the fact that the working group went to great pains in a sincere effort to develop an experimental design which would meet the requirements of all participating disciplines and yet be practically feasible and manageable. The reasons must therefore be sought beyond the mutual goodwill and skill of a number of competent scientists.

It would seem, firstly, that IBP being primarily a programme for human biologists, the bias in experimental design would naturally be towards the subject matter and methodology in which the human biologist feels most confident, i.e. genetics and physiology, and away from that which is more likely to be affected by cultural factors or the more complex genetic components of human behaviour. Unfortunately the human characteristics for which genotypes have been clearly established are purely physical, which, in modern adaptability, is not nearly as critical an issue as the determination of behavioural characteristics and learning capacity. The latter are critical because they will facilitate or inhibit adaptation to the requirements of modern technological civilization, participation in which would appear to be ultimately inevitable for all societies, and which, if properly directed and managed, would appear to hold the only prospect for human advancement and survival on earth.

Thus, for example, the study of physical work capacity in this project demonstrated differences between urban and rural experimental groups and between ethnic groups, but produced no valuable new insights into the role of nutritional factors or motivational factors as possible co-determinants of effective working capacity.

A second limiting factor in deriving a completely adequate experimental design for the IBP project is the fact that both theory and methodology are much less well developed in the fields dealing with the closely related cultural and behaviour genetics components of human adaptation to environmental change. Although much progress has been made in the field of behaviour genetics, much of the work has been hounded by the sterile aspects of the nature/nurture controversy and the social and political overtones which have been ascribed to such studies. In the case of psychology, the other behavioural science most closely involved, Biesheuvel (1968*b*) concludes that '. . . it is futile to search for genetic determinants unless we have first defined the basic behavioural entities. The phenotypical domain itself requires to be reduced to a more generally accepted order as judged by scientific rather than ideological criteria. Agreement is needed first of all about the basic dimensions . . . to which psychological observations and measurements can be reduced. Our current state of knowledge concerning the nature of intelligence illustrates the point.'

Hirsch (1967) clearly illustrates this gap in knowledge and methodology from another angle when he maintains that in contemporary behavioural science, far more attention has been paid to man's social roles than to his biological properties. In industrial psychology for example, tests are devised to select individuals who will most skilfully perform those tasks for which they are needed by industry. Because of the speed of cultural evolution, man cannot possibly have been subjected to intense natural selection for his technological skills, though man must employ the capacities he has evolved in the exercise of skills. The great challenge now before the behavioural sciences lies in the behaviour-genetic analysis of man's biological properties and the elucidation of their *modus operandi* in a socio-technological context.

There is also a third factor which contributed to the inadequacy of the experimental design of the project under discussion and which is present in

D. J. M. *Vorster*

most studies of differences between population groups whether inter- or intra-ethnic. This shortcoming is that cross-sectional samples are studied at a specific point in time whether it be school children or adults. The apparent comparability established at the time of study in respect of such environmentally influenced factors as nutrition status, educational level, socio-economic status and other demographic variables may be very misleading in that it does not take account of comparability in these and related factors during preceding periods. All human performance, physical as well as mental, is the outcome of a long period of growth and maturation. Environmental circumstances at early developmental stages have been shown to have permanent effects, both positive and negative, on later performance and this could operate differentially in subjects equated on such factors at the time of study. The IBP study reported in this paper suffered from these limitations in that it endeavoured to derive meaningful interdisciplinary results on adaptation to environmental change from a cross-sectional study, at a point in time, of adult populations that could hardly be fully equated even at the time of study.

It is my conviction that well-controlled, multi-disciplinary, longitudinal studies of different population groups, both in respect of early developmental studies and of adaptation to environmental change, are the most important means whereby science can contribute towards happy, healthy and successful co-existence of men both as individuals and as more or less differentiated ethnic groups.

Summary

A multi-disciplinary study of certain selected communities in their rural, peasant state in different parts of the Republic, and after the transition to urban industrialized life, was undertaken to throw light on psychological, anthropometrical and medical adaptations which people of different genetic constitutions have made during the course of the urbanization/industrialization process.

The two ethnic groups studied were the Venda in Vendaland and the Tshiawelo township southwest of Johannesburg and the Bapedi (Pedi) of Sekhukhuneland and the Germiston municipal township.

The conclusions reached are strictly only applicable to the defined subgroups and no generalization of these conclusions to other ethnic groups or the entire South African Bantu population can be made.

From the genetic and morphological point of view all experimental subjects would appear to have been drawn from a genetically homogeneous population. The morphological differences which were found appeared to be determined by a greater calorific intake and physical activity of the adult urban male.

In terms of nutritional status the urban groups would appear to be slightly better off than the rural groups. This, together with the higher average weight of urban males, seems to explain the significantly higher percentage of urban males in both ethnic groups who are capable of hard physical work.

On social–psychological grounds there would appear to be no absolute rural or urbanized male subgroups in the two ethnic groups studied. All samples were to a greater or lesser extent transitional. However, it was shown that both education and urbanization contribute towards improved performance in mental abilities as well as to greater differentiation in the organization of mental abilities.

Unavoidable demographic differences present in the various samples preclude any firm conclusions with regard to the respective roles of urbanization and ethnic grouping in work motivation.

I should like to acknowledge the excellent co-operation and fine team spirit which the following members of the Working Party and their staff brought to the HA project of South Africa's contribution to IBP:

Dr S. Biesheuvel, Director of Personnel, SA Breweries; Dr C. V. Bothma, Department of Bantu Administration and Development; Professor H. de Villiers, Department of Anatomy, University of the Witwatersrand; Dr J. P. du Plessis, SA Medical Research Council; Dr T. Jenkins, SA Institute of Medical Research; Dr N. F. Laubscher, National Research Institute for Mathematical Sciences, SA Council for Scientific and Industrial Research; Professor J. F. Murray, SA Institute for Medical Research; Dr H. C. Seftel, Department of Medicine, Non-European Hospital, Johannesburg; Dr C. H. Wyndham, Human Sciences Laboratory, Chamber of Mines of South Africa.

Bibliography of IBP publications

Biesheuvel, S. (1968a). Psychology and the IBP. *International Journal of Psychology*, 3, 199–207.
Biesheuvel, S. (1968b). The significance of the IBP for social sciences. *South African Journal of Science*, 64, 440–5.
de la Rey, R. P. (1969). *Socio-economic Evaluation of Rural and Urban Venda*. Pretoria: NNRI, SACSIR.
de la Rey, R. P. (1971). Socio-economic evaluation. *South African Medical Journal*, 45, 1305–14.
de Villiers, H. (1969). *IBP:HA Survey of Venda Males, July–September 1968*. Johannesburg: Department of Anatomy, University of the Witwatersrand.
de Villiers, H. (1970). A note on taste blindness in Venda males. *South African Medical Journal*, 66, 26–8.
de Villiers, H. (1971). *Morphological Variables and Genetic Markers in Urban and Rural Pedi Male Populations. IBP/HA Survey – July 1970–March 1971*. Johannesburg: Department of Anatomy, University of the Witwatersrand.

de Villiers, H. (1972). A study of morphological variables in urban and rural Venda male populations. In *Human Biology of Environmental Change*, ed. D. J. M. Vorster, pp. 110–13. London IBP.

de Villiers, H. (1976). Fingerprint and ridge counts in South African Venda males. In *Proceedings of the Anatomical Society of Southern Africa Conference, Lourenco Marques*, 1970 (in press).

du Plessis, J. P., Louw, M. E. J. & Nel, A. (1972). The principles and applications of the biochemical evaluation of the nutrition-status of populations. In *Human Biology of Environmental Change*, ed. D. J. M. Vorster, pp. 43–6. London: IBP.

du Plessis, J. P., Louw, M. E. J. & Nel, A. (1976). The principles and applications of the biochemical evaluation of the nutrition-status of populations. Typescript.

du Plessis, J. P., Louw, M. E. J., Nel, A. & Laubscher, N. F. (1972). *The Biochemical Evaluation of the Nutrition Status of Venda and Pedi Males*. Pretoria: SAMRC, NIND, and SACSIR, NRIMS.

Edgington, M. E., Hodkinson, J. & Seftel, H. C. (1971, 1972). *Disease Pattern in a South African Rural Bantu Population Including a Commentary on Comparisons with the Pattern in Urbanised Johannesburg Bantu* (1971). Johannesburg: Department of Medicine, Johannesburg Non-European Hospital and University of the Witwatersrand. And see also *Human Biology of Environmental Change* (1972), ed. D. J. M. Vorster, pp. 73–9, London: IBP.

Goosen, D. S. (1971). *Die samestelling van die Pedi-ondersoekgroep 1970–1971. S.A. en Internasionale Biologiese Program (Menslike aanpasbaarheid)*. WNNR Spesiale Verslag, PERS 158. Johannesburg: NIPN, WNNR.

Goosen, D. S. (1971). *Suid-Afrika en die Internasionale Biologiese Program (Menslike aanpasbaarheid). Die logistiek van 'n multidissiplinêre projek insake menslike aanpasbaarheid – Die Pedi-opname 1970/71*. WNNR Spesiale Verslag, PERS, 159. Johannesburg: NIPN, WNNR.

Grant, G. V. (1969a). *The Urban-Rural Scale: A Sociological Measure of Individual Urbanization*. CSIR Special Report, PERS 118. Johannesburg: NIPR, SACSIR.

Grant, G. V. (1969b). The organization of mental abilities of a Venda group in cultural transition. Ph.D. thesis, University of the Witwatersrand, Johannesburg.

Grant, G. V. (1972). The organization of intellectual abilities of an African ethnic group in cultural transition. In *Mental Tests and Cultural Adaptation*, ed. L. J. Cronbach & P. J. D. Drenth, pp. 391–400. The Hague & Paris: Mouton.

Greyson, J. A., van Graan, C. H., Rijk, J., Wyndham, C. H. & Strydom, N. B. (1971). *The Determination of the Physical Work Capacities of Urban and Rural Pedi Males*. C.O.M. Ref. Project no. HO/1/68, Research Report no. 52/61. 52/61. Johannesburg: Research Organisation, Chamber of Mines of South Africa.

Hall, S. K. P. (1971). *Motivation Among a Rural and an Urban Employed Group of Adult Pedi Males*. CSIR Special Report, PERS 154. Johannesburg: SACSIR.

Hall, S. K. P. & Harris, J. (1970). *Motivational Patterns of a Rural and an Urban Group of Adult Male Vendas. A Preliminary Report*. CSIR Special Report, PERS 124. Johannesburg: NIPR, SACSIR.

Hirsch, J. (ed.). (1967). *Behaviour Genetic Analysis*. New York: McGraw-Hill.

Jenkins, T. (1971). *Sero-Genetic Studies on the Pedi*. Johannesburg: Human Sero-Genetics Unit, SAIMR.

Jenkins, T. (1972). Some considerations in the study of Southern African populations undergoing environmental change. In *Human Biology of Environmental Change*, ed. D. J. M. Vorster, pp. 121–6. London: IBP.

Kendall, I. M. (1971). *The Organization of Mental Abilities of a Pedi Group in Cultural Transition.* CSIR Special Report, PERS 156. Johannesburg: NIPR, SACSIR.

Kendall, I. M. (1972). *A Comparative Study of the Organization of Mental Abilities of Two Watched Ethnic Groups: The Venda and the Pedi.* CSIR Special Report, PERS 171. Johannesburg: NIPR, SACSIR.

Laubscher, N. F. (1972). The Touth African multidisciplinary Venda siudp: a multivariate statistical analysis. In *Human Biologp of Enviro8mental Change,* ed. D. J. M. Vorster, pp. 199–204. London: IBP

Laubscher, N. F. & Potgieter, J. F. (1967)· *Demographic Surveys Undertaken in Sekhukhuneland and Vendaland. South Africa and International Biological Programme.* Pretoria: NRIMS and NNRI, SACSIR.

Laubscher, N. F. & Potgieter, J. F. (1969). *Selection of Samples of Venda for a Study of Human Adaptability. South Africa and International Biological Programme.* CSIR Special Report, WISK 68. Pretoria: NRIMS, SACSIR.

Louw, M. E. J., du Plessis, J. P. & Laubscher, N. F. (1971). *A Biochemical Evaluation of the Nutrition Status of Rural and Urban Pedi Males.* Pretoria: NIND, SAMRC, and NRIMS, SACSIR, and *South African Medical Journal,* **46,** 1139–42.

Lubbe, A. M. (1970). *The Nutritional Status of Rural and Urban Adult Venda Males as Determined by Dietary Surveys. International Biological Programme (Human Adaptability Section).* Pretoria: NIND, SAMRC.

Nel, A. & du Plessis, J. P. (1969). *The Biochemical Evaluation of the Nutrition Status of Urban and Rural Venda Males.* Pretoria: NNRI, SACSIR.

Nel, A., du Plessis, J. P. & Fellingham, S. A. (1971). Biochemical evaluation. *South African Medical Journal,* **45,** 1315.

Tobias, P. V. (1972). Growth and stature in Southern African populations. In *Human Biology of Environmental Change,* ed. D. J. M. Vorster, pp. 96–104. London: IBP.

van der Merwe, A. le R. (ed.) (1969). *A Comparative Study of the Nutritional Status of Rural and Urban Mono-racial Bantu Groups: 1. Venda. International Biological Programme (Human Adaptability Section).* CSIR Special Report, VOED 19. Pretoria: NNRL, SACSIR.

van der Merwe, A. le R. & Fellingham, S. A. (1971). A comparative study of rural and urban Venda males. *South African Medical Journal,* **45,** 1281–3.

van Graan, C. H., Wyndham, C. H., Strydom, N. B. & Greyson, J. S. (1972). Determination of the physical work capacities of urban and rural Venda males. In *Human Biology of Environmental Change,* ed. D. J. M. Vorster, pp. 129–31. London: IBP.

van Graan, C. H., Wyndham, C. H., Strydom, N. B., Viljoen, J. H., Greyson, J. S. & Rogers, G. G. (1969). *Determination of the Physical Work Capacities of Urban and Rural Venda Males.* C.O.M. Ref. Project no. H/1/68, Research Report no. 47/69. Johannesburg: Research Organization, Chamber of Mines of South Africa.

Vorster, D. J. M. (1968). The human adaptability programme. *South African Journal of Sciences,* **64,** 433–9.

Voster, D. J. M. (1972). *A Multidisciplinary Study of Two Ethnic Groups in Transition. Final Report to South African National Committee for International Biological Programme (Human Adaptability Section).* CSIR Special Report, PERS 167. Johannesburg: NIPR, SACSIR.

Vorster, D. J. M. (1973). Ethnic groups in transition in South Africa. *South African Journal of Science,* **69,** 267–72.

D. J. M. Vorster

Vorster, D. J. M., Steyn, D. W. & Hall, S. K. P. (1972). A study of certain psychological characteristics of a sample of adult Venda males in cultural transition. In *Human Biology of Environmental Change*, ed. D. J. M. Vorster, pp. 189–96. London: IBP.
Wyndham, C. H. (1972). Man's adaptation to heat and cold in Southern Africa. In *Human Biology of Environmental Change*, ed. D. J. M. Vorster, pp. 145–53. London: IBP.

Index

ABH (Se) secretor system, 11
  in Gujerat, 16
  in Siberia, 60, 76; under selection pressure of differentiating type, 78
ABO blood group system, 9–10
  in Africa: in calculation of genetic affinities, 198, 199; climate and, 192, 193; in Pygmy–Pygmoid and farmer pairs, 202
  in Amerindians, B absent, 28
  associations of, with goitre, 304, with leprosy, 16, and with smallpox, 16
  in Indian tribes, 15–16
  in Malaysia, 18
  in Siberia, 60, 63, 76, 79; distribution of, (A) 80, 82, (B) 80, 83, (O) 80, 84; selection pressure of stabilising type acting on, 78,79; stabilising selection, mechanism shown by frequency distribution, 88–92, 84
  in Yanomama, only O present, 113
acculturation to Western influences, in Solomon Islands, 147–8, 149, 151–2
  and blood pressure, 159, 160
  and serum cholesterol and uric acid, 157
acetylater system, in Korea, 19
acid phosphatase
  in calculating genetic affinities of African peoples, 198
  in circumpolar populations, 29
  and malarial infection, in Papua New Guinea, 304
  in Yanomama, 114, 122
adaptation to environment, 4
adenosine deaminase: search for variants of, in South American tribes, 127–9
adenylate kinase
  in calculating genetic affinities of African peoples, 198
  search for variants of, in South American tribes, 127–9
Africa
  genetic affinities of populations of centre of, for genes unresponsive to climate, 198–202
  sets of populations evolved in, 192–3
  surveys in, 19–21; of peoples of northern semi-arid zone, see Dogon, Fali, Fulani, and Kurumba; of rain forest, see rain forest
age, and blood pressure
  in African northern semi-arid zone, 260–1
  in Solomon Islands, 159–61
Ainu people, 19, 26

Åland Islands, 14
aldosterone excretion, by Yanomama, 133
Aleuts, 28, 29
  gene frequency diagrams for two populations of, settled in Komandorski Islands in 1820s, 98–101
Allen's 'rule', on length of protruding body parts at different latitudes, 244–5
altitude, and blood-factor genes in Ethiopia, 21
Amarar people (East Sudan): genetic affinities of, show Arab element, 199, 200, 201
Amerindians
  calculation of date of separation of, from Siberians, 51, 101, 103
  gene frequency polygons for Siberians and, 98–101
  genetic relations of different tribes of, 114–15
  resemblances between Siberians and, 28, 29
  of South America, 109–11; see also Yanomama tribe
Amhara people (Ethiopia), 21
  genetic affinities of, 198, 199, 200, 201
anaemia, in Papua New Guinea, 300–1
angina, in Tokelau non-migrants, 179
Arabs, Yemenite: and Yemenite Jews, 15
arid zone, Africa: stature and skull characters in, 190–1
Arid Zone Research, Negev Institute for, 224, 233
arm-folding pattern, in African northern semi-arid zone, 257–8
arm length
  in African northern semi-arid zone, 250, 251
  in Pygmy–Pygmoid and farmer pairs, 203, 204
  related to mean annual temperature, 252
asthma: in Tokelau non-migrants, 180, and in children of migrants, 182, 183
astigmatism, not found in Solomon Islands, 156
Australia, surveys in, 24, 25
Australia antigen: variable occurrence of, in Yanomama, 135
Australian Aborigines, resemblances between inland peoples of New Guinea and, 25
auto-immune diseases, and histocompatibility antigens, 11
Aymara tribe, Peru, 23

333

Index

Bantu-speaking peoples
  in Angola: hand-clasping pattern in, 257; arm-folding pattern in, 258
  spread of, 21, 189
Badik people (East Senegal), genetic affinities of, 198, 199, 200
Bergmann's 'rule', on relation between size and latitude, 244
Binga Pygmoids, 273
  genetic affinities of, 200, 201, 202, 205
Binga-Mbimu pair of hunter-farmer peoples, in African rain forest, 194
  body measurements compared, 202-6
Bira people, genetic affinities of, 205
  see also Mbuti-Bira pair
birth rates, see fertility
blood pressure
  in African northern semi-arid zone, 259-61
  in Maoris, and peoples of Pukapuka, Rarotonga, and Tokelau Islands, 166
  in Papua New Guinea, 301
  in Solomon Islands: relations with age and weight, 159-60, and with salt intake, 160-1
  in Yanomama and Cayapo tribes, 133
body measurements
  in Africa: different climatic zones, 192, 193, northern semi-arid zone, 249-56; rain forest, Pygmy-Pygmoid and farmer pairs, 202-6
  of Siberians, 56-9
  in Solomon Islands, 149, 154-5
  see also height, weight, etc.
Bolivia, surveys in, 22, 23
Bougainville Island, Papua New Guinea, 145-6
  survey in, see Solomon Islands
Brazil
  Negroes in: arm-folding pattern among, 258; hand-clasping pattern among, 257
  surveys in, 22-3, 24
breast-feeding of infants, in Tokelau migrants and non-migrants, 183
Burma, surveys in, 17
Bushmen (San), 20
  similarities between Pygmies and, 283
Bushong, Bantu farmers south of rain forest, 190
  genetic affinities of, 205
  see also Cwa-Bushong pair

cadmium, in blood of Yanomama and in US city samples, 134
Canary Islands, surveys in, 14
cardiovascular diseases, very rare in Solomon Islands, 156-7

cargo cult philosophy, 306
Caroline Islands (Micronesia), survey in, 26
Caucasoid people, in North and East Africa, 21
Cayapo tribe, Brazil, 23, 115, 120, 129
ceruloplasmin: search for variants of, in South American tribes, 128-9
chest radiography, in Solomon Islands, 153, 156
children, health of: in Tokelau migrants to New Zealand, 182, 183, 185
Chile, surveys in, 22
Chinese, in Malaysia, 18
chloride, in blood of Yanomama, 133
cholesterol in serum
  increases with acculturation in Solomon Islands, 156-7, 158
  in Tokelau migrants, 181, 184-5, and non-migrants, 175, 176
chromosome damage: extent of, in mitotic leucocytes of Yanomama, 132
Chukchi people, 28
circumpolar populations, 26-9
coconuts, in diet on Tokelau Islands, 167, 180, 184
colour blindness, not found in Siberians, 56
congenital defects, in Yanomama, 133
copper, in blood of Yanomama and in US city samples, 134
cranial abnormalities, in Neolithic and present-day Siberian populations, 95-6
  gene frequency polygon for, 96-8
Cwa (Kuba) Pygmies, genetic affinities of, 205
Cwa-Bushong pair of hunter-farmer peoples, in African rain forest, 190
  ABO groups in, 202
  body measurements compared, 202-6
Czechoslovakia, surveys in, 14

death rates
  of children in African northern semi-arid zone, 248-9
  of Yanomama, 117
demography, 3
  in African northern semi-arid zone, 247-9
  in Papua New Guinea, 293-4, 297
  of Pygmies, 276-9
  in Solomon Islands, 148, 149
  of Tokelau migants and non-migrants, 172, 185
  of Yanomama, 116-19
dendrograms
  linguistic, for Siberians, 52
  for South American tribes, 114
  for Yanomama villages, 122

334

Index

Index

potassium
  in blood of Yanomama, 133
  urinary excretion of, by Tokelau non-
    migrants, 177, 178
pottery, obtained by Pygmies from Bantu
  farmers, 188, 275
protein, in diet in Papua New Guinea, 299
Pygmies of African rain forest (hunter-
    gatherers), 20, 273–5
  ABO groups in, 202
  compared with their suzerain peoples
    (body measurements), 202–6
  demography of, 276–9
  as earliest inhabitants, now living as
    vassals to Bantu farmers, 188–90, 275
  genetic studies on, 279–80
  languages of, 276
  nutrition of, 281
  parasite infection of, 280
  stature of, 281–3, and skull characters, 191
  see also Cwa, Mbuti, Twa
Pygmoids of African rain forest (hunter-
    gatherers), defined in socio-economic
    terms, 189
  ABO groups in, 202
  compared with their suzerain peoples
    (body measurements), 202–6
  see also Binga

Quechua tribe, Peru, 23

rain forest, African
  migration of populations into, 5, 187–8
  Pygmy and Pygmoid hunter-gatherers as
    original inhabitants of, Bantu farmers
    as later arrivals, now living in vassal–
    suzerain pairs, 188–90
  stature and skull characters in, 191
  stature and weight in, 209
  surface-weight ratio for populations of,
    252–3
renin, in plasma of Yanomama, 133
respiratory disease, chronic non-specific; in
  Papua New Guinea, 301
Rhesus blood group system
  in Amhara, shows gene flow from Arabia,
    198
  in Burma, 17
  in calculating genetic affinities: of
    African peoples, 198, 199; of South
    American tribes, 114–16
  in Micronesia (Caroline Islands), 26
  in New Guinea and Australia, 25
  in Papua New Guinea, 296
  in Pygmies, 279
  in Siberians, 61, 76, 78; compared with
    Amerindians, 28

in South American tribes, 24
Romania, surveys in, 14

salt intake
  and blood pressure, 160–1
  and serum uric acid, 157–8
  in Solomon Islands, 158, 161
Sandawe people, Tanzania, 20–1
Sara people
  body measurements of, 207, 251
  fertility data for, 248
  genetic affinities of, 205, 206
  stabilising and directional selection in,
    211–13, 215
  see also Majingay
Sardinia, surveys in, 14
savanna, Africa
  stature and skull characters in, 190–1
  stature and weight in, 208
  surface-weight ratio of populations of,
    252–3
Scotland, Icelandic connections with, 28
Senoi people (Burma and Malaysia), 17, 18
serum albumin
  'private' polymorphism of, in Yanomama,
    114, 122
  search for variants of, in South American
    tribes, 128
Shangan people (South Africa), genetic af-
    finities of, 198, 199, 200, 201
shoulder width
  increases with air moisture, in Africa, 191,
    192
  in African northern semi-arid zone, 250,
    251
  in Pygmy–Pygmoid and farmer pairs, 203,
    204
Siberia: isolated indigenous populations of,
    chosen for study, 50–2; planning of
    study, 52–4; sampling of populations,
    54–5; techniques used, 55–6; conclu-
    sions, 102–3
  duration of stable equilibrium in, 95–102
  gene frequencies in: predicting variance
    of, 73–4; stationary distribution of,
    86–94; variance of, as measure of
    genetic differentiation, 74–9
  gene migrations in, 67–73
  genetic-anthropological characteristics of,
    56–63
  genogeography of, 79–80, 85–6
  sizes of indigenous populations (1959),
    63–7
skin colour
  in African northern semi-arid zone, 245
  in Sara, Oto, and Twa, 213–14
  in Solomon Islands, 146, 147

340